LIGHT ABSORPTION BY AEROSOL PARTICLES

Studies in
Geophysical Optics and Remote Sensing

Series Editor: **Adarsh Deepak**
Institute for Atmospheric Optics and Remote Sensing
Hampton, Virginia 23666

Published Volumes and Volumes in Preparation

Optical Properties of Clouds
 Peter V. Hobbs and Adarsh Deepak (Eds.)
Atmospheric Aerosols: Their Formation, Optical Properties, and Effects
 Adarsh Deepak (Ed.)
Light Absorption by Aerosol Particles
 Hermann E. Gerber and Edward E. Hindman (Eds.)

Technical Proceedings of the
First International Workshop on Light Absorption by Aerosol Particles
Held at Colorado State University, Fort Collins, Colorado
28 July to 8 August 1980

Co-Sponsored by Radiation Commission of IAMAP (IUGG), the American Meteorological Society, and the Optical Society of America. Financial support was provided by Naval Air Systems Command, National Science Foundation (Grant ATM 8005356, Division of Atmospheric Science), Office of Naval Research (Grants N0001480G0042 and N0001479C0793), Army Research Office, Naval Research Laboratory, and Colorado State University.

LIGHT ABSORPTION BY AEROSOL PARTICLES

Edited by

Hermann E. Gerber

Atmospheric Physics Branch
Naval Research Laboratory
Washington, DC 20375

Edward E. Hindman

Department of Atmospheric Science
Colorado State University
Fort Collins, Colorado 80523

SPECTRUM PRESS   1982
A Division of Science and Technology Corporation
Hampton, Virginia

Copyright © 1982, by Spectrum Press
All rights reserved.
No part of this publication may be reproduced or transmitted in any form or by any means, electronic or mechanical, including photocopy, recording, or any information storage and retrieval system, without permission in writing from the publisher.

SPECTRUM PRESS
A Division of Science and Technology Corp.
P. O. Box 7390
Hampton, Virginia 23666

Library of Congress Cataloging in Publication Data

International Workshop on Light Absorption by Aerosol Particles (lst : 1980 : Colorado State University)
Light absorption by aerosol particles.

"Technical proceedings of the First International Workshop on Light Absorption by Aerosol Particles, held at Colorado State University, Fort Collins, Colorado, 28 July to 8 August 1980"—P.
Includes bibliographical references and indexes.
1. Aerosols—Optical properties—Congresses. 2. Absorption of light—Congresses. I. Gerber, Hermann E. II. Hindman, Edward E. III. International Association of Meteorology and Atmospheric Physics. Radiation Commission. IV. American Meteorological Society. V. Optical Society of America. VI. Title

QC882.I58 1980   551.5'6   82-80728
ISBN 0-937194-00-X

Printed in the United States of America

82 83 84 85   9 8 7 6 5 4 3 2 1

CONTENTS

Preface   ix
Acknowledgments   xi
Participants   xiii
List of Symbols   xvii
Photographs of Some Participants   xix

1. KEYNOTE ADDRESS

Aerosol Research within the World Climate Research Program   1
   H.-J. Bolle

2. REVIEW

Absorption of Light by Atmospheric Aerosol Particles: Review of Instrumentation and Measurements   21
   H. E. Gerber

3. GUEST SEMINARS

New Techniques in Light Absorption Measurements   55
   Gottfried Hänel

A Laser Cavity Extinction Photometer for Measurements of Extinction Coefficient and Visual Range   65
   Robert G. Knollenberg

4. WORKSHOP AEROSOLS

Generation and Characterization of Aerosol Particles   71
   Edward E. Hindman, Randy D. Horn and William G. Finnegan

5. INSTRUMENTATION SUMMARY

Instrumentation Summary   123
   H. Gerber and E. Hindman

6. INSTRUMENTATION DESCRIPTIONS AND WORKSHOP MEASUREMENTS

CONTENTS

An Evaluation of Photoacoustic and Transmission Techniques for Monitoring Particulate Carbon Collected on Teflon Filters 129
C. A. Bennett, Jr., and R. R. Patty

The GMCC Four-Wavelength Nephelometer 149
Barry A. Bodhaine

Remote Determination of Aerosol Index of Absorption: The Diffuse-to-Direct Method 169
Michael A. Box and J. Douglas Copp

Results from University of Washington Participation in First International Workshop on Light Absorption by Aerosol Particles 173
Antony D. Clarke and Alan P. Waggoner

Light Absorption Measurements Using Integrating Plate Method 189
Stan Cowen, David S. Ensor, and Leslie E. Sparks

Optical Properties of Standard Aerosols: A Report of Measurements for the First International Workshop on Light Absorption by Aerosol Particles 197
W. G. Egan

Simultaneous Measurements of Aerosol Scattering and Extinction Coefficients in a Multi-Pass Cell 231
H. E. Gerber

Imaginary Refractive Index Measurements for Arizona Road Dust and Methylene Blue 243
James B. Gillespie

Analysis of Polar Nephelometer Data Obtained at the First International Workshop on Light Absorption by Aerosol Particles 251
Gerald W. Grams and Alessandro Coletti

Workshop Measurements of the Aerosol Absorption Coefficient with an Integrating Plate Method and an Integrating Sphere Photometer 267
Jost Heintzenberg

Spectrophone Measurements of Diesel Vehicle Particulate Material 275
S. M. Japar, J. Moore, D. K. Killinger and A. C. Szkarlat

Diffuse Reflectance and Transmission Measurements of Aerosol Absorption: Report on Results of the First International Workshop on Light Absorption by Aerosols 279
E. M. Patterson and B. T. Marshall

CONTENTS

University of Arizona Aerosol Absorption Measurements *J. A. Reagan, B. M. Herman and R. M. Schotland*	297
Photoacoustic Studies of Aerosol Optical Properties *D. M. Roessler*	301
Photoacoustic Determination of Light Absorption by Aerosols *R. Röhl, R. A. Palmer and W. A. McClenny*	307
Lawrence Berkeley Laboratory Laser Transmission Method *H. Rosen and T. Novakov*	321
Remote Sensing of Aerosol in the Free Atmosphere by Passive Optical Techniques *Glenn E. Shaw*	335
Results on Aerosol Light Absorption *Frederic E. Volz*	357
Measurement of Light Absorption by Aerosols with an Optoacoustic Detector *Wayne N. Wright, Donald H. Stedman, Leopoldo Stefanutti and Robert W. Terhune*	373

7. DATA SUMMARY

Data Summary *H. Gerber and E. Hindman*	379

8. DATA COLLATION

Data Collation *H. Gerber and E. Hindman*	387

9. WORKSHOP REVIEW

Workshop Review *S. Twomey and Donald R. Huffman*	395
AUTHOR INDEX	409
SUBJECT INDEX	415

PREFACE

This volume contains the technical proceedings of the First International Workshop on Light Absorption by Aerosol Particles held at Colorado State University, Fort Collins, Colorado, on 28 July to 8 August 1980. The workshop was co-sponsored by the Radiation Commission of IAMAP (IUGG), the American Meteorological Society, and the Optical Society of America. The workshop consisted of an instrumentation comparison where simultaneous measurements were made of the light absorption properties of laboratory-generated and ambient aerosol particles. Eighteen research groups participated by bringing to the workshop a representative collection of techniques for measuring light absorption and other physical properties of the aerosol particles. The majority of the papers in this volume describe the instrumentation brought to the workshop and tabulate the measured data. Additional papers include the keynote address by Professor H.-J. Bolle, University of Innsbruck; a review by Dr. H. Gerber, Naval Research Laboratory; seminars by Professor Gottfried Hänel, Goethe University, and by Dr. R. Knollengerg, Particle Measuring Systems, Inc.; and papers from those who were unable to attend the workshop, but whose contribution was desired in order to give a proper overview.

Much of the interest in light absorption by atmospheric aerosol particles has its origin in the concern that man's pollution of the atmosphere is affecting climate. Whether the potential change is one of cooling or heating depends in part on the magnitude of the absorption. Another area of interest is in modeling radiative transfer through the atmosphere where a description of the particles' absorption properties must be included. Although a great deal of work has been done in the last 10 years in measuring absorption and in developing new techniques, it is still not possible to assess the impact of aerosols on climate or to choose from the literature reliable and representative values of the absorption properties for modeling purposes. This state exists because of the large natural variability of light absorption by atmospheric particles and because of the historically difficult nature of absorption measurements. Of the more than one dozen techniques which have been developed to measure absorption, none are without potentially large errors. A comprehensive intercomparison between techniques for measuring light absorption by aerosol particles has not been previously held. Thus convening a workshop for establishing errors in those techniques was felt to be appropriate at this time.

PREFACE

The responsibility for specifying the scientific program of the workshop and for organizing and coordinating the workshop was given to the Organizing Committee which included the following members: R. Bird, Solar Energy Research Institute; M. Box, University of New South Wales; V. Derr, NOAA/ERL; F. Faxvog, Honeywell; H. Gerber (Co-chairman), Naval Research Laboratory; E. Hindman (Co-chairman), Colorado State University; D. Roessler, General Motors Research Laboratory; H. Rosen, Lawrence Berkeley Laboratory; R. Terhune, Ford Motor Company; and A. Waggoner, University of Washington.

The Organizing Committee established the following goals for the workshop:

1. Bring together a representative sample of the existing experimental techniques for measuring the imaginary index, absorption coefficient, and single scattering albedo of aerosols.

2. Make measurements with the techniques by exposing them simultaneously to the same aerosols.

3. Use ambient and laboratory-generated aerosol particles with well defined physical and optical characteristics, and which resemble the absorbing species found in the atmosphere, such as, soot, salt, and soil-derived particles.

4. Compare workshop measurements to determine systematic instrumentation errors.

The major goals in compiling this volume were to provide full descriptions of the experimental techniques, including cost, accuracy, and performance characteristics of the techniques; and to include a complete set of data generated at the workshop. In view of the large size of the data collection, only a partial comparison was made between the data collected with the various techniques. Additional comparisons are left to the reader. Also, no judgments were made on the accuracy and practicality of the various techniques. The reader is asked to judge for himself. Two such readers, Professor S. Twomey and Professor Donald R. Huffman, both of the University of Arizona, have kindly agreed to comment on the workshop materials as presented in Sections 1-8 of the proceedings. Their review appears in Section 9.

Hermann E. Gerber
Edward E. Hindman

ACKNOWLEDGMENTS

The editors wish to thank the other members of the Organizing Committee for their time and their valuable recommendations for the scientific program of the workshop. Appreciation is also expressed to the workshop participants, because it was their enthusiasm and cooperative spirit which greatly contributed to a successful workshop. The support provided by the Department of Atmospheric Science, CSU, T. H. Vonder Haar, Head; and by the Atmospheric Physics Branch, NRL, L. Ruhnke, Head, is acknowledged. W. Finnegan, R. Horn, C. Swain, D. Clair, and S. Laxman are thanked for their contributions in preparing and running the workshop. A. Waggoner is thanked for his helpful suggestions and work on the particle generating system, and for the loan of equipment; Naval Research Laboratory for the loan of equipment; and H.-J. Bolle for his suggestion for an Instrumentation Summary. Thermal Systems, Inc. is thanked for the loan of the fluidized-bed particle generator (TSI Model 3400), and the "Arizona Road Dust" soil particles were provided by the AC Spark Plug Division of General Motors Corporation.

Appreciation is expressed to H. Rosenwasser of Naval Air Systems Command for suggesting the workshop and providing initial funding. The majority of the financial support for the workshop was provided by the National Science Foundation (Grant ATM 8005356, Division of Atmospheric Science, R. Carrigan). Additional funds were contributed by the Office of Naval Research (Grants N0001480G0042 and N0001479C0793), Army Research Office, Naval Research Laboratory, and Colorado State University. One of us (H.G.) was supported by Naval Ocean Systems Center.

PARTICIPANTS

C. A. Bennett, Jr., Physics Department, North Carolina State University, Raleigh, North Carolina 27607

Ami Ben-Shalom, Physics Department, TECHNION-Israel Institute of Technology, Haifa, Israel

R. E. Bird, Solar Energy Research Institute, 1636 Cole Road, Golden, Colorado 80401

Barry A. Bodhaine, NOAA/ERL, Boulder, Colorado 80320

H.-J. Bolle, Institute for Meteorology and Geophysics, Schopfstrasse 41, 6020 Innsbruck, Austria

Michael A. Box, School of Physics, University of New South Wales, Kensington, NSW, Australia

Antony D. Clarke, Department of Civil Engineering, University of Washington, Seattle, Washington 98195

Alessandro Coletti, Instituto di Fisica dell'Atmosfera-C.N.R., Largo Lo Sturzo 31, 00100 Roma, Italy

J. Douglas Copp, Institute of Atmospheric Physics, University of Arizona, Tucson, Arizona 85721

Stan Cowen, Kaiser Steel Corporation, P. O. Box 217, Fontana, California 92335

R. Daniels, Atmospheric Physics Branch, Naval Research Laboratory, Washington, DC 20375

Vernon E. Derr, NOAA/ERL/WPL, Boulder, Colorado 80302

W. G. Egan, Research Department, Grumman Aerospace Corporation, Bethpage, New York 11714

Fred R. Faxvog, Honeywell, 2600 Ridgway Parkway, Minneapolis, Minnesota 55413

William G. Finnegan, Department of Atmospheric Science, Colorado State University, Fort Collins, Colorado 80523

H. E. Gerber, Atmospheric Physics Branch, Naval Research Laboratory, Washington, DC 20375

James B. Gillespie, Atmospheric Sciences Laboratory, White Sands Missile Range, New Mexico 88002

PARTICIPANTS

Gerald W. Grams, School of Geophysical Sciences, Georgia Institute of Technology, Atlanta, Georgia 30332

Gottfried Hänel, Institute for Meteorologie and Geophysics, Goethe University, Feldbergstrasse 47, D6000 Frankfurt/Main, Federal Republic of Germany

Jost Heintzenberg, Department of Meteorology, Arrhenius Laboratory, University of Stockholm, Stockholm, Sweden

B. M. Herman, Institute of Atmospheric Physics, University of Arizona, Tucson, Arizona 85721

Edward E. Hindman, Department of Atmospheric Science, Colorado State University, Fort Collins, Colorado 80523

Randy D. Horn, Department of Atmospheric Science, Colorado State University, Fort Collins, Colorado 80523

Donald R. Huffman, Department of Physics, University of Arizona, Tucson, Arizona, 85721

S. M. Japar, Research Department, Ford Motor Company, Dearborn, Michigan 48121

D. K. Killinger, Research Department, Ford Motor Company, Dearborn, Michigan 48121

Robert G. Knollenberg, Particle Measuring Systems, Inc., 1855 South 57th Court, Boulder, Colorado 80301

S. Maclean, Physics Department, York University, Downsview, Ontario, Canada M3J 1P3

W. A. McClenny, Environmental Protection Agency, Research Triangle Park, North Carolina 27711

B. T. Marshall, School of Geophysical Sciences, Georgia Institute of Technology, Atlanta, Georgia 30332

J. Moore, Research Department, Ford Motor Company, Dearborn, Michigan 48121

T. Novakov, Lawrence Berkeley Laboratory, University of California, Berkeley, California 94720

C. O. Oluwafemi, Physics Department, University of Lagos, Nigeria

R. A. Palmer, Paul M. Gross Chemistry Laboratory, Duke University, Durham, North Carolina 27706

E. M. Patterson, School of Geophysical Sciences, Georgia Institute of Technology, Atlanta, Georgia 30332

R. R. Patty, Physics Department, North Carolina State University, Raleigh, North Carolina 27607

L. F. Radke, Department of Atmospheric Sciences, University of Washington, Seattle, Washington 98195

J. A. Reagan, Department of Electrical Engineering, University of Arizona, Tucson, Arizona 85721

PARTICIPANTS

D. M. Roessler, Physics Department, General Motors Research Laboratory, Warren, Michigan 48090

R. Röhl, Paul M. Gross Chemistry Laboratory, Duke University, Durham, North Carolina 27706

H. Rosen, Lawrence Berkeley Laboratory, University of California, Berkeley, California 94720

H. Rosenwasser, Naval Air Systems Command, Code 310C, Washington, DC 20361

R. M. Schotland, Institute of Atmospheric Physics, University of Arizona, Tucson, Arizona 85721

Glenn E. Shaw, Geophysical Institute, University of Alaska, Fairbanks, Alaska 99701

Donald H. Stedman, Chemistry Department, University of Michigan, Ann Arbor, Michigan 48104

Leopoldo Stefanutti, IROE of C.N.R., Firenze, Italy

A. C. Szkarlat, Research Department, Ford Motor Company, Dearborn, Michigan 48121

Robert W. Terhune, Ford Motor Company, Dearborn, Michigan 48121

G. L. Trusty, Optical Science Division, Naval Research Laboratory, Washington, DC 20375

S. Twomey, Institute of Atmospheric Physics, University of Arizona, Tucson, Arizona 85721

Frederic E. Volz, Air Force Geophysics Laboratory, Hanscom Air Force Base, Massachusetts 01730

Alan P. Waggoner, Department of Civil Engineering, University of Washington, Seattle, Washington 98195

Wayne N. Wright, Physics Department, Kalamazoo College, Kalamazoo, Michigan 49007

C. Wyman, School of Geophysical Sciences, Georgia Institute of Technology, Atlanta, Georgia 30332

LIST OF SYMBOLS

To assist the reader, an effort has been made to use the following common symbols throughout this volume. The use of the symbols $\sigma_{e,s,a}$, $\tilde{\omega}$, δ, τ, λ, and $\varkappa$ was proposed by the Radiation Commission of IAMAP (Raschke, E., ed., 1978, *Terminology and Units of Radiation Quantities and Measurements*, NCAR, Boulder, Colorado 80303, USA).

Quantity	Symbol	Unit
Aerosol absorption coefficient	σ_a	m^{-1}
Aerosol scattering coefficient	σ_s	m^{-1}
Aerosol extinction coefficient	σ_e	m^{-1}
Single scattering albedo [$\sigma_s/(\sigma_a + \sigma_s)$]	$\tilde{\omega}$	
Optical thickness (depth)	δ	
Transmittance	τ	
Specific absorption coefficient ($\sigma_a \times 10^6/M_v$)	B_a	m^{2}g^{-1}
Refractive index of aerosol particles	n	
Real part of the index	n_1	
Imaginary part of the index	n_2	
Wavelength of light	λ	μm
Wavenumber	$\varkappa$	μm^{-1}
Particle radius	r	μm
Particle diameter	d	μm
Particle size distribution	N(r), dN/dr	no. of particles cm^{-3} μm^{-1}
Density of particle	ϱ	g cm^{-3}
Mass of particle	m	μg
Total mass of particles	M	μg
Volume of aerosol	V	liter
Mass of particle per volume of aerosol	M_v	μg m^{-3}
Flow rate of aerosol	F	liter m^{-1}
Temperature	T	
Time	t	

Workshop Participants [1st row (top) left to right]: H. J. Bolle, H. E. Gerber, G. Hänel. *[2nd row]:* R. G. Knollenberg, E. E. Hindman, W. G. Finnegan, R. D. Horn. *[3rd row]:* C. A. Bennett, Jr., R. R. Patty, B. A. Bodhaine. *[4th row]:* M. A. Box, J. D. Copp, A. P. Waggoner, A. D. Clarke, S. Cowen.

[lst row]: W. F. Egan, J. B. Gillespie, G. W. Grams, J. Heintzenberg. *[2nd row]:* S. M. Japar, E. M. Patterson, B. T. Marshall. *[3rd row]:* R. Röhl, R. A. Palmer, W. A. McClenny, H. Rosen. *[4th row]:* G. E. Shaw, F. E. Volz, D. H. Steadman, W. N. Wright, L. Stefanutti.

The following were not available at time pictures were taken: A. Coletti, B. M. Herman, D. R. Huffman, D. K. Killinger, J. Moore, T. Novakov, J. A. Reagan, D. M. Roessler, R. M. Schotland, A. C. Szkarlat, R. W. Terhune, and S. Twomey.

AEROSOL RESEARCH WITHIN THE WORLD CLIMATE RESEARCH PROGRAM

H.-J. Bolle

Institut für Meteorologie und Geophysik
Universitaet Innsbruck

Studies with climate models have demonstrated that aerosols cannot be neglected in an accurate simulation of present climate. Preliminary sensitivity tests have further demonstrated that changes in aerosol concentrations, types, and distributions, on the order of magnitude observed in nature, affect surface temperatures and atmospheric heating rates. The amplitude of such variations is comparable to the amplitude of possible variations in other parameters of the climate system. Based upon a review of the present state of knowledge, an outline is given of research priorities within the World Climate Research Program to quantitatively assess the impact of aerosols on climate.

I. INTRODUCTION

Climate generally is understood as the totality of weather events in a certain area within a specified time interval which affect the biosphere. It can be described by the mean values and the variances of all meteorological parameters during this period. However, the processes which lead to larger excursions, such as those observed in the past, or cause possible long-term trends in the climate system must be investigated for their physical reasons. For this purpose the International Council of Scientific Unions (ICSU) and the World Meteorological Organization (WMO) agreed to establish the World Climate Research Program (WCRP), which is part of the World Climate Program (Fig. l). The WCRP was founded in order to supply data and to foster the development of mathematical models for an improved understanding of the physics which determines climate. The ultimate goal of these efforts is to understand the *physical state of the earth-atmosphere system* well enough to permit climate prediction.

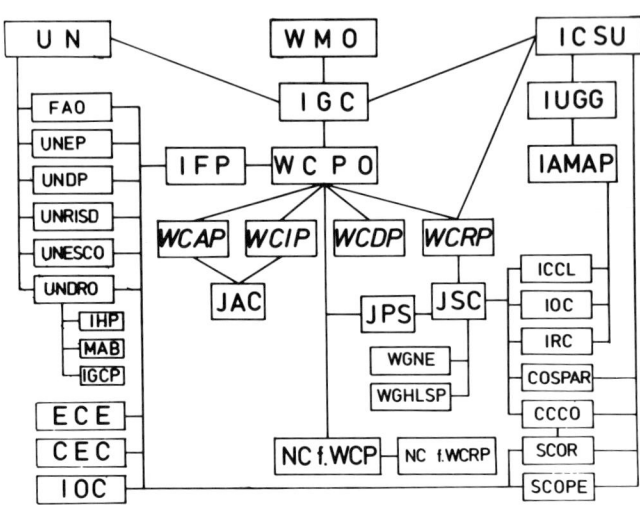

CCCO	Committee for Climate Changes and the Ocean	JPS	Joint Planning Staff
CEC	Commission of the European Community	JSC	Joint Scientific Committee (ICSU-WMO)
COSPAR	Committee on Space Research	MAB	Man and the Biosphere Program
ECE	Economic Commission for Europe	NC	National Committee
FAO	Food and Agriculture Organization	SCOPE	Scientific Committee on Problems of the Environment
IAMAP	International Association for Meteorology and Atmospheric Physics	SCOR	Scientific Committee on Ocean Research
		UN(O)	United Nations (Organization)
		UNDP	UN Development Program
ICCL	International Commission on Climate (IAMAP)	UNDRO	UN Disaster Relief Organization
ICSU	International Council of Scientific Unions	UNEP	UN Environmental Program
		UNESCO	UN Education, Scientific and Cultural Organization
IFP	Inter-Agency Focal Points		
IGC	Inter-Governmental Council		
IGCP	International Geological Correlations Program	UNRISD	UN Research Insteitute for Social Development
		WCAP	World Climate Applications Program
IHP	International Hydrological Program	WCDP	World Climate Data Program
IOC	International Ozone Commission (IAMAP)	WCDP	World Climate Data Program
		WCIP	World Climate Impact Studies Program
IOC	Inter-Governmental Oceanic Commission	WCPO	World Climate Program Office
IRC	International Radiation Commission (IAMAP)	WCRP	World Climate Research Program
IUGG	International Union of Geodesy and Geophysics	WGNE	Working Group on Numerical Experimentation
JAC	Joint Applications-Impacts Committee (now: SAC, Scientific Advisory Committee)	WGHLSP	Working Group on Hydrology and Land Surface Processes
		WMO	World Meteorological Organization

FIGURE 1. Structure of the World Climate Program. The implementation of projects is done by JSC and SAC (formerly JAC). These bodies are advised by ICSU commissions and the UN organizations. The structure is subject to modifications.

Aerosols affect weather due to their ability to act as condensation nuclei for water and ice and by their direct interaction with the radiation field of the planet. In considering the global average, the direct modulation of the radiation fluxes caused by aerosols is much smaller than the effects of other constituents such as clouds and water vapor. However, the influence of aerosols may be sufficient to be of importance. Table 1 gives an estimate of the importance of different constituents for the radiation budget components. Regionally, the relative importance of aerosols can be much greater than is reflected in Table 1 if it is taken into consideration that the optical depth of an aerosol at 0.55 μm

TABLE 1. Estimated Relative Importance of Different Atmospheric Constituents for the Earth's Global Mean Radiation Budget.[a]

Percent of shortwave radiation (0.3-3 μm)			Percent of longwave radiation (3-100 μm)	
Solar extraterrestrial flux at 1 AU	100		Emitted IR flux	100
Reflected sunlight	30		Clouds	47-50
Aerosols		1	Gases	26-27
Surface		4-5	H_2O	16-18
Gases		5-6	CO_2	7-8
Clouds		19	O_3, CH_4	2
Absorption of solar radiation	70		Surface	23-24
Surface	43-45		Aerosols	1-2
Open oceans		30-31		
Open land		9		
Ice and snow		4-5		
Gases	18-20			
H_2O		12		
O_3, O_2		5-6		
CO_2		1		
Others		< 1		
Clouds		4-5		
Aerosols		3-5		

[a]*Estimated from numbers reported by Kondratyev, K. Y.,* Meteorological Investigations by Means of Rockets and Satellites, *Leningrad (1962); London, J., and Sasamori, T., Radiation budget of the atmosphere,* Space Research IX, *pp.639-649, Akademieverlag, Berlin (1971); Robinson, N. (ed.),* Solar Radiation, *Elsevier, Amsterdam (1966); Sasamori, T., London, J., and Hoyt, D. V., Radiation budget of the southern hemisphere,* Met. Monographs 13, *9-23 (1972); and Schulze, R., Strahlenklima der Erde, Dr. Dietrich Steinkopff Verlag, Darmstadt (1970).*

wavelength can vary between about 0.05 or less in a clean maritime atmosphere to 1 or more during strong desert dust outbreaks and in urban-industrial atmospheres.

Climate research attempts to determine how much of the heating or cooling of the earth and its atmosphere is due to aerosols, and how much this heating or cooling would be changed if the amount or type of aerosol in the atmosphere changes. The thermal effect on the climate system of any atmospheric constituent is quantitatively determined by the heating rate, which can be computed from the energy divergence div $\vec{\phi}$ of the net radiative flux $\vec{\phi}$ by

$$\frac{dT}{dt} = \frac{1}{\varrho_{air} c_p} \text{ div } \vec{\phi} \qquad (1)$$

where
- T = temperature (K)
- t = time
- ϱ_{air} = air density
- c_p = specific heat capacity at constant pressure (1.005 kJ/kg K).

The average heating rate of the atmosphere is on the order of -1 K per day. This radiative cooling is compensated for by the release of sensible and latent heat from the surface. The heat is transported into the atmosphere by turbulent mixing. Sensible heat principally warms the planetary boundary layer, while the latent heat is released in the cloud layer.

In order to study the effects of aerosols on the heat budget, one has to consider two different processes which can be separated by wavelength. Solar radiation has short wavelengths, from the ultraviolet up to about 2.5 μm. In this spectral range aerosols absorb and scatter the solar radiation, which generally reduces the amount of solar energy available at the surface for heating and evaporation. A purely scattering aerosol enhances albedo and will only decrease the amount of solar radiation reaching the ground. The result would be cooling of the lower troposphere. As the aerosol starts to absorb short-wave radiation its albedo will be reduced, and additional heat is then transferred directly to the atmosphere. The amount of radiant energy which can thus be deposited in the atmosphere increases with increasing surface albedo, since more of the electromagnetic energy travels through the atmosphere twice. Eventually the heat gained by absorption in the atmosphere will exceed the cooling due to the enhanced albedo at the surface, and the net effect for the whole system may be a warming. At wavelengths beyond 3 μm scattering becomes less important, and the infrared emission of the earth's surface and of atmospheric constituents plays the dominant role for the radiation budget. Aerosols which absorb at these longer wavelengths will therefore contribute mainly to the greenhouse effect of the atmosphere and compensate for the decrease in solar energy reaching the ground. The resulting net effect depends upon the relative magnitude of the optical properties of the aerosols in these two wavelength regions.

Research on the effects of aerosols on climate should therefore be directed toward the accurate determination of the flux divergences:

$$\text{div } \bar{\phi} = -\frac{d}{dz} | \{\phi^+_{SW} - \phi^-_{SW}\} + \{\phi^+_{LW} - \phi^-_{LW}\} | \qquad (2)$$

The subscripts SW and LW denote short-wave and long-wave radiation, respectively, and the superscripts denote direction toward the earth's center (-) and toward zenith (+).

The fluxes can in principle be computed from the equation of radiative transfer, which for a plane-parallel atmosphere can be written

$$\mu \frac{\partial L(\delta;\theta;\phi)}{\partial \delta} = L(\delta;\theta;\phi) - J(\delta;\theta;\phi) \qquad (3)$$

where
- L = radiance (in Wm^{-1}sr^{-1})
- J = source function
- ð = exp($-\delta/\mu$) transmittance
- δ = optical depth
- μ = cos θ
- θ = zenith angle
- φ = azimuth angle

The solution of this equation requires information on the scattering and absorption coefficients (which, together with the aerosol mass distribution are necessary to compute ð) as well as on the directional distribution of the scattered radiation (the phase function), the temperature, and the extinction coefficients which enter the source function. Integration of the radiance over all directions gives the net flux. Numerical methods to compute the fluxes in turbid atmospheres have recently been reviewed by Lenoble (1) and Fouquart et al. (2), and among the many monographs on this topic one of the most recent is that of van de Hulst (3).

With the ability to handle radiative transfer in a turbid atmosphere to the desired accuracy reasonably established, the next and even more important question arises: Is the aerosol a critical parameter? A critical parameter is a quantity which varies in space and time in such a way that it can either regionally or globally disturb the energy budget of the climate system and change the condition in the biosphere if no other compensating processes occur at the same time.

Atmospheric temperature is commonly used as a measure of whether or not a climate parameter can be considered critical. While this is not the only criteron for climate change in nature, temperature is such a fundamental quantity in both atmospheric dynamics and the biosphere that for practical as well as physical reasons it is the most convenient measure. However, it must be noted that compensating processes may cause the temperature to remain essentially constant while other parameters such as cloudiness, surface albedo, humidity, or ozone concentration change. Such a situation could also affect life, e.g., by a change in the radiation necessary for photosynthesis.

II. AEROSOL PARAMETERS FOR CLIMATE MODELING

Methods of evaluating the equation of radiative transfer for a turbid atmosphere have been developed during the last few decades to such a level that almost any prescribed realistic accuracy can be achieved, provided the input parameters are precisely known and computer time is no problem. In climate research, however, a rigorous treatment of aerosol scattering is very often not possible since aerosol must be treated as one constituent among many others, and a modeling of all of these parameters with the best possible refinement would exceed any available computer capacity.

Methods of computing aerosol effects must therefore be simplified without losing too much in accuracy. The procedure of arriving at parameterized solutions for the radiative transfer must be looked at as a controlled degradation of the precise methods.

The radiative properties of any individual aerosol may be computed from the basic set of parameters which includes the size distribution N(r), the particle shape, and the complex index of refraction n($\varkappa$) where r is the radius of an individual particle and $\varkappa$ is the wave number. From these basic parameters the following quantities can be derived by applying Mie theory: albedo for single scattering $\tilde{\omega}$, scattering phase function (or indicatrix) p which describes the spatial distribution of the scattered radiance, and the scattering coefficient per unit mass, σ_m, or per unit length, σ_s, from which the optical depth $\delta = \int \sigma_m dm = \int \sigma_s ds$ can be computed. Instead of the complete scattering phase function, the anisotropy factor

$$g = <\cos \theta> = \frac{1}{2} \int_{-1}^{+1} \mu p(\theta) d\mu \qquad (4)$$

is often used, which is an average cosine of the foreward-scattering angle θ weighted by the phase function.

The geographical and temporal variability of aerosol concentration and composition is very large, and thus far cannot be handled adequately in climate models. For sensitivity studies, therefore, carefully selected standard aerosol models must be used which reflect the natural conditions sufficiently well. Such models generally define for each layer of the atmosphere (boundary layer, free troposphere, lower stratosphere, upper stratosphere, and mesosphere) the optical depth δ (or the linear extinction coefficient and the nature of the particulate material). The following types of aerosols are distinguished: mineral dust, oceanic haze, soot, aerosol generated by gas-to-particle conversion, water solubles, and volcanic ash. From these basic materials aerosol mixtures like those listed in Table 2 can be constructed.

TABLE 2. Approximate Composition of Standard Aerosol Mixtures in Percentages (4,5)

Aerosol	Mineral dust	Oceanic haze	Soot	Gas-to-particle or water soluble	Volcanic ash
Continental	70		1	29	
Maritime		95.0		5	
Urban-industrial	20		10	70	
Stratosphere					
Undisturbed				100	
Volcanic					100

In standard aerosol models which are under construction for use with climate models, the size distributions are specified for each basic aerosol type (5). Radiation properties will be presented in tabular form for 61 wavelengths between 0.2- and 40-μm. In these tables the ratios of the spectral attenuation coefficients (extinction, scattering, absorption) to the extinction coefficient at 550 μm are presented, together with the asymmetry factor and the albedo for single scattering. The relative attenuation coefficients must be scaled by the user according to the optical depth adopted at 550 μm for each layer. These models assume that the relative humidity does not approach the limit of about 70% where some aerosols start to swell. Models for this case have been developed by Hänel and Bullrich (6).

At present most climate studies are made with globally averaged aerosol models such as those defined by Deirmendjian (7) and Toon and Pollack (8). Extensive computations on the role of aerosols have been carried out by Braslau and Dave (9) for different Lambertian surface albedos and sun angles. These authors evaluated the equation of radiative transfer numerically by Fourier series expansion of the scattering phase function and radiance. Ohring (10) used these data in order to introduce aerosol effects into a zonally averaged model. A two-stream approximation was developed by Sagan and Pollack (11) and was subsequently used by Charlock and Sellers (12), Wang and Domoto (13), and Reck (14). Table 3 contains the model data used by some of these authors.

TABLE 3. Numerical Values of Tropospheric Aerosol Parameters Adopted by Different Authors

Quantity	Braslau & Dave (9), Ohring (10)	Reck (14)	Toon & Pollack (18), Charlock & Sellers (12)	Carlson & Benjamin (15), Carlson & Caverly (16)	Levin, Joseph & Mekler (17)	Yamamoto & Tanaka (18)	Shettle & Fenn (19)
$n_1(0.55\mu m)$	1.5	1.5		1.54		1.50;0.0-0.1	1.38-1.75^f
$n_2(0.55\mu m)$	0;0.01	0.1		0.008-$0.003^{a,e}$	0.005^c-0.03^d		$4.3 \cdot 10^{-9}$-0.44^f
$N(r)$	modified γ	ar^{-4}	$\approx r^{-(\beta+1)}$		Cr^{β}	$C \cdot 10^4$, $0.03\mu m<r<0.01\mu m$ Cr^{-4}, $0.1\mu m<r<10\mu m$	log-normal
$r_{min}(\mu m)$	0.001	0.05		0.1	$\beta=1.4 \ldots 2.6$ 0.15		10^{-3}
$r_{max}(\mu m)$	7.0	10.0		10.0	10.0		10^2
$N_{tot}(cm^{-3})$	1.97×10^7 (8.24×10^7)						
$\delta(0.55\mu m)$	0.1	0.065;0.325	0.125	1.0^b	$0.2 \ldots 3.17$	$\beta_0 a^{-1}$, $0.05 \leq \beta_0 < 0.375$	adjustable
$\delta(vis,0.55\mu m)$		0.90	0.75-1.0	0.71-0.86			0.792;0.897;0.988
$\delta(IR,10\mu m)$		0.28					0.245;0.486;0.677
$\omega_{crit}(0.55\mu m)$		0.75	0.81	≈ 0.9			
$<\cos\theta>$		0.64	0.7	0.85-0.78(0.73)			0.613;0.638;0.744
$\sigma_e(vis)km^{-1}$	not appl.	0.065;0.325		0.003-0.5			normalized
$\sigma_e(IR)/\sigma_e(vis)$		0.108					
ϱ_s(surface albedo)	0.07;0.3;0.6		0.139	0.05;0.23			0;0.05;0.15
Remarks	Rigorous integration, Legendre polynomials	One-dimensional radiative convective models (20)	Global mean model	Delta-Eddington approximation	Observations under different conditions (desert)	Computations based upon principle of invariance	Urban, rural, maritime models, adjustable to visibility

[a]Difficult to compare because of alternative choice of wavelength bands. [b]$\delta_{Raleigh} = 0.0088[\lambda\cdot(4.15 + 0.2/(\Delta p/1013)]$ (p=pressure in mb), $0.45<\delta_{Mie}<1.2$. [c]Sharav, Israel. [d]Background. [e]Patterson et al. (21) report a value of 0.0055 (300nm:0.025, 600nm:0.0038) for Saharan aerosol. [f]For components.

III. RESULTS OF SENSITIVITY STUDIES

1. *Tropospheric Aerosol*

In the troposphere, aerosols are concentrated near the surface in the planetary boundary layer. Their residence time is affected by the weather processes and will be on the order of 1 week in regions where they are frequently washed out by precipitation and a few weeks elsewhere. For this reason strong regional differences in aerosol concentration and type can be expected. The globally averaged optical depth due to aerosols is estimated to be 0.123 (8) to 0.155 (22), in comparison with 0.145 due to molecular scattering. Sellers (22) estimated that an optical depth of 0.023 is generated by human activities. He found that a doubling of the anthropogenic contribution to δ from 0.023 to 0.046 would decrease the surface temperature by 0.5 K or $\Delta t/\Delta \delta = -22$ K.

Since then more detailed aerosol models have been explored. It has been demonstrated that the imaginary part of the refractive index and the optical depth δ are more critical parameters than the phase function and the mean particle radius $\bar{r}$ [e.g., Chylek and Coakley (23,24)]. Hansen et al. (25) found a critical value $\tilde{\omega} = 0.85$ above which cooling and below which warming occurs. In some regions of the world, especially with strong industrial concentrations, warming can therefore be expected while in other regions even with rather heavy aerosol loads the aerosol may cool if the imaginary part of its refractive index in the shortwave range is low. According to Hansen et al. (25) the temperature effect may range between 3 K ($\tilde{\omega} = 0.6$) and -1 K ($\tilde{\omega} = 0.95$).

Ohring (10) found that complete neglect of aerosol in a zonally averaged model would introduce an error of 1 K in the computed surface temperatures, and that an increase in the optical aerosol depth to the present value would decrease the surface temperature by

$$\Delta T/\Delta \delta = -11.6 \text{ K} \tag{5}$$

Additional consideration of cloud albedo feedback may reduce the temperature effect (26). Ohring's work also indicated that in an atmosphere with constant relative humidity the effect on surface temperature is stronger than in an atmosphere with fixed absolute humidity. On the other hand, the ice-albedo feedback enhances the surface temperature effect. The temperature effect is also latitude dependent, decreasing with increasing latitude. In the case of absorbing aerosols the polar albedo decreases further with increasing aerosol concentration but the temperatures in the atmosphere continue to decrease. This is interpreted to be a result of a dynamic coupling between low and high latitudes.

Charlock and Sellers (12) computed the aerosol effect with a time-dependent monthly as well as annual model. By adding aerosols of the type described by Toon and Pollack (8) with an optical depth of 0.125, the surface temperature is reduced by 1.6 K or $dT/d\delta = -12.8$ K, a value which is close to that found by Ohring (10). For a strongly absorbing aerosol ($\tilde{\omega} = 0.75$) and a surface albedo of 0.139, Sellers' annual model gives $dT/d\delta = 15.4$ K at a latitude of 40° to 50°N. No temperature change occurs for $\tilde{\omega} \cong 0.81$. The dependence on surface albedo has been studied by Reck (14,17,18) who found a transition range between a surface albedo of 0.37 (for an extinction coefficient >10 km^{-1} at 550 nm) and 0.6 (for an extinction coefficient >1 km^{-1} at 550 nm) where the surface is cooled for the lower albedo and heated for the higher albedo values respectively. The system remains in thermal balance at the division line.

A comparison of computations with (13,29) and without (10,18) infrared aerosol absorptivity shows that the main effect of tropospheric aerosols is due to modulation of the short-wave radiation fluxes, a result which was first spelled out by Rasool and Schneider (29). The bulk of the aerosol is concentrated in the lower atmosphere where the temperatures are not much different from surface temperatures, especially if one considers daily averages with the nocturnal inversions. Over deserts, however, silicate material can be transported high up into the troposphere and cause an albedo for single scattering three- to five-times smaller than under undisturbed conditions (17). In these cases, the infrared radiation properties gain in importance and the heating due to absorption of solar radiation is partly compensated for by the infrared emission. But in these cases as well, an increase in atmospheric heating still seems to prevail. The deficit in solar energy reaching the ground is partly compensated for by the long-wave opacity and its greenhouse effect.

In cases of dust outbreaks from the Sahara over the Atlantic as studied by Carlson and Benjamin (15) for the GATE area, aerosols were injected up to tropospheric levels around 500 mb. Infrared cooling rates of up to -3 K/day resulted in the center of the dust cloud, which had an optical depth (550 nm) of around 3.0. In these cases as well, the net effect averaged from 1000 mb to 500 mb and over 24 hours produced a warming of about 1 K due to the combined effects of net absorption of solar radiation in the dust layer and an additional greenhouse effect in a layer between 1000 mb and 900 mb due to the infrared aerosol properties. If the same aerosol were over the desert, the greenhouse effect would be converted into cooling near the surface, but net heating would still be maintained through the whole layer.

2. Stratospheric Aerosol

Stratospheric aerosol is generated by both gas-to-particle conversions and volcanic events. It therefore consists of two components, a permanent component and another component which is injected locally, spreads globally by turbulent mixing, and has a residence time of a few years. Thus the scale of aerosols in the stratosphere is quite different from that observed in the troposphere. On the other hand, the optical properties of the stratospheric aerosols seem to be more uniform than in the troposphere. However, this may be an inaccurate impression with respect to the volcanic component, since there are indications that each volcano may generate its own specific aerosol. The climate effect can again be computed from a data set of $\tilde{\omega}$, $<\cos\theta>$, $\bar{r}$, $\delta_{(vis)}$ and the albedo of the underlying levels which is very close to the planetary albedo ϱ_p. Typical values for the optical aerosol parameters are given in Table 4 (19).

TABLE 4. Stratospheric Aerosol Models

Authors	Shettle & Fenn (19) Harshvardhan (30)	Toon & Pollack (8)
Size distribution N(r)	$324r\ exp(-18r)$ $[5461.33r\ exp(-16\sqrt{r})]^a$	$A\ exp\{-[ln^2(r/\bar{r}\ )]/[2\ ln^2 \sigma]\}$ $\sigma = 2(1.8)^a$
Mean radius		$0.035;\ (0.1\text{-}1)^a$
Extinction coefficient	$1.1 \cdot 10^{-4}$	
Albedo of single scattering $\tilde{\omega}$	$1(0.947)^a$	1
Asymmetry factor g	$0.726(0.698)^a$	≈ 0.7
Real part of refractive index n_1	$1.43(1.50)^a$	b
Imaginary part of refractive index n_2	$10^{-8}(8 \cdot 10^{-3})^a$	b
Optical depth δ		$0.005;\ (0.02\ldots >0.1)^a$

[a] volcanic perturbations.
[b] Palmer and Williams (31).

The heating rates in the stratosphere turn out to be very sensitive to a change of $\tilde{\omega}$ (32,33). Values of 1- to 1.5-K/day have been computed at 17 km for $\tilde{\omega}$ between 0.6 and 0.8. In the troposphere a stratospheric aerosol layer can generate a warming (on the order of 0.1 K/day) if $\tilde{\omega} < 0.98$, or a small cooling effect if $\tilde{\omega}$ approaches 1.0; the cooling is greatest if there is no infrared opacity.

Due to stratospheric background aerosols, the spherical albedo of the earth can only be changed on the order of 10^{-3}. Above $\tilde{\omega} = 0.94$ aerosols will increase the earth's albedo and below this critical value it will be decreased. For climate research it is therefore extremely important to accurately determine the albedo for single scattering in the range above 0.9.

After strong volcanic eruptions the stratosphere may maintain an optical depth of 0.1 and even 0.3 for a period of 1 or 2 years (32). Charlock and Sellers (12) and Hansen et al. (34) computed the effect of a volcanic dust layer with time-dependent models for the time following the Agung eruption in 1962 (data listed also in Table 4). A depression in the tropical surface temperatures on the order of 0.4 to 0.55 K was computed for a period of about 2 to 3 years, which is in rather good agreement with an observed depression of 0.5 K for March 1963 and lasting from mid-1964 to 1966. In the stratosphere a warming of up to 6 degrees has been observed over Australia from autumn 1963 to spring 1964 which agrees with the models that give the warming maximum in late autumn 1963, but also give a steeper decrease to "normal" conditions than has been measured.

Harshvardhan (30) investigated the stratospheric aerosol layer and found the strongest effects in polar regions and very small effects at latitudes < 50°. His conclusion is that perturbations on the order of $\delta \approx 0.1$ which occur after volcanic eruptions may induce significant climatic changes due to the change of the equator-to-pole radiative gradient.

During periods of high plutonic activity the aerosol concentration in the stratosphere may be maintained in a highly disturbed state over long periods of time, on the order of decades. Model simulations indicate that temperature excursions can be expected in the troposphere which are on the same order of magnitude as those generated by the increase of CO_2 during the last hundred years, i.e., a few tenths of a degree (33). If such an effect were to also occur in the higher latitudes or globally, it would be on the order of magnitude expected for a little ice age.

IV. FEEDBACK PROCESSES WITH AEROSOL INVOLVEMENT

The aerosol-temperature effect cannot be viewed as isolated from the behavior of the whole climate system, but little is known about the magnitude of possible feedback mechanisms. As mentioned earlier, another major characteristic of aerosol particles is their ability to function as condensation nuclei. In addition, aerosols may be imbedded in clouds in between the cloud droplets, thus again affecting the optical properties of the clouds (35). Another cloud-related effect is the dissolution of clouds in areas of desert-aerosol outbreaks over the oceans, a phenomenon which can often be observed in satellite images. This effect may cause the planetary albedo to decrease in the areas affected by the desert aerosols rather than increase as is assumed in some simulation studies.

The combined effects of climate-sensitive parameters on the temperature field may result in a change of horizontal (latitudinal) temperature gradients which would change the general circulation and heat transports. Also the stability of the atmosphere depends *inter alia* on the radiation effects of aerosols. Such processes may lead to altered precipitation patterns and wind speeds. Joseph (36) has demonstrated such effects with a general circulation model. If arid zones are affected in such a way that their climate becomes dryer and from time to time more intense storms develop, then the amount of aerosols may also increase over these regions and be transported over large distances, thus enhancing the aerosol effect. A look at Mars demonstrates the extreme possibilities which exist on a dry planet. At the present time feedback processes of these types have not been simulated, and it is pure speculation that they may have impact on the stability of climate.

Anthropogenic pollution is another mechanism which, because of its permanence, may already have affected climate in certain regions. Globally seen, its effect still can only be small because of the short residence time of these aerosols, which are not transported high up into the atmosphere. There is, however, a man-induced feedback mechanism in arid zones which should be mentioned here. Due to extensive overgrazing and agriculture, erosion is increased in these regions, which again increases the probability that dust is raised from the ground and mixed up into the middle troposphere, thus adding to the aerosol load. The surface albedo change connected with this erosion will again act as a feedback mechanism by reducing the tendency for cooling of the whole system, which has an increasing surface albedo (27,28).

V. FUTURE RESEARCH ASPECTS

Due to the smallness of the signals generated by the aerosol-radiation interaction, it is difficult to extract from operational meteorological measurements the impact of this interaction on climate. Since, in addition, the signs of the effect may be opposite in different parts of the world, it is even more problematic to empirically determine a global mean effect. Simulation models will therefore also play a dominant role in the future assessment of the importance of aerosols for climate. For these mathematical-physical models, efficient algorithms for the treatment of radiative transfer due to aerosols must be developed further and their accuracies have to be determined by comparison with more elaborate computation schemes. Pollack and Cuzzi (37) recently attempted to model the effect of nonspherical particles in a semi-empirical approach. Nonspherical particles cause a larger increase in the planetary albedo than spherical equivalents.

In order to apply the radiative transfer codes in climate models, radiative aerosol properties and the global distribution of aerosols with these properties must be known. One can speculate that in the future very advanced models will be able to generate their own aerosols as they generate clouds. However, this would first require that several problems be solved. First, the generation of aerosols over land and over oceans as a function of wind shear must be parameterized. There are now a number of approaches toward solving this problem over deserts, stemming especially from research on the generation of Martian duststorms (38). This is one example of how planetary research can guide the solution of terrestrial problems. On earth this problem is, however, interconnected with the seasonal variations of soil moisture which affect adhesion, and this in turn is interrelated with general circulation. Second, the transports of these aerosols and their interaction with atmospheric humidity must be computed. Third, the very important gas-to-particle conversion requires the implementation of physico-chemical reactions in the models (39,40). Fourth, plants are sources of aerosols due to the evaporation of volatile compounds such as terpenes, and, of course, also due to wood fires. Fifth, the injection of aerosols by volcanic eruptions and from anthropogenic sources must be introduced into the models, which requires a detailed knowledge of the mechanism of the injection processes. So far even the numbers for the aerosol source strengths are only crudely known (Table 5).

TABLE 5. *Estimated Relative Rates for Aerosol Production Processes in Percentages[1]*

Processes	a	b	c	d
Sulfates (from H_2S and SO_2, gas-to-particle conversion, including volcanic and man-made)	35	27-15	29-18	41-37
Ammonium salts	--	8-10	9-12	41-37
Sea-salt spray	33	31-11	33-14	23-27
Soil and rock debris	16	10-19	11-23	14-16
Organic volatiles from plants, forest fires, agricultural burning, and hydrocarbons from natural decay	6	9-17	9-16	9-12
Volcanic dust	2	3- 6	3- 7	7- 1
Nitrates (from NO)	6	9-18	3- 2	3
Man-made particulates (heating, industry, engine exhausts)	2	3- 6	3- 8	3- 4

[1] *After (a) Peterson and Junge (41); (b) SMIC (42); (c) Twomey (43); and (d) Dittberner (44).*

Contributions are expected from the experimentalists which support the construction of the models and the parameterizations involved. There are four fields of necessary activities which have to be continued and stressed: (a) measurements of the optical properties of natural aerosols as needed for the radiative transfer computations; (b) investigations of the radiative effects of aerosols imbedded in clouds; (c) validation in field experiments of theoretically derived radiation effects, such as heating rates; and (d) global assessment of the distribution of aerosol concentrations, inventory of specific aerosol types, and construction of advanced aerosol models for use in the climate simulations. The foregoing discussions have indicated which of the optical parameters are needed.

The instruments for the necessary experiments are now highly developed. For measurement at different places in the world, however, a certain standardization and intercalibration is needed in order to arrive at consistent data sets. For this reason workshops, such as the First International Workshop on Light Absorption by Aerosol Particles organized by H. Gerber, Naval Research Laboratory, and E. Hindman, Colorado State University, at Ft. Collins in August 1980, are absolutely necessary and extremely valuable.

Experimental validation of numerically simulated effects will only be possible at certain locations. The validation requires a variety of simultaneous and complementary measurements of aerosol parameters and radiation fluxes in order to achieve an "overdetermined" system which allows fundamental conclusions to be reached on the validity of the assumptions and algorithms. Attempts in this direction have been made in the Soviet CEANEX experiment (45), during GATE (15), and in the southwestern United States (46), and more efforts are needed. It must be reiterated that experimental and operational monitoring stations have to be installed in order to obtain representative samples all over the world and to assess possible changes in aerosol types and trends in their concentrations in the most efficient way.

It is most likely that for the time being the monitoring of stratospheric aerosols will remain in a much more advanced state than the monitoring of tropospheric aerosols. Satellite measurements are providing a detailed picture of the distribution of aerosols in and above the upper troposphere (47). Comparison and validation programs with coordinated satellite, lidar, and balloon investigations have provided the basis for a quantitative evaluation of the data (48,49,50). Satellite measurements may continue to be the only unique data source to monitor aerosols over long enough periods of time and with the internal consistency necessary to detect variations in the climate system as far as aerosols and related temperatures in the stratosphere are concerned. The continuation of this monitoring system is therefore a major requirement for climate research.

To implement a complementary monitoring system with adequate coverage in the troposphere is extremely difficult. It seems to be necessary to improve on the present techniques and to fill in the existing gaps in such a network by special experimental campaigns made at selected times in selected locations (51). It will be the task of the Joint Scientific Committee for the World Climate Research Program to implement such activities according to specifications emerging from the scientific community.

ACKNOWLEDGMENTS

This contribution is an updated verson of an address presented at the First International Workshop on Light Absorption by Aerosol Particles. Some facts have been incorporated which emerged from more recent publications or were presented at the CAS-Radiation Commission meeting of experts on aerosols and climate in Geneva, November 1980. Some of the results discussed here by J. Joseph, M. P. McCormick, and J. B. Pollack are still in the process of publication (see references). The careful typing of the manuscript by Miss Barbara Schmidinger is gratefully acknowledged.

REFERENCES

1. Lenoble, J., ed., *Standard Procedures to Compute Atmospheric Radiative Transfer in a Scattering Atmosphere*, IAMAP-Radiation Commission, Boulder (1977).
2. Fouquart, Y., Irvine, W. M., and Lenoble, J., eds. *Standard Procedures to Compute Atmospheric Radiative Transfer in a Scattering Atmosphere, Vol. II*, IAMAP-Radiation Commission, Boulder (1980).
3. Van de Hulst, H. C., *Multiple Light Scattering, Vols. I and II*, Academic Press, New York (1980).
4. *Report on Aerosol and Climate*, WMO-ICSU, Geneva (1981).
5. McClatchey, R., ed., *A Cloudless Standard Atmosphere for Radiation Computations*, Radiation Commission-IAMAP, Boulder, 1981.
6. Hänel, G., and Bullrich, K., Physico-chemical property models of tropospheric aerosol particles, *Beitr. z. Phys. d. Atm. 51*, 129-138 (1978).
7. Deirmendjian, D., *Electromagnetic Scattering on Spherical Polydispersions*, Elsevier, New York (1969).
8. Toon, O., and Pollack, J., A global average model of atmospheric aerosols for radiative transfer calculations, *J. Appl. Met. 15*, 225-246 (1976).
9. Braslau, N., and Dave, J. V., Effect of aerosols on the transfer of solar energy through realistic model atmospheres. Part II: Partly-absorbing aerosols, *J. Appl. Met. 12*, 616-619 (1973).
10. Ohring, G., The effect of aerosols on the temperatures of a zonal average climate model, *PAGEOPH 117*, 851-864 (1979).
11. Sagan, C., and Pollack, J. B., Anisotropic nonconservative scattering and the clouds of Venus, *J. Geophys. Res. 72*, 469-477 (1967).
12. Charlock, T. P., and Sellers, W. D., Aerosol effects on climate calculation with time-dependent and steady-state radiative convective models, *J. Atmos. Sci. 37*, 1327-1341 (1980).
13. Wang, W.-C., and Domoto, G. A., The radiative effect of aerosols in the earth's atmosphere, *J. Appl. Met. 13*, 521-534 (1974).

14. Reck, R. A., Influence of airborne particles on the earth's radiation balance, *GARP Publ. Series No. 22*, 947-973 (1979).
15. Carlson, T. N., and Benjamin, S. G., Radiative heating rates for Saharan dust, *J. Atmos. Sci. 37*, 193-213 (1980).
16. Carlson, T. N., and Caverly, R. S., Radiative characteristics of Saharan dust at solar wavelengths, *J. Geophys. Res. 82*, 3141-3152 (1977).
17. Levin, Z., Joseph, J. H., and Mekler, Y., Properties of Sharav (Khamsin) dust—comparison of optical and direct sampling data, *J. Atmos. Sci. 37*, 881-891 (1980).
18. Yamamoto, G., and Tanaka, M., Increase of global albedo due to air pollution, *J. Atmos. Sci. 29*, 1405-1412 (1971).
19. Shettle, E. P., and Fenn, R. W., Models of the atmospheric aerosols and their optical properties, *AGARD Conf. Proc. No. 183, Optical Propagation in the Atmosphere*, 45 pp. (1976).
20. Manabe, S., and Wetherald, R. T., Thermal equilibrium of the atmospere with a given distribution of relative humidity, *J. Atmos. Sci. 24*, 241-259 (1967).
21. Patterson, E. M., Gillette, D. A., and Stockton, B. H., Complex index of refraction between 300 and 700 nm for Saharan aerosols, *J. Geophys. Res. 82*, 3153-3160 (1977).
22. Sellers, W. D., A new global climatic model, *J. Appl. Met. 12*, 241-254 (1973).
23. Chylek, J. P., and Coakley, J. A., Aerosols and climate, *Science, 183*, 75-77 (1974).
24. Coakley, J. P., and Chylek, P., The two-stream approximation in radiative transfer, including the angle of incident radiation, *J. Atmos. Sci. 32*, 409-418 (1975).
25. Hansen, J. E., Lacis, A., Lee, P., and Wang, W., Climate effects of atmospheric aerosols, *Conf. on Aerosols: Urban and Rural Characteristics Source and Transport Studies*, New York Academy of Science (1979).
26. Temkin, R. L., and Snell, F. M., An annually zonally averaged hemispherical climate model with diffuse cloudiness feedback, *J. Atmos. Sci. 33*, 1671-1635 (1976).
27. Reck, R., Aerosols in the atmosphere: Calculation of the critical absorption backscatter ratio, *Science, 186*, 1034-1036 (1974).
28. Reck, R., Influence of surface albedo on the change in the atmospheric radiation balance due to aerosols, *Atmos. Environ. 8*, 823-833 (1974).
29. Rasool, S. I., and Schneider, S. H., Atmospheric carbon dioxide and aerosols: Effects of large increases on global climate, *Science, 29*, 138-141 (1971).
30. Harshvardhan, Perturbation of the zonal radiation balance by a stratospheric aerosol layer, *J. Atmos. Sci. 36*, 1274-1285 (1979).

31. Palmer, K. F., and Williams, D., Optical constants of sulfuric acid; application for the clouds of Venus?, *Appl. Optics, 14*, 208-219 (1975).
32. Pollack, J. B., Toon, O. B., Sagan, C., Summers, A., Balduin, B., and VanCamp, W., Stratospheric aerosols and climate change, *Nature, 263*, 551-555 (1976).
33. Pollack, J. B., Toon, O. B., and Wiedman, D., Radiative properties of the background stratospheric aerosols and implications for perturbed conditions, *Geophys. Res. Lett. 8*, 26-28 (1981).
34. Hansen, J. E., Wang, W. C., and Lacis, A. A., Mount Agung provides test of a global climate perturbation, *Science, 199*, 1065-1068 (1978).
35. Twomey, S., The influence of pollution on the shortwave albedo of clouds, *J. Atmos. Sci. 34*, 1149-1152 (1977).
36. Joseph, J. H., The effect of desert aerosol on a model of the general circulation, *Proc. of the Symposium on Radiation in the Atmosphere*, (H.-J. Bolle, ed.), Science Press, 487-492 (1977).
37. Pollack, J. B., and Cuzzi, J. N., Scattering by nonspherical particles of size comparable to a wavelength: A new semi-empirical theory and its application to tropospheric aerosols, *J. Atmos. Sci. 37*, 868-881 (1980).
38. Maegley, W. J., Saltation and Martian sandstorms, *Rev. Geophys. and Space Phys. 14*, 135-142 (1976).
39. Turco, R. P., Hamill, P., Toon, O. B., Witten, R. C., and Kiang, C. S., A one-dimensional model describing aerosol formatiion and evolution in the stratosphere: I: Physical processes and mathematical analogs, *J. Atmos. Sci. 36*, 699-717 (1979).
40. Friend, J. P., Leifer, R., and Triehon, M., On the formation of stratospheric aerosols, *J. Atmos. Sci. 30*, 465-479 (1973).
41. Peterson, J. T., and Junge, C. E., Sources of particulate matter in the atmosphere, *Man's Impact on the Climate* (W. H. Matthews, W. W. Kellogg, and G. D. Robinson, eds.), pp. 310-320, MIT Press, Cambridge, Massachusetts (1971).
42. SMIC, *Inadvertent Climate Modifications*, MIT Press, Cambridge, Massachusetts (1981).
43. Twomey, S., *Atmospheric Aerosols*, Elsevier Sci. Publ., Amsterdam (1977).
44. Dittberner, G. J., Climatic change: Volcanoes, man-made pollution and carbon dioxide, *IEEE Trans. on Geoscience Electronics, GE-16*, 50-61 (1978).
45. Kondratyev, K. Y., The complete atmospheric energetics experiment, *GARP Publ. Series No. 12*, Geneva (1978).
46. DeLuisi, J. J., Furukawa, P. M., Gillette, D. A., Schuster, B. G., Charlson, R. J., Porch, W. M., Fegley, R. W., Herman, B. M., Rabinoff, R. A., Fwitty, J. T., and Weiman, J. A., Results of a comprehensive atmospheric aerosol-radiation experiment in the southwestern United

States. Part I: Size distribution, extinction optical depth and vertical profiles of aerosols suspended in the atmosphere, *J. Appl. Met. 15*, 441-454 (1976); Part II: Radiation flux measurements and theoretical interpretation, *J. Appl. Met. 15*, 455-463 (1976).
47. McCormick, M. P., Hamill, P., Pepin, T. J., Chu, W. C., Swissler, T. J., and McMaster, L. R., Satellite studies of the stratospheric aerosol, *Bull. Am. Met. Soc. 60*, 1038-1046 (1979).
48. Russell, P. B., Swissler, T. J., McCormick, M. P., Chu, W. P., Livingston, J. M., and Pepin, T. J., Satellite and correlative measurements of the stratospheric aerosol: I. An optical model for data conversions, *J. Atmos. Sci. 38*, 1279-1294 (1981).
49. Russell, P. B., McCormick, M. P., Swissler, T. J., Chu, W. P., Livingston, J. M., Fuller, W. H. Rosen, J. M., Hofmann, D. J., McMaster, L. R., Woods, D. C., and Pepin, T. J., Satellite and correlative measurements of the stratospheric aerosol: II. Comparison of measurements made by SAM II, dentzondes, and an airborne lidar, *J. Atmos. Sci. 38*, 1295-1312 (1981).
50. *Global Monitoring of the Environment for Selected Atmospheric Constituents,* Environmental Data and Information Service, National Climatic Center, Asheville (1978). [A joint publication of the World Meteorological Organization, the Environmental Protection Agency, and the US Department of Commerce/NOAA, in cooperation with the United Nations Environment Program.]
51. Pollack, J. B., McCormick, M. P., eds. Aircraft and spacecraft measurements of stratospheric aerosols and their implications, *Geophys. Res. Lett. 8*, 1-28 (1981).

ABSORPTION OF LIGHT BY ATMOSPHERIC AEROSOL PARTICLES: REVIEW OF INSTRUMENTATION AND MEASUREMENTS

H. E. Gerber

Naval Research Laboratory
Washington, DC

A review is given of the technology for measuring light absorption by aerosol particles in the visible spectrum and in the infrared. Measurement techniques are described and evaluated for usefulness and accuracy. Atmospheric measurements made with the techniques are briefly summarized, and recommendations are given for improving instrumentation and the knowledge of light abosrption by atmospheric aerosols.

[1]*This work was supported by Naval Ocean Systems Center, Project No. ZF59551002.*

I. INTRODUCTION

Light absorption by atmospheric aerosol particles has generated substantial interest for several reasons. The pollution of the atmosphere with aerosol particles is one of the mechanisms by which man may be changing global climate. Whether the change is one of cooling or warming depends in part on the ratio of light absorption to scattering by the particles (e.g., see 1-3). Direct atmospheric measurements have shown that this absorption cannot be neglected, since it can range from 5% to 20% of the solar constant (4-7). Evidence that man is contributing to the absorption comes from observations that urban aerosols absorb visible light most strongly (8,9), and from the identification of graphitic-like carbon (soot), a by-product of man's burning of fossil fuels, as a major and ubiquitous absorber of light (e.g., 10). Another potential effect is the absorption of infrared radiation by suspended soil particles (11) which may become more important because of excessive land use in semiarid regions. Other environmental influences include the effect of absorbing particles on atmospheric visual range (12), and the behavior of soot particles as catalysts for SO_2 oxidation (13). Much interest also exists in modeling radiative transfer in the atmosphere. The preceding optical effects as well as many others are modeled with Mie theory, which requires knowledge of the particles' absorption and other optical properties (e.g., see 14,15).

Historically, Waldram (16) appears to have been first in demonstrating the importance of light absorption by aerosol particles. However, it was not until about 10 years ago, with the appearance of the SCEP (17) and SMIC (18) reports which drew attention to the possible relationship between pollution and climate, that a large amount of interest developed in this field. Since then a great deal of work has been done and important knowledge has been gained. Still, major uncertainties remain; whether anthropogenic aerosols influence climate is still not known, modeling absorption with Mie theory includes undetermined errors, and measurements of light absorption with various techniques are often of unknown accuracy. These and other remaining problems are a result of factors described succinctly by Toon et al. (19). They note that atmospheric aerosol particles are complex optically because of their chemical inhomogeneity, and also because the particles vary in space and time. This causes substantial difficulty in measuring absorption and other optical properties of aerosols and in calculating optical effects in the atmosphere.

This review concentrates on describing the various types of techniques for measuring light absorption by aerosol particles. An attempt is made to judge the accuracy of each type. A selection of imaginary-index values measured with the techniques is shown for one wavelength of light in the visible, and brief discussions are given of the wavelength dependence of absorbers and of the composition of absorbing particles. Also discussed are the relationships between various light absorption properties of the particles, as well as the meaning of "effective" optical properties.

II. LIGHT ABSORPTION PROPERTIES

Light absorption properties often used in the literature are n_2, n_2/ϱ, σ_a, σ_a/M_v, and $\tilde{\omega}$, where the first two are properties of the aerosol particles and the others are properties of the aerosols (particles and atmosphere).

In Mie theory the imaginary index n_2 determines the damping of the light wave incident on the particle. The value of n_2 depends only on the chemical composition of the particles, and it is related to the absorption coefficient k of bulk material with the same composition by

$$n_2 = \frac{k\lambda}{4\pi} \tag{1}$$

where λ is the wavelength of light and k is found from the exponential light-attenuation law (Bouger's, Beer's, or Lambert's law):

$$i/i_o = \exp(-kx) \tag{2}$$

where i_o is the light intensity incident on material of thickness x, and i is the transmitted intensity.

The complex refractive index is given by

$$n = n_1 - in_2 \tag{3}$$

where $i = -1^{1/2}$, and n_1 is the real index which corresponds to the velocity of the light wave in the particle. [An alternate but seldom applied approach is to use the complex dielectric function for the particles instead of n, which proves useful in locating resonant absorption bands (20).]

Normalizing n_2 by the density ϱ of the particle results in the mass absorption coefficient n_2/ϱ.

The aerosol absorption coefficient σ_a is the fraction of energy in a parallel beam of light absorbed per unit length of aerosol, and the total fractional loss of light per unit aerosol length is the extinction coefficient

$$\sigma_e = \sigma_a + \sigma_s \tag{4}$$

where σ_s is the aerosol scattering coefficient.

The transmittance of light over an atmospheric path of length L under the conditions of negligible multiple and molecular scattering also has the form of the exponential attenuation law:

$$I/I_o = \exp(-\sigma_e L) \tag{5}$$

When all scattering is negligible or has been subtracted from the extinction, Eq. (4) reduces to

$$I/I_o = \exp(-\sigma_a L) \quad (6)$$

Normalizing σ_a to the mass M of aerosol particles per volume V of atmosphere gives the specific absorption coefficient σ_a/M_v, where the specific mass $M_v = M/V$.

The single-scattering albedo $\tilde{\omega}$ is equal to $\sigma_s/\sigma_e = 1 - \sigma_a/\sigma_e$.

The aerosol coefficients and indices are related with Mie theory. For example

$$\sigma_a = \int \pi r^2 Q_a(r,\lambda,n) \, N(r) \, dr \quad (7)$$

where r is the particle radius, N(r) is the particle size distribution, and $Q_a(r,\lambda,n)$ is the Mie efficiency factor for absorption. Expressions similar to Eq. (7) exist for σ_s and σ_e which also require n as inputs. The same holds for the equations which describe the angular dependence of scattered light. The Mie expressions are normally applied by assuming that aerosol particles consist of homogeneous spheres. [An introduction to the use of Mie theory is given in (21) and a thorough treatment appears in (22).]

III. MEASUREMENT TECHNIQUES

Techniques of measuring light absorption properties of aerosol and aerosol particles can be divided into four categories, as shown in Fig. 1. Opposite each sketch is a partial list of the investigators who have either contributed to the development of the technique or have made use of it.

1. *Particle Sample*

In this category a sample of ambient aerosol particles is collected by filtration or impaction and analyzed in the laboratory by a variety of means to obtain σ_a and/or n_2. Volz (23-25) and others (26-29) applied the potassium bromide (KBr) pellet technique to particle samples [this technique is commonly used in analytical chemistry for infrared spectroscopy of powders (30)]. In this technique the particle sample is mixed with finely ground KBr and compressed under high pressure to form a transparent disk or pellet through which light transmission is measured as a function of wavelength. Volz (24,25) used Eqs. (1) and (2) to determine n_2.

Fischer (8,31,32) measured light transmission through a layer of aerosol particles of thickness x collected by impaction on a glass substrate. He used wavelengths of light covering the solar spectrum, and he also extended his technique into the infrared by using an AgCl substrate (33). He applied Eq. (2)

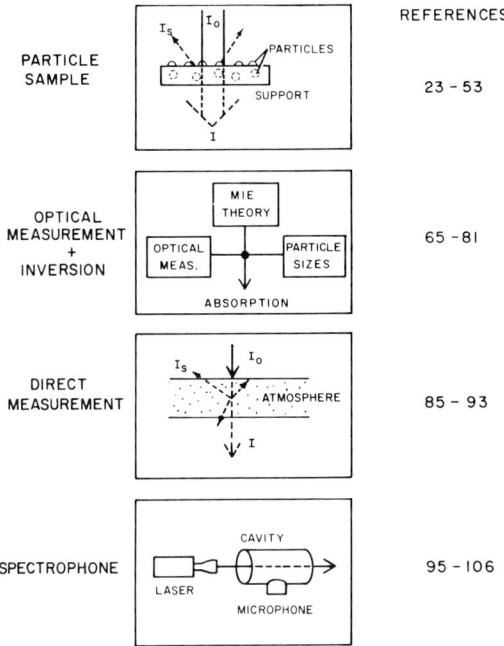

FIGURE 1. Sketches of the four major experimental categories of measuring light absorption by aerosol particles (see text for details).

after the transmission measurement was corrected for the light scattered by the particle layer which was measured in an integrating sphere. The difficult measurement of the layer thickness x was not necessary, because x was replaced by the expression

$$x = M/\varrho A \qquad (8)$$

where M, the mean density ϱ of the particles in the layer, and the area A of the deposit were measurable quantities. Combining Eqs. (1), (2) and (8) gave values of n_2/ϱ. A similar use of an integrating sphere was reported for measurement of $\tilde{\omega}$ on various laboratory-produced particles (34).

Lin et al. (35) developed the "integrating plate method," which resembled Fischer's technique in that light transmission was measured through a particle deposit. It differed in that only a tenuous particle deposit was collected on a transparent Nuclepore filter, and a light-diffusing opal-glass plate was used instead of the integrating sphere. They also approached the analysis of the transmission measurements differently by assuming that the absorption coefficient of the particles on the filter deposit was equivalent to σ_a of the atmosphere. This was consistent with the use of Eq. (6) to calculate σ_a from filter transmission measurements, where

$$L = Ft/A = V/A \tag{9}$$

where F is the filter flow rate, t is the sampling time, A is the filter deposit area, and V is the aerosol volume. The practical nature of this technique has resulted in numerous measurements (9,35,36), and its use simultaneously with an integrating nephelometer (37) gave values of $\tilde{\omega}$ (9,35).

Variations of the preceding techniques include (a) using an integrating sphere similar to the one of Fischer (8) for which corrections were made for the partially specular reflection of the inner wall, and where Eq. (6) was applied to the particle layer to determine σ_a (38); (b) using a tenuous filter deposit of particles with an integrating sphere (39); and (c) measuring absorption by particles collected on a filter in such a way as to minimize the interference from light scattering (40).

Another variation, termed the "laser transmission method", was developed by Rosen (10) and has been used extensively (41-44). The method consists of measuring light transmission through particles collected on light-diffusing filters (e.g., Millipore and quartz-fiber filters). The filter side coated with particles is placed facing the incident light beam (HeNe laser) and a wide-angle lens on the other side of the filter focuses the diffusively transmitted light onto a sensor. Eq. (6) is used to determine σ_a.

An entirely different approach for determining the absorption in the visible and near infrared uses the Kubelka-Munk theory for diffuse reflectance of light by a layer of powder. This phenomenological theory is well known and has been described in detail (45-47). It was applied to collections of aerosol particles by Lindberg and Laude (48), Patterson et al. (49), and others (29,50-53). The theory relates the reflectance from the powder to the light absorption and scattering within the powder. The absorption is given in terms of the Kubelka-Munk absorption coefficient k_M which is approximately equal to k from which n_2 is found [Eq. (1)]. In practice a sample of particles is usually diluted with a nonabsorbing material such as $BaSO_4$, since the theory works best for weakly absorbing powders, and the mixture better resembles the required isotropically scattering medium. The size of the aerosol particles has some effect on the relationship between k_M and k, which usually differs by not more than a factor of 2 (48).

The particle sample techniques as a whole have an important advantage over other techniques in that they are the most practical and inexpensive to use. However, the price for this convenience has been the difficulty in interpreting their measurements. One problem is the hygroscopic nature of atmospheric particles which causes them to rapidly increase in size at greater than approximately 65% relative humidity. None of the described techniques can be used for particle deposits which are part liquid, therefore nearly all reported measurements are for dried particle samples. It is known, however, that the absorption properties of atmospheric particles are strongly dependent on the liquid content of the particles; hence the particle sample techniques give erroneous results unless corrections are applied for the loss of water. Hänel (54) [Eq. (13)] gives a correction for n_2 in the visible as a function of relative humidity. It is common to find in the literature no mention of relative humidity with absorption measurements. In those cases values of n_2 can be much too large, and $\tilde{\omega}$ too small.

Other potential difficulties include multiple scattering among the particles on the deposit which enhances absorption. Lin et al. (35) estimated that this effect was negligible for their arrangement; however, it may become important for a denser deposit used in Fischer's technique (8,31,32,38). Andre et al. (38), on the other hand, suggested that the strong inhomogeneity caused by soot in the aerosol samples reduces the multiple-scattering influence, as demonstrated by the agreement between their measured and calculated values of σ_a. Scattering of light by the particles, substrates, and light diffusers must be properly accounted for in all techniques in order to use Eqs. (2) or (6); this scattered light is not clearly accounted for in descriptions of some of the techniques (e.g., 10,35). Misinterpretation of the measurements is possible when the particles are illuminated with diffuse light, since the definition of σ_a requires parallel incident light.

Greater problems exist for measurements in the infrared than in the visible as summarized by Jennings et al. (55). Large uncertainties are possible in the values of n_2 and σ_a determined with the KBr technique (24) and with other transmission methods (32,56). This is because σ_a and n_2 depend more strongly on n_1 which varies more than in the visible, and because the size distribution of the particles is hard to specify since the grinding process for the KBr technique changes the distribution as well as other physical properties of the particles (57). Still, some evidence exists that useful measurements are possible when the particles meet the condition that $r \ll \lambda$ (55,58).

Bergstrom (59), Toon et al. (19), and Jennings et al. (55) pointed out that values of n_2 derived by Eq. (1) from measurements of k (e.g., 8,24,48,56) can be ambiguous, since Eq. (1) is only strictly valid for a plane light wave in a medium with uniform optical properties, whereas atmospheric particles are inhomogeneous to an unknown degree. The possible dependence of the inhomogeneity on particle size and location in thicker particle deposits is an additional complication. They thus concluded that the use of n_2 obtained from Eq. (1) led to large uncertainties in Mie calculations of aerosol optical properties.

Some information is available in the literature to estimate the magnitude of the errors in the particle sample techniques. The integrating plate technique (35) and the laser transmission method (10) were used simultaneously for ambient measurements (36). The latter gave values of σ_a which were larger than the former by a factor of about 2.5. The laser transmission method was also compared to a particle spectrophone (see Section III-4), and nearly perfect agreement was found (41). On the other hand, an *in situ* spectrophone was compared with the integrating plate method (35), which gave values larger by a factor of 1.3 or 2.4 depending on whether the side of the filter with the particle deposit faced the light source or the opal glass plate (60). Lindberg and Laude (48) gave a factor of 2 as the uncertainty in the value of the bulk absorption coefficient k measured with the diffuse-reflectance technique. This suggests that errors in measured values of σ_a or k will generally be less than about a factor of 2 to 3 in the visible.

Errors in n_2 appear to be larger. Bergstrom (59) estimates from model calculations that particle inhomogeneity results in values of n_2 which cause uncertainties in aerosol optical properties of about a factor of 5 to 10. Patterson (53) compares the variability in measurements of n_2 in the visible for soil particles analyzed with various techniques (9,29,49,61,62); the differences are somewhat less than the smaller factor given by Bergstrom (59), perhaps because the particles have chemical similarities. In the infrared a similar comparison (53) with different techniques (25,33,56,63) for soil particles shows a variability of about twice that in the visible. Those results suggest that measurements of n_2 can be in error by a factor of about 5 to 10, although, increased inhomogeneity of the particles may well increase that factor.

2. *Optical Measurement and Inversion*

In this second group of techniques the absorption properties are not measured directly, but are inferred from a combination of other optical measurements on the aerosols, information on the size distribution of the particles, and Mie theory. The techniques attempt to solve for n_2 which is found in the integral that sums the optical effects of these particles according to Mie theory. A mathematical inversion of the integral is necessary to find n_2 (see 64).

The particular optical measurement of the aerosols and the means of determining the particle size distribution vary with the investigator. Eiden (65,66) measured the ellipticity of sunlight polarized by the aerosols, and he assumed various particle size distributions. Grams et al. (62,67) measured the scattered-light phase function of linearly polarized light with a polar nephelometer for airborne soil particles sized with an impactor, and Hansen (68) and Hansen and Evans (69) inverted several elements of the scattering matrix measured with a polar nephelometer to determine n_2 as well as the particle size distribution. The polar nephelometer approach was also used for various types of single particles suspended electrostatically; the measured phase functions were compared to tabulated functions to determine n (70-72). Ward et al. (73) used bistatic laser scattering, aureole measurements, and one model size distribution. Carlson and Caverly (61) inverted model radiative transfer calculations to fit measured spectral optical depths for different atmospheric pathlengths and measured size distributions to arrive at values of σ_a and n_2 for Saharan

REVIEW OF INSTRUMENTATION AND MEASUREMENTS 29

dust. The sensitivity of the ratio of direct to diffusively scattered solar radiation to light absorption was noted by Robinson (74), and was developed theoretically by Herman et al. (75) and King and Herman (76) into the "diffuse-direct radiation method," in which observed ratios and particle size distributions are inverted to yield n_2. Ambient measurements of this ratio were made and inverted with assumed size distributions (77,78) and with distributions themselves inverted from spectral optical depths (79). Spinhirne et al. (80) used multizenith-angle lidar measurements of the extinction-to-backscatter ratio of atmospheric aerosols, and Reagan et al. (81) added a bistatic lidar in similar measurements; particle size distributions were found by inverting spectral optical depths with the method of King et al. (82).

Except for the nephelometers which give point estimates of n_2, the techniques in this section give remote measurements of n_2 which are the mean values of that parameter over a path in the atmosphere. Given that optical phenomena over long distances are often of importance, and that local sources and atmospheric inhomogeneities make the extrapolation of the point measurements difficult, these techniques have a definite advantage.

Since Mie theory is involved in the inversions for these techniques it becomes necessary to specify the shape and uniformity of the particles. In most of the methods described, the assumption was made that the atmospheric particles consisted of chemically homogeneous spheres. Because nonspherical particles are common in the atmosphere and behave optically different than spheres (e.g., 83,84), and particle inhomogeneity is the norm, the results obtained with the techniques described in this section are open to question. The omission of particle shape and inhomogeneity from Mie theory has not been purposeful because of their intractable influence on the theory and their unknown distribution in the atmosphere. In view of those difficulties n_2 has been termed an "effective imaginary index" (68,75,81), which is defined as the index of the hypothetical spherical homogeneous particles which gives a match between the modeled performance of a technique and the atmospheric optical measurement made with the technique. The applicability of the effective imaginary index to radiative transfer calculations has not been made clear in the literature.

3. *Direct Measurement*

The techniques in this category fall into two subgroups. The first uses aerosol extinction and scattering to yield σ_a and $\tilde{\omega}$ [Eq. (4)], while the second, termed the "flux-divergence method," directly measures the radiation budget in an atmospheric layer. In the latter the net radiative flux (the difference between the upward and downward hemispheric fluxes) is measured at two levels which define the atmospheric layer. The difference in the net flux between the levels results in the total absorption in the layer, and by subtracting molecular absorption from the total the contribution by the particles is obtained.

Extinction and scattering were measured in an aerosol-filled sphere with an omnidirectional light at its center and nonreflecting inner walls (85); this method required corrections for small particles (86). Similar but improved methods were developed by Ebersole (87) for the infrared and by Gerber (88) for the visible where transmission measurements were made through an aerosol-filled cell while an integrating nephelometer placed in a cell wall simultanteously determined σ_a; inversions gave n_2. Direct measurements of σ_a and σ_s included those of Waldram (16) who used aircraft telephotometer observations of a ground target to get σ_e and vertical profiling with a polar nephelometer to get σ_s. Russell et al. (89) made direct measurements of $\tilde{\omega}$ from an instrumented tower, and inverted optical depths to obtain size distributions which were used with $\tilde{\omega}$ to deduce a value of n_2 weighted for the solar spectrum. Major (90) found a mean absorption by particles of 15% of the solar constant from comparison of surface radiometer measurements and albedo data from the Nimbus-3 satellite.

The *in situ* flux divergence measurements were preceded by the measurements of Robinson (74) who measured diffuse and direct solar fluxes on the ground and used the phase functions measured by Waldram (16) to estimate absorption. In the direct measurements, radiometers mounted on aircraft were used to make vertical profiles of the net hemispheric fluxes to obtain the total energy absorbed by the particles (4-7,77,91-93).

The techniques in this section have the advantage of giving *in situ* values of aerosol light absorption; however, they all require differencing large and sometimes nearly equivalent quantities to arrive at absorption, so that large errors can result. In cases where absorption is significant, reasonable accuracies were obtained (e.g., see 7,94).

4. Spectrophone

The techniques in this section rely on the heating of aerosol particles by their absorption of pulsed light energy. The periodic expansion of the gas around the particles due to heat conduction from the particles generates a sound wave detected with a sensitive microphone. The amplitude of the sound wave is proportional to the energy absorbed by the particles which for dry particles is related to σ_a [a detailed derivation of the photoacoustic response for aerosols is given by Roessler and Faxvog (95)]. The laser has often been used as the light source (96-99), and a broadband source simulating solar radiation has also been used (100). While the preceding techniques provide *in situ* measurements by passing the aerosols through the acoustic cavity of the spectrophone, other approaches have used the spectrophone principal for particles dispersed in KBr (56) and collected on a filter (41).

REVIEW OF INSTRUMENTATION AND MEASUREMENTS 31

Uses of spectrophones have included measurements in the infrared on various combustion particles (95,98,99), diesel particles (101,102), and atmospheric dusts (56,103). Atmospheric measurements in the visible appear to have been limited to those of Foot (100). The measurements of Bruce et al. (103) at a CO_2-laser wavelength are of special interest, because they compare the spectrophone value of σ_a with Mie calculations of σ_a using dust with measured size distributions and identified absorption peaks from the analysis of the dust's composition; the difference between measured and observed values of σ_a was a factor of approximately 5. They also used the CO_2 spectrophone with simultaneous transmission measurements through smoke to determine σ_s which is difficult to measure for particles in the infrared (104).

Large light energies are required to obtain measurements when particles are moderate or weak absorbers. This raises the possibility that for *in situ* spectrophone techniques the volatile components lost by heating the particles cause errors in the interpretation of the acoustic signal (97,105). This possibility has been exploited by Campillo et al. (106) who deduced the infrared light absorption by moist particles by measuring the amplitude modulation of visible light scattered by the particles which are irradiated with modulated CO_2-laser light.

IV. IMAGINARY INDEX IN VISIBLE SPECTRUM

Many attempts have been made to measure n_2 in estimating the influence of aerosol particles on climatic change. A selection of n_2 values measured with the techniques described in the preceding section is given in Fig. 2 for ambient particles for a wavelength of light near 550 nm.

Immediately apparent from Fig. 2 (and Table 1) is the variability in n_2, which is more than two orders of magnitude, causing difficulty in choosing representative values for modeling. The value of $\tilde{\omega}$ suggested by Hansen et al. (111) to separate global heating from cooling is the critical value $\tilde{\omega}_o = 0.85$, which for typical atmospheric particle size distributions gives a value of $n_2 \approx 0.017$ [Eq. (16)]. In Fig. 2 $n_2 = 0.017$ separates the total number of n_2 values into two nearly equal groups. This demonstrates the problem of assessing the effect on climate, over which there has been much disagreement in the past 10 years [e.g., Jaenicke (112) and Bach (113)].

The choice of representative values of n_2 is complicated by the difficulty of separating the two factors which contribute to the variability. The first factor is the natural variation of absorbing particles in the atmosphere, as demonstrated by the vertical lines about some of the data points in Fig. 2, and the second is measurement error which may contribute a substantial variation to n_2, as shown in Section III.

In an attempt to specify values of n_2 which are useful for modeling, the values of n_2 were arranged into several broad categories as shown in Fig. 2. The urban values were numerous enough to permit their separation into several other categories: Values of $n_2 < 0.01$ (Nos. 1-5) were plotted first in

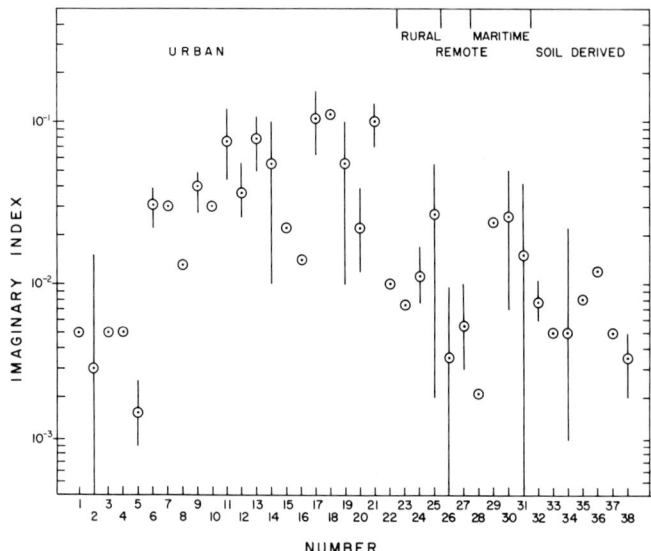

FIGURE 2. *Imaginary part of the refractive index of atmospheric aerosol particles for visible light with a wavelength near 550 nm. The points give the mean, and the vertical lines give the range of the measurements, except in one case (No. 2) where they give the measurement uncertainty (see Table 1 for a numerical listing of the indices, and further explanation).*

Fig. 2 and were not included in the calculation of an average value $\bar{n}_2$ because Nos. 1-4 required a different interpretation (see Section VII), and 5 obtained with a spectrophone by Foot (100) appeared unrealistically small. Numbers 8-13 were obtained with filter transmission measurements, Nos. 14-18 with diffuse reflectance, and the rest with miscellaneous techniques. The largest urban n_2 values are generally for measurements using the particle sample techniques (Nos. 8-18). A possible explanation for these large values may be that most resulted from measurements made on dried particles which under average atmospheric conditions consisted partly of water and thus had lower values of n_2. A decrease in n_2 to compensate for the lost water was estimated from Eq. (13) to be a factor of 2 for a relative humidity of 70% (54). This correction was applied to all urban n_2 values from No. 8 to No. 18 with the exception of those where the correction was already included (No. 10) and those (Nos. 16 and 18) from normally dry areas. Given those changes the average urban value of $\bar{n}_2$ is 0.036 ($\bar{\omega} \approx 0.75$). Correcting to RH = 0% gives $\bar{n}_2 \approx 0.05$; averaging all urban

Table 1. Key to Figure 2.

No.	Reference	Index	Comments
1	Ward et al. (73)	0.005	Bistatic lidar, aureole; Gainesville
2	Spinhirne et al. (80)	0.003	Multizenith angle lidar; Tucson
3	Reagan et al. (81)	0.005	Bistatic-monostatic lidar; Tucson
4	Hansen (68)	0.005	Polar nephelometer; Tucson
5	Foot (100)	0.0017^a	Spectrophone; London suburb
6	King (79)	0.0306	Diffuse-direct; Tucson
7	DeLuisi et al. (78)	0.03	Diffuse-direct; Denver "brown cloud"
8	Fischer (8)	0.013	Particle layer; Mainz; RH = 35%
9	Lin et al. (35)	0.04	Filter and integrating plate; NYC
10	Hänel (54)	0.03	Particle layer; Nainz; RH = 70%
11	Weiss et al. (9)	0.076	Filter and integrating plate; US
12	Rosen et al. (43)	0.036^b	Filter and laser transmission; Anaheim
13	Sadler et al. (36)	0.078^b	Filter and integrating plate; Seattle
14	Andre et al. (38)	0.055	Particle layer; Frankfurt
15	Patterson (107)	0.022	Diffuse reflectance; Denver "brown cloud"
16	Lindberg (108)	0.014	Diffuse reflectance; Yuma
17	Lindberg et al. (109)	0.104	Diffuse reflectance; three European cities
18	Levin and Lindberg (29)	0.110	Diffuse reflectnce; Tel Aviv
19	Eiden (65)	0.055	Sunlight polarization; Mainz; RH = 70%
20	Murai et al. (7)	0.022^c	Flux divergence; Tokyo
21	Gerber (94)	0.10^a	σ_e and σ_s; Rota and Athens
22	Russell et al. (89)	0.01	$\tilde{\omega}$ and size distribution; San Francisco
23	Fischer (8)	0.0074	Particle layer; Jungfraujoch
24	DeLuisi et al. (77)	0.011	Flux divergence; southwestern US
25	Weiss et al. (9)	0.027	Filter and integrating plate; US
26	Weiss et al. (9)	0.0035^a	Filter and integrating plate; southwestern US & Mauna Loa
27	Rahn (personal communication)	0.0055^a	Various methods; Arctic
28	Fischer (8)	0.0021	Particle layer; Mace Head; RH = 35%
29	Hänel (54)	0.024	Particle layer; N. Atlantic, RH = 70%
30	Gerber (94)	0.026^a	σ_e and σ_s; N. Atlantic near Europe
31	Gerber (94)	0.015^a	σ_e and σ_s; Mediterranean
32	Lindberg and Laude (48)	0.0077	Diffuse reflectance; White Sands
33	Kondratyev et al. (6)	0.005^d	Flux divergence: Central Asia
34	Grams et al. (62)	0.005	Polar nephelometer; southwestern US
35	Patterson et al. (49)	0.008	Diffuse reflectance; Saharan dust
36	Carlson and Caverly (61)	0.012	Spectral optical depths; Saharan dust
37	Lindberg (108)	0.005	Diffuse reflectance; Negev desert (high visibility)
38	Patterson (52)	0.0035	Diffuse reflectance; Mt. St. Helens ash

[a] Estimated from Eq. (16) with measured values of $\tilde{\omega}$.
[b] Estimated from Eq. (10) where the ratio (0.15) of particulate carbon mass to the total mass M_v is from Macias et al. (110).
[c] Measured values of δ gave σ_e, and values of total light absorption in atmospheric layers gave σ_a, from which $\tilde{\omega}$ was calculated and used with Eq. (16) to get n_2. The measured absorption was decreased by 30% to account for the absorption of diffuse light [a value estimated from Herman et al. (75)] and by an additional 10% to account for the upwelling radiation.
[d] Downward hemispheric fluxes as a function of altitude were used to estimate δ and σ_e, the absorbed light was used for σ_a, and Eq. (16) was used for n_2. Since δ is underestimated and σ_a is overestimated by about the same amount due to the contribution of diffusively scattered light, $\tilde{\omega}$ remains about the same.

values shown in Fig. 2 gives $n_2 = 0.038$. For the other regions in Fig. 2 the averages (including humidity corrections to RH = 70% where appropriate) are 0.011 for rural, 0.0036 for remote, 0.17 for maritime, and 0.0071 for soil-derived (No. 38 not included). The maritime value $\bar{n}_2$ is too large for background aerosol, since the n_2 values for Nos. 30 and 31, and probably also for No. 29, were influenced by aerosols from continental areas.

These n_2 values are rough estimates at best, since the lack of data prevents proper weighting by geographical area; and the natural variability of n_2 values about $\bar{n}_2$ is at least one order of magnitude. Other factors to consider are that most n_2 values were obtained from measurements near the ground in Europe or the United States and that most were measured at one wavelength and at an unknown RH.

V. COMPOSITION AND WAVELENGTH DEPENDENCE OF ABSORBERS

1. *Visible Spectrum*

Absorbers in the visible can be separated into four main groups, each of which has a different effect on light absorption: The strongest absorber consists of the soot fraction of carbonaceous particles which results from the combustion of fossil fuels (see reviews in Refs. 114,115), soil or crustal particles absorb light moderately, salt particles form the bulk of the aerosol particle mass but have negligible absorption, and water particles are also nonabsorbers. An additional group of slightly absorbing particles for which little is known consists of natural and anthropogenic organic material other than soot. These various materials have different source regions; e.g., soot originates in populated areas, soil particles in arid regions, NaCl over oceans, and $(NH_4)_2SO_4$ in areas that produce SO_2. The materials are mixed in a complicated way by atmospheric processes which create aerosols with a large variability in their absorption properties.

Atmospheric soot particles are generated primarily by diesel vehicles, with residential wood fires and gas- and oil-fired furnaces also contributing (116,117). These particles contain an absorbing fraction identified by Raman spectroscopy (see Fig. 3), and by solvent extraction and heat treatment of particle samples as a graphitic form of carbon. The soot particles generated during the combustion process consist of primary carbon particles a few hundreths of a micron in diameter which coagulate into branched chain aggregates a few tenths of a micron in diameter [see Whitby (118) for a review]. Electron microscope analyses of particles in polluted atmospheres showed carbonaceous particles predominantly of the aggregate form (119,120).

As a first approximation soot particles can be considered to act as purely light-absorbing particles. Although the typical soot aggregate size of approximately 0.2 μm is found in the accumulation mode (118) where other particles

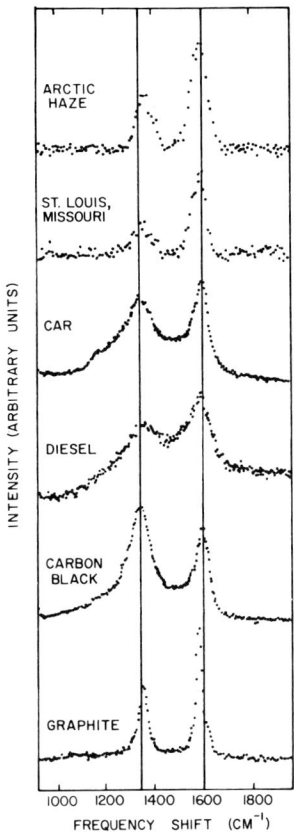

FIGURE 3. Raman spectra of Arctic haze and urban particulates, various source emissions, carbon black, and polycrystalline graphite. All demonstrate the two intense Raman modes characteristic of graphitic carbon [from Rosen et al. (44), by courtesy of the authors].

scatter light significantly, the much smaller size of the primary particles gives an optical equivalent size of the aggregate which is smaller than the physical aggregate size. The optical size depends on the degree of aggregation; Dalzell et al. (121) found the optical size to be approximately one third the size of highly aggregated soot particles, while Roessler and Faxvog (122) found that the optical size was closer to the size of the primary particles for slightly aggregated soot. Further evidence on the predominantly absorbing nature of soot is found in the measurements of the specific-extinction (122) and -absorption (60,102) coefficients (diesel and other soot particles) which consistently showed values between 8 and 10 $m^2 \ g^{-1}$. These values agree with Mie calculations of the specific extinction coefficient when the mean optical size (carbon spheres) is less than 0.1 μm (122). In view of those factors the value of $\tilde{\omega} \approx 0.5$ given by Twitty and Weinman (123) for atmospheric soot particles is probably too large. A more likely value of ≈ 0.2 was estimated from scattering and absorption cross sections given by Faxvog and Roessler (124).

The wavelength dependence of soot is found in absorption measurements on urban particles which contain substantial amounts of soot. In Fig. 4, taken from a review of optics of crustal aerosols by Patterson (53), a Denver aerosol sample shows n_2 approximately independent of wavelength. Measurements by Andre et al. (38) near Frankfurt show a similar independence for the small-particle fraction and unactivated particles. Those observations are consistent with the work of Twitty and Weinman (123) where n_2 was nearly independent of λ for a variety of graphitic materials.

The lower curve in Fig. 4, for soil-derived particles in Denver, is typical in shape and magnitude of others for crustal particles from a variety of global locations (53). Iron oxides are thought to be the absorbing component in these particles, which give the red-brown coloration to particle deposits from arid and semiarid regions.

The transparent nature of water and of salt particles commonly found in the atmosphere was established with laboratory measurements on pure bulk samples of those substances. Huffman (20) and Toon et al. (19) summarize techniques for such measurements, and a comprehensive listing of the pertinent n_2 values is available in the literature.

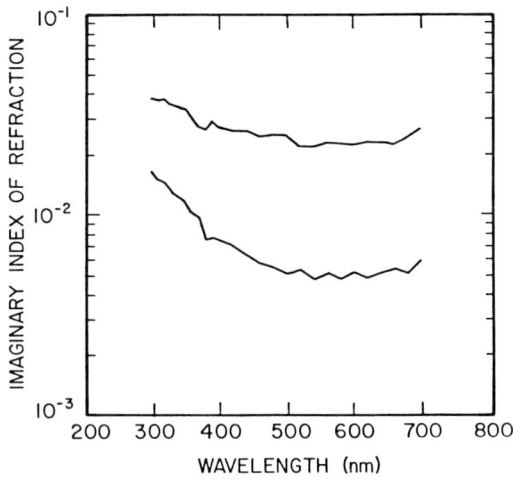

FIGURE 4. Spectral variation of the imaginary index of Denver aerosol particles measured with the diffuse-reflectance technique. The upper curve represents aerosol containing a large amount of soot, and the lower curve is for particles of crustal origin [from Patterson (53), by courtesy of author].

2. Infrared

Infrared transmission spectra reveal numerous absorption bands, most of which have been identified using measured spectra of pure bulk materials. Some of the bands, found in the 3- to 5-μm and 8- to 12-μm atmospheric windows, are strong enough to play an important role in the global longwave radiation balance (125). Strong absorbers in polluted areas are $(NH_4)_2SO_4$ and organic particles (31), while elsewhere soil-derived particles predominate. Numerous efforts have been made to identify the infrared absorption bands of soil-derived particles (e.g., Refs. 24,27-29,58,126). The work of Hoidale and Blanco (27), which identified the major absorbing constituents of typical desert dust as quartz, clay minerals (kaolinite, illite, and montmorillonite), and calcite, has been applied in many subsequent efforts (see Fig. 5). Carlson and Benjamin (127) summarize the published values of n_2 in the infrared and visible for crustal aerosol.

VI. RELATIONSHIPS BETWEEN n_2, σ_a, σ_s, $\tilde{\omega}$

Waggoner et al. (128) derived the approximation formula

$$n_2 = \frac{\sigma_a \varrho}{M_v f(n_1, \lambda)} \tag{10}$$

by evaluating Eq. (7) for typical atmospheric particle size distributions using the assumption of homogeneous spherical particles. This equation is consistent with Mie theory which predicts an approximately linear relationship between a particle's absorption cross section and the product of n_2 times the particle's volume, under the condition that the Mie size parameter, $\alpha = 2\pi r/\lambda$, is sufficiently small.

Numerical values for $\lambda/f(n_1)$ were calculated by Lin et al. (35); and an analytical expression

$$f(n_1, \lambda) = \frac{12\pi n_1}{\lambda(n_1^2 + 2)} \tag{11}$$

was derived by Hänel and Dlugi (129), who also showed that Eq. (10) predicted values of n_2 with an error of less than 40% as compared to complete Mie calculations with Eq. (7) (for $\alpha \lesssim 10$ and $\lambda = 550$ nm).

Weiss et al. (9) gave a slightly different version of $f(n_1, \lambda)$:

$$f(n_1, \lambda) = \frac{36\pi n_1}{\lambda(n_1^2 + 2)^2} \tag{12}$$

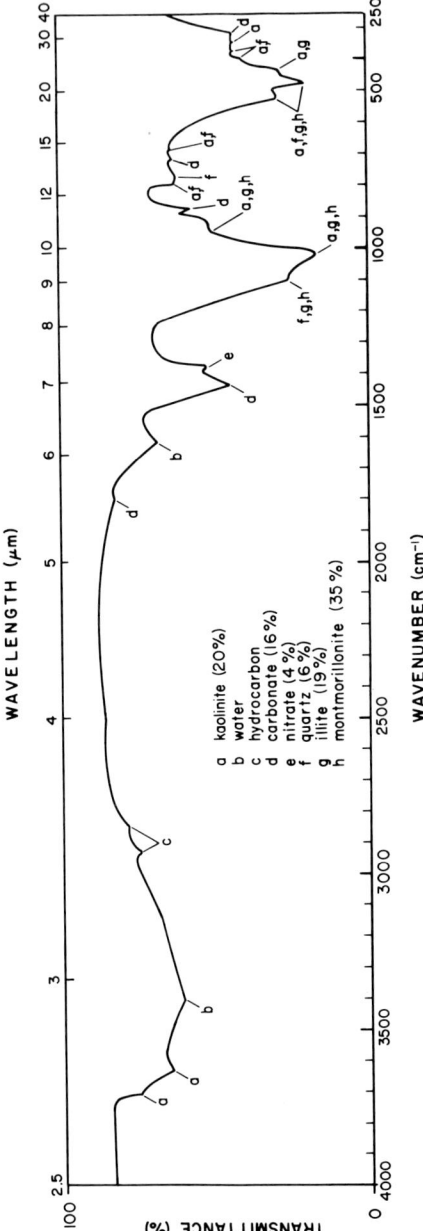

FIGURE 5. Typical infrared transmittance spectrum of atmospheric dust from an interior desert basin in Southwest United States measured with the potassium-bromide-pellet technique [from Hoidale and Blanco (27), by courtesy of the authors].

Hänel (54) gave the formula

$$n_2' = n_w + (n_2 - n_w) [1 + f\bar{\varrho}\bar{\mu}/(1 - f)]^{-1} \qquad (13)$$

to correct for the decrease of the imaginary index for increasing relative humidity f; n_2' is the corrected index, n_w is the imaginary index of water, and $\bar{\mu}$ is the linear mass increase coefficient which is tabulated (54) for aerosols of various regions. Values of n_2 and ϱ in Eq. (10) can be used with Eq. (13) under the condition that n_2 and ϱ correspond to dried particles.

Combining Eqs. (6), and (9) to (11) gives

$$n_2 = -\frac{\ln(I/I_0)\lambda\varrho A(n_1^2 + 2)}{12\pi M} \qquad (14)$$

which has been used with measurements of σ_a obtained with the integrating plate method of Lin et al. (35) to estimate n_2, and combining Eqs. (1), (2), and (8) gives

$$n_2 = -\frac{\ln(i/i_0)\lambda\varrho A}{4\pi M} \qquad (15)$$

for the technique used by Fischer (8) and others to determine k and n_2. Given that filter light transmission is measured in both techniques so that $I = i$, $I_0 = i_0$, and $\sigma_a L = kx$ [Eqs. (2) and (6)], Eqs. (14) and (15) differ only by a factor of $(n_1^2 + 2)/3$ which is between 1.25 and 1.5 for atmospheric particles in the visible. This leads to the conclusions that both techniques measure light absorption in essentially the same way, that the assumption by Fischer (8) of an isotropically homogeneous absorbing medium for the particle layer on the filter is nearly the same as considering the collected particles as discrete absorbers and homogeneous spheres (35) except for the above factor, and that literature values of the absorption coefficient and n_2 for both techniques should be comparable.

The relationship between n_2 and $\tilde{\omega}$ was estimated by calculating each parameter with Mie theory (homogeneous spheres, $\lambda = 632.8$ nm) for a large number of ambient particle size distributions [Gerber (94)]. The calculations shown in Fig. 6 agree well with measured values of $\tilde{\omega}$ for which n_2 was estimated with Mie theory for remote, rural, urban, and maritime aerosols. The curves in Fig. 6 demonstrate that $1 - \tilde{\omega}$ is a strong function of n_2 and relatively independent of n_1 and the particle size distribution, i.e.,

$$1 - \tilde{\omega} \approx f(n_2) \qquad (16)$$

where $f(n_2) = 1.47(n_2 + 0.001)^{0.152} - 0.627$ for the solid curve in Fig. 6 and for $0.005 < n_2 < 0.2$. Jennings et al. (130) reached similar conclusions concerning

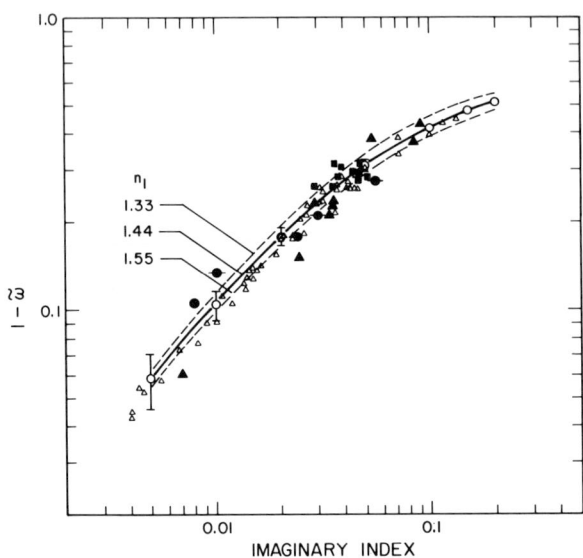

FIGURE 6. The single scattering albedo $\tilde{\omega}$ of ambient aerosols as a function of the aerosol particles' effective imaginary index n_2, real index n_1, and size distributions. The curves give mean relationships between $\tilde{\omega}$ and n_2 calculated with Mie theory ($\lambda = 632.8$ nm) for a large number of observed size distributions; and the vertical lines give twice the standard deviation of the variation in the curves due to differences in the distributions (94). The data points are from Gerber (94) ($\triangle$) for maritime aerosols near Europe ($\lambda = 632.8$ nm, $n_1 = 1.55$), from Lin et al. (35) ($\blacksquare$) for urban aerosols, from Weiss et al. (9) ($\blacktriangle$) for urban, rural, and remote aerosols (average values), and from Hänel (54) for urban ($\bullet\!\!\!-$) and maritime ($\bullet$) aerosols. The three sets of data points for Hänel (54) with increasing values of the imaginary index are for relative humidities of 90%, 70%, and 0%, respectively.

$\tilde{\omega}$ and n_2 by using atmospheric model size distributions in their calculations. The results in Fig. 6 suggest that measurements of σ_a with the particle sample and other techniques can be used with Eqs. (10) and (16) to obtain $\tilde{\omega}$ directly with a reasonable degree of accuracy for many atmospheric conditions. It is of interest to note that measurements of the total carbon content of aerosols in urban and possibly also rural areas can also be related to $\tilde{\omega}$ (with less precision, however), because Rosen et al. (43) and Sadler et al. (36) found a strong correlation between σ_a and the carbon content.

The curves in Fig. 6 also give information on the relationship between σ_s and M_v. This relationship is significant because both quantities are easily measured and the ratio σ_s/M_v for ambient aerosols has been shown to be relatively constant (e.g., 131,132). The ratio is found by combining $\sigma_e = \sigma_a + \sigma_s$ with Eqs. (10) and (16):

$$\frac{\sigma_s}{M_v} \approx \frac{n_2 f(n_1,\lambda)}{\varrho} [1/f(n_2) - 1] \qquad (17)$$

Evaluating Eq. (17) for $n_2 = 0.02$, $n_1 = 1.44$, $\lambda = 550$ nm, and $\varrho = 1.5$ g cm^{-3} gives $\sigma_s/M_v = 1.38$ m^2 g^{-1}. This value is less than the nearly constant value of $\sigma_s/M_v = 3.13$ m^2 g^{-1} found by Waggoner and Weiss (132) in many different locations. The difference may be due to the fact that Eq. (17) corresponds to all particles in ambient size distributions, whereas Waggoner and Weiss (132) made their measurements on the dried small particle fraction (r < 1.25 μm).

The relationships between the optical parameters for visible light discussed in this section do not hold well in the infrared, as can be seen from Jennings et al. (55,130). However, under certain limited conditions, relationships of the nature of Eqs. (1) and (10) may give useful results (53), because the alternative, which consists of identifying the absorbing species and applying bulk optical constants, measuring size distributions, and using Mie theory, does not appear to give better results (103).

VII. EFFECTIVE OPTICAL PROPERTIES

Each time the assumption is made that aerosol particles consist of homogeneous spheres, n_2 must be termed an effective imaginary index because of the inhomogeneity of ambient particles. An important question must be answered: Does one effective n_2 cover all possibilities or must more be defined for different types of theoretical calculations or measurement techniques? A first attempt to answer that question is made in the following discussion.

One such effective n_2 can be defined when n_2 and σ_a are related as in Eqs. (7) or (10). Under the condition that n_2 is determined with those equations from measurements of σ_a, n_2 is simply a parameterization of σ_a and can be used in Mie calculations when σ_a is involved. This n_2 can be termed the effective imaginary index for absorption n_2^a. The only requirement on n_2^a is that it has a monotonic relationship with σ_a, which is the case in the visible (130). The imaginary index given in Fig. 6 is also n_2^a, because it results from the application of Mie theory to measurements of σ_a or $\tilde{\omega}$. The σ_e part of $\tilde{\omega}$ is approximately independent of n_2 for all reasonable size distributions and values of n_2 found in the atmosphere. This well known property of σ_e in the visible is clearly demonstrated by Jennings et al. (130). Thus the criticism by Bergstrom (59) and others of measurements obtained with the particle sample techniques (Section III-1) must exclude those instances when n_2^a is estimated with those techniques and is used in Mie calculations involving σ_a and $\tilde{\omega}$ in the visible.

However, n_2^a cannot be used in calculations of the scattering properties of the particles, such as the angular dependence of the scattered light (phase function). Given that soot particles, which are the strongest absorbers in the atmosphere in the visible, scatter only a small fraction of the total scattered light (Section V-1), the shape of the relative phase function must be dominated by scattering from the much larger number of other particles in the atmosphere which consist of material with relatively low or negligible values of n_2. Thus, to a first approximation, phase functions in the atmosphere will appear to be for transparent particles. This will apply best in unpolluted areas and in source regions of soot (urban areas) where the soot particles are probably physically separate from the nonabsorbing particles; following the homogenizing processes of particle coagulation and cycling through clouds the soot particles will be part of complex mixtures which behave optically differently (133). For calculations or measurements dealing with the relative phase function, an effective imaginary index for scattering n_2^s can be defined which reflects the much lower values of n_2 for soil, salt, and water particles. Values of n_2^s cannot be used to calculate σ_a or $\tilde{\omega}$ unless the soot content is negligible, in which case $n_2^a \approx n_2^s$. Generally, $n_2^s < n_2^a$ in rural areas, and $n_2^s \ll n_2^a$ in urban areas.

The insensitivity of the relative phase function to strongly absorbing soot particles provides a possible explanation for the low values of n_2 given by some investigators (68,73,80,81) for urban aerosols (see Fig. 2). Hansen (68) measured phase functions so that he in effect determined n_2^s. The others used lidars to measure a portion of the phase function (as well as other quantities) to deduce n_2. It is only possible to speculate that the lidar techniques also measured n_2^s, and that they also demonstrated in their measurements that the urban phase function corresponds to nearly transparent particles. Perhaps the lidar techniques will require their own definition of an effective n_2. The evaluation of lidar techniques with aerosols containing soot has been suggested (81).

A simple qualitative argument can be given to demonstrate that the diffuse-direct method (75) gives n_2^a approximately. The starting point is a curve for the diffuse-direct ratio S/F given by Herman et al. (75) for a small solar zenith angle ($\theta = 15°$) and surface albedo (A = 0.1), and a constant value of the atmospheric optical depth ($\delta = 0.25$) [see Fig. 3 in (75)]. Assuming that S/F is correct for $n_2 = 0$, soot with negligible scattering is added to the atmosphere and S/F is recalculated for n_2^a. Since the curve in (75) is for a constant δ, the addition of soot will increase σ_a at the expense of σ_s, and $\tilde{\omega}$ is determined. The type of scattering particles can remain the same, but their numbers must be decreased to keep δ constant. With the approximation that the diffuse light is attenuated by the same amount as the direct light (which is a reasonable first approximation, since most of the diffuse light comes from scattering angles less than 45°), the addition of soot results in an S/F curve for n_2^a [n_2^a is determined from $\tilde{\omega}$ in Fig. 6] which is nearly the same as the one given in Fig. 3 of (75) for homogeneous spheres.

REVIEW OF INSTRUMENTATION AND MEASUREMENTS

VIII. CONCLUSIONS AND SUGGESTED FURTHER WORK

All techniques for measuring light absorption by aerosol particles were found to have potentially large errors which, however, were smaller than the natural variability of the light absorption properties of the aerosols, so that useful measurements can presently be made. The errors were estimated to be a factor of 2 or 3 in the visible and about three times larger in the infrared.

The particle sample techniques are the most practical and inexpensive to use, and they give useful measurements of σ_a, $\tilde{\omega}$, and n_2^a, provided that a humidity correction is made and that the light scattered by the particles and the substrates is properly accounted for. The assumption that the deposit of collected particles behaves as an isotropic homogeneous absorbing medium does not greatly influence the accuracy of those measurements if multiple-scattering effects are negligible, nor of the use of n_2^a for Mie theory calculations of σ_a and $\tilde{\omega}$. Remote sensing techniques appear to underestimate urban values of n_2^a because they use inversions of the scattered-light phase function which is nearly independent of the atmospheric soot content.

Research is suggested in the following areas to improve measurement techniques:

1. The contribution of light scattered by the particle deposit and substrates in the particle sample techniques is not well understood and must be explained.

2. The accuracy of absorption properties inferred from infrared transmission measurements on particle samples should be better established.

3. The optical measurement and inversion techniques should be reevaluated as a function of particle inhomogeneity caused by soot.

4. An atmospheric comparison between the particle sample, optical measurement and inversion, and direct measurement techniques is recommended.

5. Desirable goals in instrumentation development are to arrive at simple *in situ* and satellite techniques for measuring light absorption.

Important progress was made in the last 10 years in the identification of the composition, optical properties, and source regions of light-absorbing particles. Soot contributed by man and crustal particles are the most important species in the troposphere. Areas in which additional progress is required include the following:

1. Due to the natural variability of absorbing particles and the lack and sporadic nature of light absorption measurements, it is difficult to assess the role of absorbers in large-scale radiative transfer calculations. A representative global measurement network is necessary.

2. Except for the work of Waldram (16) and Murai et al. (7), the vertical distribution of absorbing particles has not been investigated. The contribution of strong absorbers generated at the earth's surface to stratospheric aerosols is unknown.

3. Work is required on the size dependence of absorbing particles, which was shown to be of importance by Lindberg and Gillespie (134), and Andre et al. (38).

4. The spatial distribution and life cycle of optically important soot particles is not well known.

5. The physical relationship of soot particles to the more numerous nonabsorbing atmospheric particles must be determined since the optics of the mixture depends on that relationship (133). Careful electron microscopy on ambient particles is needed.

6. Measurements of the light-absorption properties of ambient aerosols should be accompanied at the very least by measurements of σ_s, M_v, and relative humidity.

ACKNOWLEDGMENTS

This paper is an expanded version of a presentation given at the General Motors symposium "Particulate Carbon: Atmospheric Life Cycle," 12-14 October 1980, Warren, Michigan. Helpful discussions with Gottfried Hänel on the particle sample techniques and with Rudi Husar on atmospheric soot are gratefullly acknowledged.

REFERENCES

1. Ensor, D., Porch, W., Pilat, M., and Charlson, R., Influence of the atmospheric aerosol on albedo, *J. Appl. Met. 10*, 1303-1306 (1971).
2. Mitchell, J. M., The effect of aerosols on climate with special reference to temperature near the earth's surface, *J. Appl. Met. 10*, 703-714 (1971).
3. Chylek, P., and Coakley, J., Aerosols and climate, *Science, 183*, 75-77 (1974).
4. Roach, W. T., Some aircraft observations of fluxes of solar radiation in the atmosphere, *Quart. J. Roy. Met. Soc. 87*, 346-363 (1961).
5. Drummond, A. J., and Robinson, G. D., Some measurements of the attenuation of solar radiation during BOMEX, *Appl. Opt. 13*, 487-492 (1974).
6. Kondratyev, K. Y., Vassilyev, O. B., Grishechkin, V. S., and Ivlev, L. S., Spectral radiative flux divergence and its variability in the troposphere in the 0.4-2.4 μm region, *Appl. Opt. 13*, 478-486 (1974).
7. Murai, K., Kobayashi, M., Yamauchi, T., and Goto, R., The absorption of solar radiation in the lower atmosphere, *Papers Met. Geophys. 27*, 21-32 (1976).
8. Fischer, K., Mass absorption coefficient of natural aerosol particles in the 0.4-2.4 μm wavelength interval, *Beit. Phys. Atmos. 46*, 89-100 (1973).

9. Weiss, R. E., Waggoner, A. P., Charlson, R. J., Thorsell, D. L., Hall, J. S., and Riley, L. A., Studies of the optical, physical, and chemical properties of light absorbing aerosols, *Proc. Conf. Carbonaceous Particles in the Atmosphere* (T. Novakov, ed.), pp. 257-262, Report LDL-9037, Lawrence Berkeley Laboratory, Berkeley, California (1979).
10. Rosen, H., Hansen, A. D. A., Gundel, L., and Novakov, T., Identification of the optically absorbing component in urban aerosols, *Appl. Opt. 17*, 3859-3861 (1978).
11. Russell, P. B., and Grams, G. W., Application of soil dust optical properties in analytical models of climate change, *J. Appl. Met. 14*, 1037-1043 (1975).
12. Roessler, D. M., and Faxvog, F. R., Visibility in absorbing aerosols, *Atmos. Environ. 15*, 151-155 (1981).
13. Novakov, T., Chang, S. C., and Harker, A. B., Sulfates as pollution particulates: Catalytic formation on carbon (soot) particles, *Science, 186*, 259-261 (1974).
14. Toon, O. B., and Pollack, J. B., A global average model of atmospheric aerosols for radiative transfer calculations, *J. Appl. Met. 15*, 225-246 (1976).
15. Shettle, E. P., and Fenn, R. W., Models for the aerosols of the lower atmosphere and the effects of humidity variations on their optical properties, *Environmental Research Paper No. 676,* AFGL-TR-79-0214, Air Force Geophysics Laboratory, Hanscom AFB, Massachusetts, 94 pages (1979).
16. Waldram, J. M., Measurements of the photometric properties of the upper atmosphere, *Quart. J. Roy. Met. Soc. 71,* 319-336 (1945).
17. *Study of Critical Environmental Problems, Man's Impact on the Global Environment (SCEP),* MIT Press, Cambridge, Massachusetts, 319 pp (1970).
18. *Study of Man's Impact on Climate (SMIC),* MIT Press, Cambridge, Massachusetts, 308 pp (1971).
19. Toon, O. B., Pollack, J. B., and Khare, B. N., The optical constants of several atmospheric aerosol species: Ammonium sulphate, aluminum oxide, and sodium chloride, *J. Geophys. Res. 81,* 5733-5748 (1976).
20. Huffman, D. R., Intersteller grains. The interaction of light with a small-particle system, *Adv. Phys. 26,* 129-230 (1977).
21. Fenn, R. W., Optical properties of aerosols, *Handbook on Aerosols* (R. Dennis, ed.), pp 66-92, Report TID-26608, U.S. Energy Research and Development Administration (1976).
22. van deHulst, H. C., *Light Scattering by Small Particles,* Wiley, New York, 470 pp (1957).
23. Volz, F. E., Infrared refractive index of atmospheric aerosol substances, *Appl. Opt. 11,* 255-259 (1972).

24. Volz, F. E., Infrared absorption by atmospheric aerosol substances, *J. Geophys. Res.* 77, 1017-1031 (1972).
25. Volz, F. E., Infrared optical constants of ammonium sulphate, Saharan dust, volcanic pumice, and fly ash, *Appl. Opt.* 12, 564-568 (1973).
26. Blanco, A. J., and Hoidale, G. B., Microspectrophometric technique for obtaining the infrared spectrum of microgram quantities of atmospheric dust, *Atmos. Environ.* 2, 327-330 (1969).
27. Hoidale, G. B., and Blanco, A. J., Infrared absorption spectra of atmospheric dust over an interior desert basin, *Pure Appl. Geophys.* 74, 151-164 (1969).
28. Flanigan, D. F., and DeLong, H. P., Spectral absorption characteristics of the major components of dust clouds, *Appl. Opt.* 10, 51-57 (1971).
29. Levin, Z., and Lindberg, J. D., Size distribution, chemical composition, and optical properties of urban and desert aerosols in Israel, *J. Geophys. Res.* 84, 6941-6950 (1979).
30. Kendall, D. N., *Applied Infrared Spectroscopy*, Reinhold, New York, 560 pp (1966).
31. Fischer, K., Measurement of absorption of visible radiation by aerosol particles, *Beit. Phys. Atmos.* 46, 89-100 (1973).
32. Fischer, K., Mass absorption indices of various types of natural aerosol particles in the infrared, *Appl. Opt.* 14, 2851-2856 (1975).
33. Fischer, K., The optical constants of atmospheric aerosol particles in the 7.5-12 μm region, *Tellus,* 3, 266-274 (1976).
34. Glushko, V. N., Livshits, G. SH., and Tashenov, B. T., Investigation of true light absorption by aerosols in the 0.4-2.4 μm region, *Isz. Atmos. Ocean. Phys.* 8, 236-237 (1972).
35. Lin, C.-I., Baker, M., and Charlson, R. J., Absorption coefficient of atmospheric aerosol: A method of measurement, *Appl. Opt.* 12, 1356-1363 (1973),
36. Sadler, M., Charlson, R. J., Rosen, H., and Novakov, T., An intercomparison of the integrating plate and the laser transmission methods for determination of aerosol absorption coefficients, *Atmos. Environ.* 15, 1265-1268 (1981).
37. Charlson, R. J., Horvath, H., and Pueschel, R. F., The direct measurement of atmospheric light scattering coefficient for studies of visibility and pollution, *Atmos. Environ.* 1, 469-478 (1967).
38. Andre, K., Dlugi, R., and Schnatz, G., Absorption of visible radiation by atmospheric aerosol particles, fog and cloud water residues, *J. Atmos. Sci.* 38, 141-155 (1981).
39. Heintzenberg, J., Measurement of light absorption and elemental carbon in atmospheric aerosol samples from remote locations, *Particulate Carbon: Atmospheric Life Cycle* (G. T. Wolff and R. L. Klimisch, eds.), Plenum Press, New York (1981).

40. Twomey, S., Direct visual photometric technique for estimating absorption in collected aerosol samples, *Appl. Opt. 19,* 1740-1741 (1980).
41. Yasa, Z., Amer, N. M., Rosen, H., Hansen, A. D. A., and Novakov, T., Photoacoustic investigation of urban aerosol particles, *Appl. Opt. 18,* 2528-2530 (1979).
42. Rosen, H., Hansen, A. D. A., Gundel, L., and Novakov, T., Identification of the graphitic carbon component of source and ambient particulates by Raman spectroscopy and an optical attenuation technique, *Proc. Conf. Carbonaceous Particles in the Atmosphere* (T. Novakov, ed.) pp 49-55, Report LDL-9037, Lawrence Berkeley Laboratory, Berkeley, California (1979).
43. Rosen, H., Hansen, A. D. A., Dod, R. L., and Novakov, T., Soot in urban atmospheres: Determination by an optical absorption technique, *Science, 208,* 741-744 (1980).
44. Rosen, H., Novakov, T., and Bodhaine, B. A., Soot in the Arctic, *Atmos. Environ. 15,* 1371-1374 (1981).
45. Wendlandt, W. W., and Hecht, G. H., *Reflectance Spectroscopy,* Interscience, New York, 298 pp (1966).
46. Kortum, G., *Reflectance Spectroscopy,* Springer-Verlag, New York, 366 pp (1969).
47. Egan, W. G., and Hilgeman, T. W., *Optical Properties of Inhomogeneous Materials,* Academic Press, New York, 235 pp (1979).
48. Lindberg, J. D., and Laude, L. S., Measurement of the absorption coefficient of atmospheric dust, *Appl. Opt. 13,* 1923-1927 (1974).
49. Patterson, E. M., Gillette, D. A., and Stockton, B. H., Complex index of refraction between 300 and 700 nm for Saharan aerosols, *J. Geophys. Res. 82,* 3153-3160 (1977).
50. Lindberg, J. D., Absorption coefficient of atmospheric dust and other strongly absorbing powders: An Improvement on the method of measurement, *Appl. Opt. 14,* 2813-2815 (1975).
51. Levin, Z., Joseph, J. H., and Mekler, Y., Properties of Saharan (Khamsin) dust—comparison of optical and direct sampling data, *J. Atmos. Sci. 37,* 882-891 (1980).
52. Patterson, E. M., Measurements of the imaginary part of the refractive index between 300 and 700 nanometers for Mount St. Helens ash, *Science, 211,* 836-838 (1981).
53. Patterson, E. M., Optical properties of the crustal aerosol: Relation to chemical and physical characteristics, *J. Geophys. Res. 86,* 3236-3246 (1981).
54. Hänel, G., The properties of atmospheric aerosol particles as functions of the relative humidity at thermodynamic equilibrium with surrounding moist air, *Adv. Geophys. 19,* 73-188 (1976).

55. Jennings, S. G., Pinnick, R. G., and Gillespie, J. B., Relation between absorption coefficient and imaginary index of atmospheric aerosol constituents, *Appl. Opt. 18,* 1368-1371 (1979).
56. Schleusener, S. A., Lindberg, J. D., and White, K. O., Differential spectrophone measurements of the absorption of laser energy by atmospheric dust, *Appl. Opt. 14,* 2564-2565 (1975).
57. Duyckaerts, G., The infrared analysis of solid substances, a review, *Analyst, 84,* 201-214 (1959).
58. Carlon, H. R., Contributions of particle absorption to mass extinction coefficients (0.55-14 μm) of soil-derived atmospheric dusts, *Appl. Opt. 19,* 690-693 (1980).
59. Bergstrom, R. W., Comments on the estimation of aerosol absorption coefficients in the atmosphere, *Beit. Phys. Atmos. 46,* 198-202 (1973).
60. Szkarlat, A. C., and Japar, S. M., Light absorption by airborne aerosols: Comparison of integrating plate and spectrophone techniques, *Appl. Opt. 20,* 1151-1155 (1981).
61. Carlson, T. N., and Caverly, R. S., Radiative characteristics of Saharan dust at solar wavelengths, *J. Geophys. Res. 82,* 3141-3152 (1977).
62. Grams, G. W., Blifford, I. H., Jr., Gillette, D. A., and Russell, P. B., Complex index of refraction of airborne soil particles, *J. Appl. Met. 13,* 459-471 (1974).
63. Pollack, J. B., Toon, O. B., and Khare, B. N., Optical properties of some terrestrial rocks and glasses, *Icarus, 19,* 372-389 (1973).
64. Twomey, S., *Introduction to the Mathematics of Inversion in Remote Sensing and Indirect Measurements,* Elsevier, New York, 243 pp (1977).
65. Eiden, R., The elliptical polarization of light scattered by a volume of atmospheric air, *Appl. Opt. 5,* 569-575 (1966).
66. Eiden, R., Determination of the complex index of refraction of spherical aerosol particles, *Appl. Opt. 10,* 749-754 (1971).
67. Grams, G. W., Dascher, A. J., and Wyman, C. M., Laser polar nephelometer for airborne measurements of aerosol optical properties, *Opt. Eng. 14,* 85-90 (1975).
68. Hansen, M. Z., Atmospheric particulate analysis using angular light scattering, *Appl. Opt. 19,* 3441-3448 (1980).
69. Hansen, M. Z., and Evans, W. H., Polar nephelometer for atmospheric particulate studies, *Appl. Opt. 19,* 3389-3395 (1980).
70. Philips, D. T., and Wyatt, P. J., Single-particulate light-scattering measurement: Photochemical aerosols and atmospheric particulates, *Appl. Opt. 11,* 2082-2087 (1972).
71. Wyatt, P. J., Some chemical, physical, and optical properties of fly ash particles, *Appl. Opt. 19,* 975-983 (1980).
72. Pluchino, A. B., Goldberg, S. S., Dowling, J. M., and Randall, C. M., Refractive-index measurements of single micron-sized carbon particles, *Appl. Opt. 19,* 3370-3372 (1980).

73. Ward, G., Cushing, K. M., McPeters, R. D., and Green, A. E. S., Atmospheric aerosol index of refraction and size-altitude distribution from bistatic laser scattering and solar aureole measurements, *Appl. Opt. 12,* 2585-2592 (1973).
74. Robinson, G. D., Absorption of solar radiation by atmospheric aerosol as revealed by measurements at the ground, *Arch. Met. Geoph. Biokl. B9,* 19-40 (1962).
75. Herman, B. M., Browning, R. S., and DeLuisi, J. J., Determination of the effective imaginary term of the complex refractive index of atmospheric dust by remote sensing: The diffuse-direct radiation method, *J. Atmos. Sci. 32,* 918-925 (1975).
76. King, M. D., and Herman, B. J., Determination of the ground albedo and the index of absorption of atmospheric particulates by remote sensing. Part I: Theory, *J. Atmos. Sci. 36,* 163-173 (1979).
77. DeLuisi, J. J., Furakawa, P. M., Gillette, D. A., Schuster, B. G., Charlson, R. J., Portch, W. M., Fegley, R. W., Herman, B. M., Rabinoff, R. A., Twitty, J. T., and Weiman, J. A., Results of a comprehensive atmospheric aerosol-radiation experiment in the Southwestern United States. Part II: Radiation flux measurements and theoretical interpretation, *J. Appl. Met. 15,* 455-463 (1976).
78. DeLuisi, J. J., Bonelli, J. E., and Sheldon, C. E., Spectral absorption of solar radiation by the Denver brown (pollution) cloud, *Atmos. Environ. 11,* 829-836 (1977).
79. King, M. D., Determination of the ground albedo and the index of absorption of atmospheric particulates by remote sensing, Part II: Application, *J. Atmos. Sci. 36,* 1072-1083 (1979).
80. Spinhirne, J. D., Reagan, J. A., and Herman, B. M., Vertical distribution of aerosol extinction cross section and inference of aerosol imaginary index in the troposphere by lidar technique, *J. Appl. Met. 19,* 426-438 (1980).
81. Reagan, J. A., Byrne, D. M., King, M. D., Spinhirne, J. D., and Herman, B. M., Determination of the complex refractive index and size distribution of atmospheric particulates from bistatic-monostatic lidar and solar radiometer measurements, *J. Geophys. Res. 85,* 1591-1599 (1980).
82. King, M. D., Byrne, D. M., Herman, B. M., and Reagan, J. A., Aerosol size distributions obtained by inversion of spectral optical depth measurements, *J. Atmos. Sci. 35,* 2153-2167 (1978).
83. Perry, R. J., Hunt, A. J., and Huffman, D. R., Experimental determinations of Mueller scattering matrices for nonspherical particles, *Appl. Opt. 17,* 2700-2710 (1978).
84. Pollack, J. B., and Cuzzi, J. N., Scattering by nonspherical particles of size comparable to a wavelength: A new semi-empirical theory and its application to tropospheric aerosols, *J. Atmos. Sci. 37,* 868-881 (1979).

85. Wootten, N. W., A method for the measurement of true light absorption in aerosols, *Brit. J. Appl. Phys. 13,* 406-408 (1962).
86. Brinkworth, B. J., Measurement of true light absorption in aerosols, *Brit. J. Appl. Phys. 16,* 1901-1906 (1965).
87. Ebersole, J. F., Experimental investigation of battlefield aerosols pertinent to high energy laser propagation, *Report ARI-RR-115,* Aerodyne Research Inc., Bedford, Massachusetts, 25 pp (1977).
88. Gerber, H. E., Portable cell for simultaneously measuring the coefficients of light scattering and extinction for ambient aerosols, *Appl. Opt. 18,* 110-115 (1979).
89. Russell, P. B., Livingston, J. M., and Uthe, E. E., Aerosol-induced albedo change: Measurement and modelling of an incident, *J. Atmos. Sci. 36,* 1587-1608 (1979).
90. Major, G., Effect of gases, aerosols and clouds on atmospheric absorption of solar radiation, *Beit. Phys. Atmos. 49,* 216-221 (1976).
91. Reynolds, D. W., Vonder Harr, T. H., and Cox, S. T., The effect of solar radiation absorption in the tropical troposphere, *J. Appl. Met. 14,* 433-444 (1975).
92. Pueschel, R. F., and Kuhn, P. M., Infrared absorption of tropospheric aerosols: Urban and rural aerosols of Phoenix, Arizona, *J. Geophys. Res. 80,* 2960-2962 (1975).
93. Robinson, G. D., Some determinations of atmospheric absorption by measurement of solar radiation from aircraft and at the surface, *Quart. J. Roy. Met. Soc. 92,* 263-269 (1966).
94. Gerber, H. E., Absorption of 632.8 nm radiation by maritime aerosols near Europe, *J. Atmos. Sci. 36,* 2502-2512 (1966).
95. Roessler, D. M., and Faxvog, F. R., Optoacoustic measurement of optical absorption in acetylene smoke, *J. Opt. Soc. Amer. 69,* 1699-1704 (1979).
96. Bruce, C. W., and Pinnick, R. G., In situ measurements of aerosol absorption with a resonant cw laser spectrophone, *J. Appl. Opt. 16,* 1762-1765 (1977).
97. Terhune, R. W., and Anderson, J. E., Spectrophone measurements of the absorption of visible light by aerosols in the atmosphere, *Opt. Lett. 1,* 70-72 (1977).
98. Treux, T. J., and Anderson, J. E., Mass monotoring of carbonaceous aerosols with a spectrophone, *Atmos. Environ. 13,* 507-509 (1979).
99. Japar, S. M., and Killinger, D. K., Photoacoustic and absorption spectrum of airbrone carbon particulate using a tunable dye laser, *Chem. Phys. Lett. 66,* 207-209 (1979).
100. Foot, J. S., Spectrophone measurements of the absorption of solar radiation by aerosol, *Quart, J. Roy. Met. Soc. 105,* 275-283 (1979).
101. Faxvog, F. R., and Roessler, D. M., Optoacoustic measurements of diesel particulate emissions, *J. Appl. Phys. 50,* 7880-7882 (1979).

102. Japar, S. M., and Szkarlat, A. C., Measurement of diesel vehicle exhaust particulate using photoacoustic spectroscopy, *Combust. Flame Technol. 24,* 215-219 (1981).
103. Bruce, C. W., Yee, Y. P., Hinds, B. D., Brewer, R. J., Minjares, J., and Pinnick, R. G., Initial field measurements of atmospheric absorption at 9-11 μm wavelengths, *J. Appl. Met. 19,* 997-1004 (1980).
104. Bruce, C. W., Yee, Y. P., and Jennings, S. G., In situ measurement of the ratio of aerosol absorption to extinction coefficient, *Appl. Opt. 19,* 1893-1894 (1980).
105. Baker, M. B., Energy absorption by volatile atmospheric aerosol particles, *Atmos. Environ. 10,* 241-248 (1976).
106. Campillo, A. J., Dodge, C. J., and Lin, H.-B., Aerosol particle absorption specstroscopy by photothermal modulation of Mie scattered light, *Appl. Opt. 20,* 3100-3103 (1981).
107. Patterson, E. M., Optical properties of urban aerosols containing carbonaceous material, *Proc. Conf. Carbonaceous Particles in the Atmosphere* (T. Novakov, ed.), pp 247-251, Report LDL-9037, Lawrence Berkeley Laboratory, Berkeley, California (1979).
108. Lindberg, J. D., The composition and optical absorption coefficient of atmospheric particulate matter, *Opt. Quant. Electron. 7,* 131-139 (1975).
109. Lindberg, J. D., Gillespie, J. B., and Hinds, B. D., Measurements of refractive indices of atmospheric particulate matter from a variety of geographical locations, *Proc. Intern. Symp. Radiation in the Atmos., Garmisch-Partenkirchen* (H.-J. Bolle, ed.), pp 102-104, Science Press, New York (1976).
110. Macias, E. S., Delumyea, R., Chu, L.-C., Appleman, H. R., Radcliffe, C. D., and Stanley, L., The determination, specification, and behavior of particulate carbon, *Proc. Conf. Carbonaceous Particles in the Atmosphere* (T. Novakov, ed.), pp. 70-78, Report LDL-9037, Lawrence Berkeley Laboratory, Berkeley, California (1979).
111. Hansen, J. E., Lacis, A. A., Lee, P., and Wang, W. C., Climate effects of atmospheric aerosols, *Aerosols: Anthropogenic and Natural, Sources and Transport* (T. Kneit and P. Lioy, eds.), pp. 575-587, New York Academy of Sciences, New York (1979).
112. Jaenicke, R., Atmospheric aerosols and global climate, *J. Aerosol Sci. 11,* 577-588 (1980).
113. Bach, W., Global air pollution and climatic change, *Rev. Geoph. Space Phys. 14,* 429-474 (1976).
114. Novakov, T., ed., *Proc. Conf. Carbonaceous Particles in the Atmosphere,* Report LDL-9037, Lawrence Berkeley Laboratory, Berkeley, California, 283 pp (1979).
115. Wolff, G. T., and Klimish, R. L., eds., *Particulate Carbon: Atmospheric Life Cycle,* Plenum Press, New York (1981).

116. Muhlbaier, J. L., and Williams, R. L., Fireplaces, furnaces and vehicles as emission sources of particulate carbon, *Particulate Carbon: Atmospheric Life Cycle* (G. T. Wolff and R. L. Klimish, eds.), Plenum Press, New York (1981).
117. Nolan, J. L., Measurement of light absorption aerosols from combustion sources, *Proc. Conf. Carbonaceous Particles in the Atmosphere* (T. Novakov, ed.), pp 265-269, Report LDL-9037, Lawrence Berkeley Laboratory, Berkeley, California (1979).
118. Whitby, K. T., Size distribution and physical properties of combustion aerosols, *Proc. Conf. Carbonaceous Particles in the Atmosphere* (T. Novakov, ed.), pp 201-208, Report LDL-9037, Lawrence Berkeley Laboratory, Berkeley, California (1979).
119. Cartwright, J., Nagelschmidt, G., and Skidmore, J. W., The study of air pollution with the electron microscope, *Quart. J. Roy. Met. Soc. 82,* 82-86 (1959).
120. Russell, P. A., Carbonaceous particulates in the atmosphere: Illumination by electron microscopy, *Proc. Conf. Carbonaceous Particles in the Atmosphere* (T. Novakov, ed.), pp 133-140, Report LDL-9037, Lawrence Berkeley Laboratory, Berkeley, California (1979).
121. Dalzell, W. H., Williams, G. C., and Hottel, H. C., A light-scattering method for soot concentration measurements, *Combust. Flame, 14,* 161-170 (1970).
122. Roessler, D. M., and Faxvog, F. R., Optical properties of agglomerated acetylene smoke particles at 0.5145-μm and 10.6-μm wavelengths, *J. Opt. Soc. Am. 70,* 230-235 (1980).
123. Twitty, J. T., and Weinman, J. A., Radiative properties of carbonaceous aerosols, *J. Appl. Met. 10,* 725-731 (1971).
124. Faxvog, F. R., and Roessler, D. M., Carbon aerosol visibility vs particle size distribution, *Appl. Opt. 17,* 2612-2616 (1978).
125. Grassl, H., Influence of different absorbers in the window region on radiative cooling (and on surface temperature determination), *Beit. Phys. Atmos. 47,* 1-12 (1974).
126. Toon, O. B., Pollack, J. B., and Sagan, C., Physical properties of the particles composing the Martian dust storm of 1971-1972, *Icarus, 30,* 663-696 (1977).
127. Carlson, T. N., and Benjamin, S. G., Radiation heating rates for Saharan dust, *J. Atmos. Sci. 37,* 193-213 (1980).
128. Waggoner, A. P., Baker, M. B., and Charlson, R. J., Optical absorption by atmospheric aerosols, *Appl. Opt. 12,* 896 (1973).
129. Hänel, G., and Dlugi, R., Approximation for the absorption coefficient of airborne atmospheric aerosol particles in terms of measurable bulk properties, *Tellus, 29,* 75-82 (1977).
130. Jennings, S. G., Pinnick, R. G., and Auvermann, H. J., Effects of particulate complex refractive index and particle size distribution variations

on atmospheric extinction and absorption for visible through middle IR wavelengths, *Appl. Opt. 17,* 3922-3929 (1978).
131. Charlson, R. J., Atmospheric visibility related to aerosol mass concentration, *Environ. Sci. Tech. 3,* 913-918 (1969).
132. Waggoner, A. P., and Weiss, R. E., Comparison of fine particle mass concentration and light scattering extinction, *Atmos. Environ. 14,* 623-626 (1980).
133. Ackerman, T. P., and Toon, O. B., The absorption of visible radiation in atmospheres containing mixtures of absorbing and non-absorbing particles, *Appl. Opt. 20,* 3661-3667 (1981).
134. Lindberg, J. D., and Gillespie, J. B., Relationship between particle size and imaginary refractive index in atmospheric dust, *Appl. Opt. 16,* 2628-2630 (1977).

NEW TECHNIQUES IN LIGHT ABSORPTION MEASUREMENTS

Gottfried Hänel

Institute of Meteorology and Geophysics
University Frankfurt, FR Germany

A new radiometer is described which simplifies the measurement of radiation supply of solar wavelengths. Two methods for measuring the radiant energy absorbed by aerosol particles are described. A photometric technique is used for particles collected on filters, and a calorimetric technique is used for in situ *measurements. Data collected with the radiometer and the light-absorption techniques yield the heating rate of the atmosphere due to light absorption by the particles. Example measurements show substantial atmospheric termperature increases due to absorption, especially in industrial regions.*

I. INTRODUCTION

The annual mean incoming solar radiation at the top of the atmosphere and the earth's surface is fairly well known. However, estimates given by different authors of the amounts of shortwave radiation (0.29 μm to 3.5 μm wavelength) absorbed in the cloud-free atmosphere and in clouds differ considerably. The relevant numbers are 8% to 20% of the incoming solar radiation for the first effect and 3% to 11% for the second one. Because of these uncertainties, *in situ* measurements of the absorption of shortwave radiation by gases and particles are desired. In the following, measuring techniques are described allowing *in situ* measurements of the amount of shortwave radiation absorbed by aerosol and cloud particles. It is the aim of future measurements with this instrumentation to get an insight into the heating of the atmosphere due to absorption of solar radiation by particles in different weather situations and at different sites.

II. THEORY

It was the aim of this work to measure *in situ* the time rate $\dot{Q}_a$ of electromagnetic energy which is transformed into internal energy by absorption of shortwave radiation in aerosol or cloud particles. To get $\dot{Q}_a$, the energy budget of radiation processes in a small volume V of air has to be combined with the equation of radiative transfer yielding

$$\dot{Q}_a = V \bar{\sigma}_a \int_{4\pi\omega_o} N(\omega) d\omega \qquad (1)$$

where $\bar{\sigma}_a$ = mean absorption coefficient of the particles for shortwave radiation, $N(\omega)$ = radiance of shortwave radiation, and ω = solid angle (ω_o = unit solid angle). Equation (1) is valid when the energy density of shortwave radiation in the volume remains constant in the time interval considered. For illustration, $\dot{Q}_a$ is expressed as the heating rate of the air $(dT/dt)_a$ caused by the absorption of shortwave radiation in particles yielding

$$\left(\frac{dT}{dt} \right)_a = \frac{\bar{\sigma}_a}{\varrho_l c} \int_{4\pi\omega_o} N(\omega) d\omega \qquad (2)$$

where T = temperature, t = time, ϱ_l = density of air, and c_p = specific heat of air at constant pressure. In general, the absorbed radiative energy first causes an increase in the internal energy of the particles. After that this energy is partly used to heat the atmosphere and to evaporate water from the particles. Thus, $(dT/dt)_a$ is the real heating rate of the atmosphere due to absorption of shortwave radiation within the particles only when the water on the particles is in equilibrium with the water vapor in the atmosphere. Under conditions of changing relative humidity in the cloud-free atmosphere and within clouds a term must be added to Eq. (2) which considers the energy release during condensation of water vapor on the particles or the energy needed to evaporate water from the particles.

From Eq. (1), the way to determine *in situ* $\dot{Q}_a$ or $(dT/dt)_a$ is to measure the mean absorption coefficient $\bar{\sigma}_a$ of the particules in the shortwave region and the integrated shortwave radiation $\int_{4\pi\omega_0}$, i.e. the total shortwave radiation flowing from all directions toward the point of measurement. In the following, this integral will be called shortwave radiation supply.

III. MEASURING TECHNIQUES

1. *Shortwave Radiation Supply*

The perfect measurement of the shortwave radiation supply would involve pointing a photometer successively to all directions and integrating the readout of N. However, this technique is too time-consuming and too expensive for simultaneous measurements at different places on a routine basis. Therefore, a simplified measuring technique applicable for routine measurements has been developed. With this instrument, the flux densities of shortwave radiation coming from six equal ranges of the solid angle are measured separately by thermopiles covered with a quartz disk (Fig. 1). The sum of these six ranges of the solid angle is equal to 4π sr. The six flux densities measured are corrected for (a) the error (about -3.5%) arising from the fact that radiation flowing from an edge of a funnel to one of the sensors can also be determined partly by the sensor detecting the neighboring range of the solid angle, and (b) the error (about $25 \pm 5\%$) coming from the cosine response of the sensors for that part of incoming radiation not arriving perpendicularly with respect to the planes of the sensors.

To avoid large cosine errors during sunshine, the solar radiant flux density is measured separately with an Angstrom pyrheliometer, and the output of the instrument is corrected accordingly. The calibration errors are $\pm 1\%$ for the pyrheliometer and $\pm 2\%$ for the integrating radiometer. The total error by which the shortwave circum-global radiation can be measured is smaller than $\pm 7.2\%$ during sunshine when the shortwave albedo of the ground $\alpha_G = 0.4$, is smaller than $\pm 6.1\%$ during sunshine and when $\alpha_G = 0.2$, and is smaller than $\pm 8.6\%$ when there is no sunshine and $\alpha_G = 0.2$. These errors have been computed considering the calibration errors of the instruments, the distributions of the shortwave sky radiation, and the reflected shortwave radiation from walls and from the ground according to the literature and present calculations.

At present the integrating radiometer is being modified slightly by introducing additional light-diffusing surfaces to reduce considerably the cosine error.

The maximum error of the modified version will be in the order of 3% to 5%, and additional pyrheliometer measurements will be unnecessary. Moreover the geometry of the radiometer will allow measurements from an aircraft. Thus the variation of the shortwave radiation supply with height can be measured in the future. This is important, because the atmospheric heating by absorption of solar radiation by particles is dependent on height.

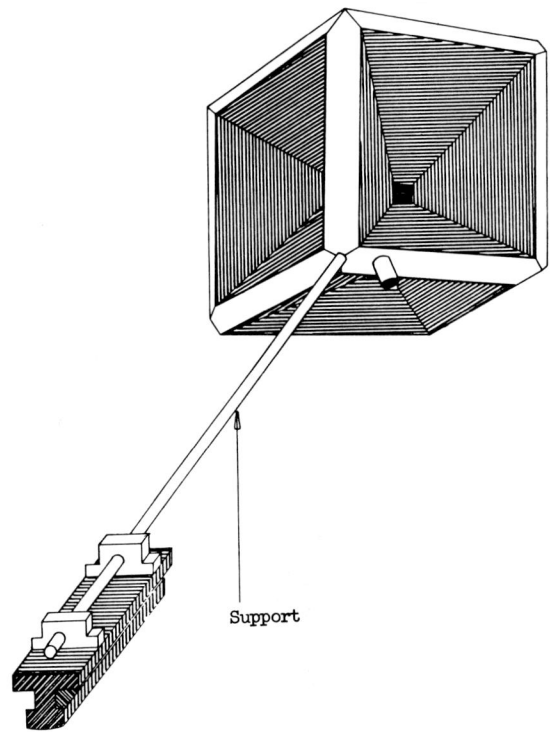

FIGURE 1. *Schematic diagram of the integrating radiometer.*

2. *Absorption Measurements*

2.1 *Photometric method for samples of particles on filters*

For the photometric absorption measurement, particles from the atmosphere are collected on a filter. In the apparatus, the particle-loaded filter is placed in front of a light diffuser so that the particles are directly irradiated with shortwave radiation from a lamp emitting radiation close to the solar radiation spectrum (Fig. 2). The flux density of the incoming radiation and the angular dependence of the radiation scattered by the particle/filter/light-diffuser system is measured with a thermopile covered with a quartz plate. For comparison purposes this measurement is repeated for the filter/light-diffuser system alone. Considering (a) the energy budget equations for both systems (i.e. the sum of forward-scattered, backward-scattered, and absorbed energies equals the energy input), and (b) the multiple reflections of radiation between the particles and the filter/light-diffuser system, the part of the flux density of the incoming parallel radiation which is absorbed by the particles is equal to

$$\frac{F_{apo}}{F_o} = 1 - \frac{F_R F_{VS} + F_o F_V}{F_{RS} F_V + F_o F_{VS}} \tag{3}$$

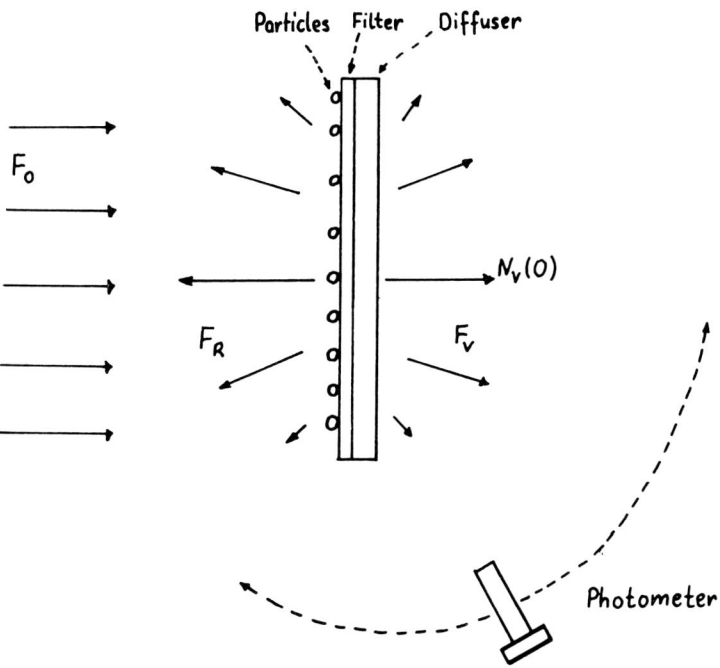

FIGURE 2. Schematic diagram of the photometric absorption measurement.

where F_{apo} = flux density of that part of incoming parallel radiation absorbed by the particles, F_o = flux density of incoming parallel radiation, F_V, F_R = flux density of the forward-scattered and backward-scattered radiation by the particle/filter/diffuser system, and F_{VS}, F_{RS} = flux density of the forward-scattered and backward-scattered radiation by the filter/diffuser system.

According to Lin et al. (1), the absorption of radiation by the particles on the filter is equalized to the absorption of radiation by the same particles in airborne state according to

$$1 - \frac{F_{apo}}{F_o} = \exp\left(-\bar{\sigma}_{ap}\frac{V}{F}\right) \tag{4}$$

where V is the volume of air sucked through the filter, F is the particle-loaded area of the filter, and V/F is the length of the pipe of air sucked through the filter during sampling. When the filter/diffuser system absorbs most of the incident radiation, $F_R \cdot F_{VS} \ll F_o \cdot F_V$ and $F_V \cdot F_{RS} \ll F_o \cdot F_{VS}$, and consequently $1 - F_{apo}/F_o = F_V/F_{VS}$. When, in addition to that, the forward scattered radiation is in both cases directly proportional to the radiance N(O) in the forward direction, then

$$1 - F_{apo}/F_o = N_V(O)/N_{VS}(O) \tag{5}$$

This formulation of $1 - F_{apo}/F_o$ is used by Lin et al. (1). Our present results show that there are cases where Eq. (5) gives negative absorption coefficients. This has not been found using Eq. (3). Therefore it is one of the goals of this work to study experimentally the conditions for which Eqs. (3) and (5) give the same results.

The mean absorption coefficient of the particles must be corrected after it is obtained according to Eqs. (3) and (4) because the wavelength dependencies of the shortwave radiation emitted by the lamp and the shortwave circum-global radiation under real atmospheric conditions are slightly different. The correction is

$$\bar{\sigma}_a(\text{atm}) = (0.955 \pm 0.010)\,\bar{\sigma}_a(\text{lamp}) \tag{6}$$

This small correction is due to the comparatively small wavelength dependence of the spectral absorption coefficient, especially in the visible region of the solar spectrum [e.g., Fischer (2)].

2.2 *Calorimetric method for airborne particles*

The experimental setup for the calorimetric method mainly consists of two closed chambers connected by a differential pressure meter of high sensitivity (Fig. 3). Into one of these chambers (the measuring chamber) an absorber plus dry air are sucked, while only dry air is sucked into the second one (the comparison chamber). When these chambers are illuminated at the same time by parallel shortwave radiation, electromagnetic energy is transformed into internal energy within the absorber. When this energy is transferred to the surrounding dry air, a temperature and, consequently, a pressure increase in the measuring chamber takes place. This pressure increase is directly proportional to the absorbed flux density. The relevant formula is

$$1 - \frac{F_{apo}}{F_o} = 1 - \frac{c_v V \Delta P_a}{RF_o} = \exp(-\bar{\sigma}_a l) \qquad (7)$$

where $V = 524$ cm^3 (volume of measuring chamber), $l = 18$ cm (length of measuring chamber), c_v = specific heat of the air, R = specific gas constant of the air, and ΔP_a = pressure increase in the measuring chamber due to the absorption of shortwave radiation within the particles. When a mean absorption coefficient for the airborne particles within the chamber has been

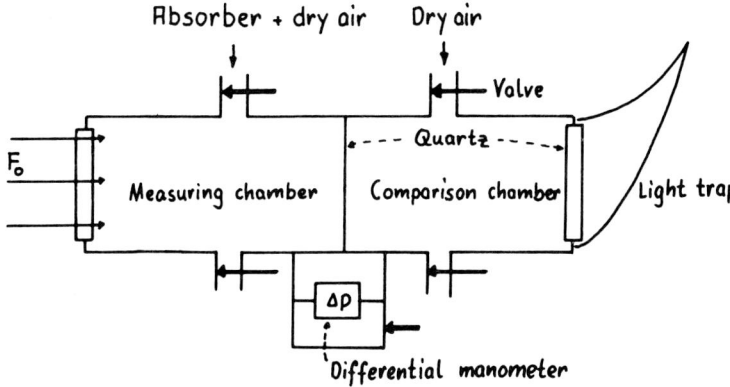

FIGURE 3. Schematic diagram of the calorimetric absorption measurement.

obtained according to Eq. (7), it must be corrected for real atmospheric conditions according to Eq. (6). Up to now, the mean shortwave absorption coefficients have been measured for airborne aerosol particles by the calorimetric method appear to be too small. This is due to difficulties which are primarily caused by the evaporation of very small amounts of water from the particles after the absorption of radiation has taken place, and by heat exchange between the air in the chamber and the walls.

IV. RESULTS OF MEASUREMENTS

In Fig. 4 some measurements of the shortwave radiation supply are compiled. These measurements have been taken in the city of Frankfurt and on the top of Kleiner Feldberg (about 800 m above mean sea level) in the Taunus mountains near Frankfurt. The results show that the shortwave radiation supply may be much larger in a city than in a natural environment. This is due to the reflection of light from bright walls. On the average, the shortwave radiation supply is on the order of the solar constant. The maximum values measured in Frankfurt are about 50% larger than the solar constant. Above snow and between clouds much larger values than these maximums should be expected.

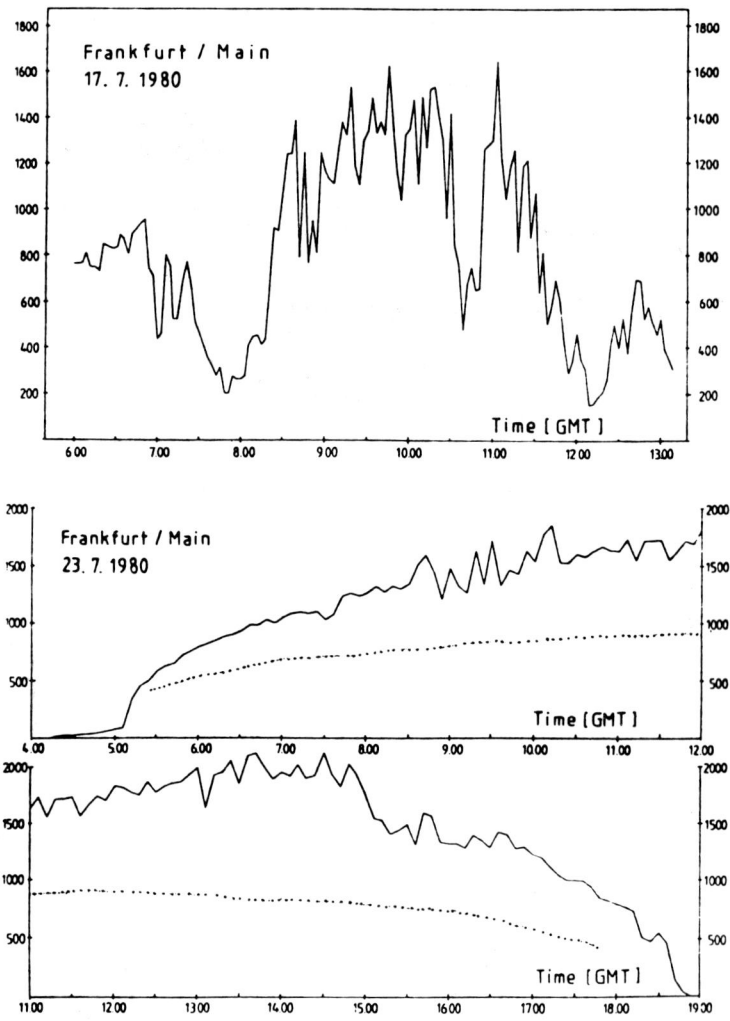

FIGURE 4. Shortwave radiation supply $\int_{4\pi\,\omega_0} N\,d\omega$ in W/m². Solar constant = 1353 W/m²; solid line = $\int_{4\pi\,\omega_0} N\,d\omega$; dotted line = flux density of direct solar radiation.

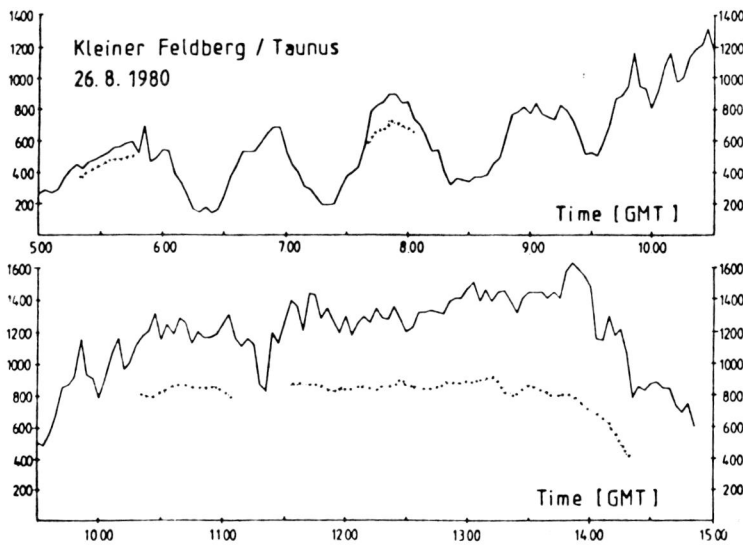

FIGURE 4. Continued.

Mean shortwave absorption coefficients in the city of Frankfurt, together with the pertinent heating rates for the shortwave radiation supply of 1000 W/cm^2 and 2000 W/cm^2, are compiled in Table 1. The absorption coefficients have been measured with the photometric method. Their values range between about 10^{-7} cm^{-1} and 10^{-6} cm^{-1}. The corresponding heating rates at 1000 W/m^2

TABLE 1. *Mean Shortwave Absorption Coefficients $\bar{\sigma}_a$ and Corresponding Heating Rates of the Atmosphere in Frankfurt/Main; V/F = Equivalent Length of Absorption Path in the Atmosphere.*

Date/time	V/F (km)	$\bar{\sigma}_a$ (cm^{-1})	Heating rates in °K/h for shortwave radiation supply of	
			1000 W/m^2	2000 W/m^2
2/13/80--1500-1700	1.96	$1.98 \times 10^{-7} \pm 16\%$	0.058	0.120
11/7/80--1000-1400	3.89	$7.91 \times 10^{-7} \pm 9\%$	0.230	0.460
11/7/80--1410-1805	2.98	$1.34 \times 10^{-7} \pm 17\%$	0.039	0.078
11/10/80--1450-2215	5.30	$1.67 \times 10^{-7} \pm 10\%$	0.049	0.098

range between 0.94 K/day and 5.53 K/day. They can be regarded as approximate mean values during the day for situations with clear skies and scattered cloudiness. The heating rates at 2000 W/m^2 are maximum values during clear-sky situations. These first results show that the absorption of solar radiation in aerosol particles is a process of high climatological importance, especially in industrial regions.

ACKNOWLEDGMENT

Travel expenses to the First International Workshop on Light Absorption by Aerosol Particles were partially funded by NSF Grant ATM 8005356.

REFERENCE

1. Lin, Chin-I, Baker, M., and Charlson, R. J., Absorption coefficient of atmospheric aerosol: a method of measurement, *Appl. Opt. 12*, 1356-1363 (1973).
2. Fischer, K., Mass absorption coefficient of natural aerosol particles in the 0.4-2.4 μm wavelength interval, *Contrib. Atmos. Phys. 46*, 89-100 (1973).

A LASER CAVITY EXTINCTION PHOTOMETER FOR MEASUREMENTS OF EXTINCTION COEFFICIENT AND VISUAL RANGE

Robert G. Knollenberg

Particle Measuring Systems Inc.
Boulder, Colorado

An instrument has been designed and fabricated to measure the visual range and extinction coefficient of the atmosphere using a very short attenuation path. The prototype instrument operates over a nominal 0.5- to 40-km visual range. The device uses a high "Q" He-Ne laser cavity operating as a nonlinear amplifier of very small insertion losses. The design employs vacuum and molecular gas reference measurements in addition to the aerosol-laden air for internal referencing and a self-calibration feature.

I. INTRODUCTION

The measurement of the visual range has been of concern to the atmospheric science community and the public in general for many years. Currently the visual range (or visibility) is generally reported by trained observers viewing high-contrast features or objects at known distances. At selected locations, such as at airports, the visual range is measured using transmissiometers and reported as runway visual range (RVR). These devices, in general, require a light source and receiver pair spaced at 152 m from which visibilities can be measured down to Category II minima. [Category II is a Federal Aviation Administration (FAA) designation for operations in the air transport category requiring special precision approach avionics.] Obviously the sensitivity of, as well as need for, such a device diminishes as visibilities become on the order of a few kilometers. Transmissiometers with larger path spacings (on the order of several kilometers) are also used routinely for atmospheric research on a host of problems where visibilities are of prime interest. A problem arises in trying to compare such long-path measurements with point measurements. The latter has been noted routinely when Particle Measuring Systems Inc. (PMS) has compared computed visual ranges using particle size spectrometer data with those measured with transmissiometers. The particle size spectrometer performs measurements at essentially one point, whereas the transmissiometer is integrating over a lengthy path. In certain cases this lengthy path corresponds to an appropriate "ideal" length considering such factors as the actual visual range, signal-to-noise ratio, and attenuation; however, due to alignment and siting it is not a simple matter to adjust the path length. Attempts have been made to use folded paths of a variety of sorts, but here again the alignment problem is compounded.

During 1980 PMS developed an instrument technique that can substitute for a long-path or folded-path transmissiometer. The technique is called a laser cavity extinction photometer and uses an extended high Q laser cavity to develop the desired sensitivity over a very short path. Laser cavities are strongly sensitive to small internal losses (1,2).

II. LASER CAVITY EXTINCTION PHOTOMETER

The first prototype instrument developed at PMS used a helium-neon (He-Ne) laser. A diagram of this instrument is shown in Fig. 1. The physical construction is rather simple and includes a hybrid laser plasma tube fitted within a large tube, one end of which can be opened or closed to the ambient air by a sliding cover. The principle of operation is as follows: the cavity is alternately exposed to ambient air containing visual-range-reducing aerosol and to air without aerosol, and the laser output power is measured as an extinction value. Detector output is calibrated in terms of extinction coefficient and/or visual range and is unique to the laser cavity design. To provide a self-calibration feature, a third measurement point can be introduced by removing all gas as well as aerosol. Ratioing the contents of the evacuated cavity with those of aerosol-free air yields an extinction value that is purely due to molecular scattering and is a function only of atmospheric pressure. Since the latter nominally varies only 1% to 2% from high- to low-pressure regions, the self-calibration feature is realized. A correction needs only to be applied for altitude.

In actual operation the instrument head is connected through solenoid-operated valves to a vacuum pump and a filtered air source. The sample cover is also operated using a solenoid mechanism. The sample modes are controlled via a microprocessor controller which also computes visual range and adjusts calibration. The altitude correction is manually set and input to the controller.

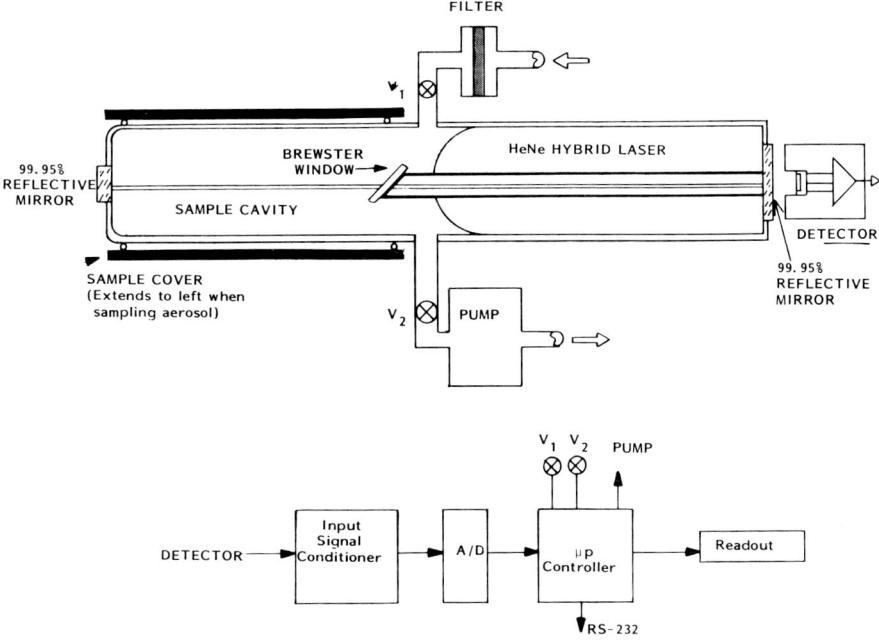

FIGURE 1. Diagram of the prototype cavity extinction photometer.

It is worthwhile to discuss the cavity parameters in greater detail in order to clearly understand the origin of the device's sensitivity. First it should be pointed out that the gain within a nominal He-Ne laser is on the order to 1% to 2%. Thus one typically has an output mirror whose transmission (T) is nominally 1% if maximum output power is to be realized. It should be apparent that single pass losses greater than the single pass gain would extinguish laser action. Thus the output power would drop from 100% to 0% with something on the order of 1% to 2% total loss. Roughly, then, this cavity would have 100 times the sensitivity of a standard transmissiometer of equal path length or would have the same sensitivity with 1% of the path length. However, as noted in Fig. 1, this laser has mirrors with much higher reflectivities (99.95%) since little output power is needed. If the laser gain remains the same, most of the energy that would normally be transmitted is reflected internally. The internal energy density increases, and so does its sensitivity to the uncontrolled losses being measured. The ratio of the mirror reflectivity to all other losses is called Q (Q = R/all losses) and is a relative figure of merit of its sensitivity. Assuming that in Fig. 1 the Brewster window is lossless and the cavity is evacuated, the Q is approximately equal to reflectivity/transmission = 99.95/0.05 = 1999. Thus the equivalent path length of many standard transmissiometers can be equalled in inches to feet of sample cavity length.

The output laser power is a nonlinear function of cavity losses. In general this creates a calibration curve with greatest sensitivity when losses are small. This is clearly shown in Fig. 2 which is the calibration curve for the prototype device. A change in visual range from 40 km to 4 km results in less than 10% of the total allowable cavity loss, but the power reduces by more than 60%. On the other hand, the final 10% of output power is lost by the reduction of visual range from 1 km to 100 m.

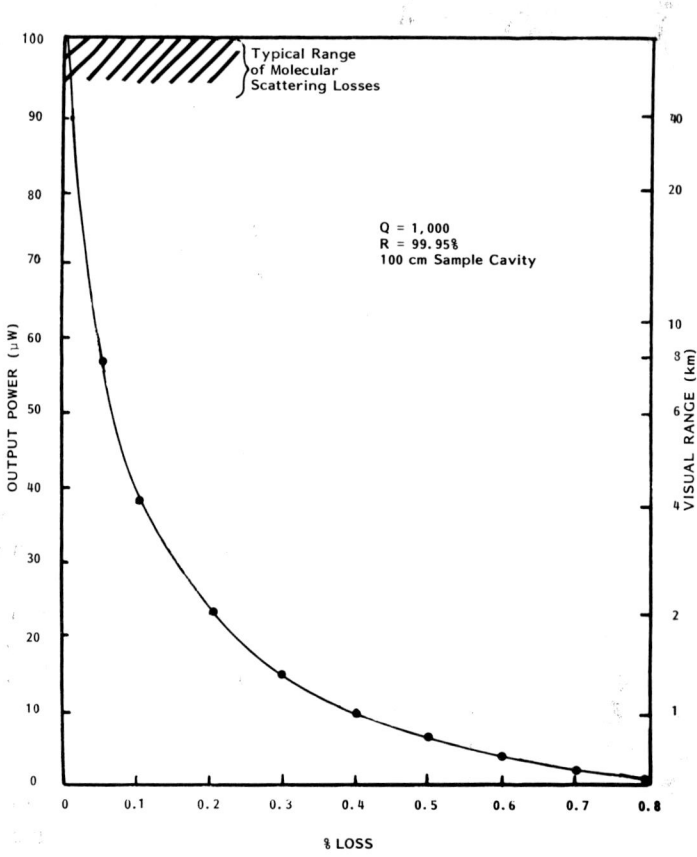

FIGURE 2. *Visual range and output power versus percent cavity loss. The prototype uses a 100-cm sample cavity length. With current state-of-the-art high reflectivity mirrors, the sample cavity length can be reduced to less than 30 cm with a Q of 2,000 to 5,000.*

It is easy to see that the sensitivity can be modified by various parameters. The most convenient parameter to vary is cavity Q using mirror reflectivity. Cavity length should be kept below 1 m for ease of mounting and mechanical stability. With highest-quality mirrors and loss-free Brewster windows, Q's as high as 5,000 can be realized which would provide a measurement range of 200- to 300-km. Laser gain is another parameter that can be adjusted. A higher gain would permit operation at lower visibilities.

III. CONCLUSIONS

From the preceding discussion it is clear that laser cavity extinction photometers can be used in place of long-path transmissiometers. Their utility primarily lies in the convenience or desirability of point measurement capability. They would nominally be used for measurements of visual ranges greater than 1 km. Although ultimately they can be designed for almost any range of values, their primary limitation is in precipitation where the cavity losses are no longer represented by a viewed particle ensemble but by discrete particle events. The cavity itself must be regularly serviced and its critical optical surfaces kept clean. Continuing efforts are being concentrated on making the device as serviceable and service-free as practical.

REFERENCES

1. Schuster, B. G., and Knollenberg, R., Detection and sizing of small particles in an open cavity gas laser, *Appl. Opt. 11,* 1515-1520 (1972).
2. Schleusener, S. A., and Read, A. A., Variable Brewster angle flat used as a gas laser gain control, *Rev. Sci. Instrum. 37,* 287-289 (1966).

GENERATION AND CHARACTERIZATION OF AEROSOL PARTICLES[1]

Edward E. Hindman[2]
Randy D. Horn
William G. Finnegan

Cloud Simulation and Aerosol Laboratory
Department of Atmospheric Science
Colorado State University
Fort Collins, Colorado

Aerosol particles with a wide range of light absorption properties were generated, transported, and characterized to permit meaningful intercomparisons of all the major types of light absorption instruments. The particles were generated in concentrations of $\sim$ 1 mg m^{-3} and 50 µg m^{-3} for periods up to 2 hours. The particle characteristics ranged from highly absorbing carbonaceous to nearly transparent ammonium sulfate, from submicron ammonium sulfate to supermicron Arizona road dust, and from spherical ammonium sulfate, to chain-aggregate carbonaceous, to irregular ambient particles.

---

[1]The workshop was funded by NSF Grant ATM 8005356 and ONR Grants N0001480G0042 (funds provided by Naval Air System Command 310C and the Army Research Office) and N0001479C0793.

[2]On leave to the Research Institute of Colorado, Fort Collins, Colorado.

I. INTRODUCTION

Wide variations exist in the reported values of the light absorption properties for aerosol particles in the free atmosphere (1). This variability is attributed to the natural atmospheric variability of absorbing particles and to measurement uncertainties. Consequently, a workshop was proposed at which all the major techniques for measuring light absorption by aerosol particles would simultaneously sample the same well-characterized particle population to separate the natural variability from instrumentation error and to establish the amount of measurement uncertainty.

The Cloud Simulation and Aerosol Laboratory of Colorado State University (CSU) was chosen as the site of the workshop because a number of national and international workshops have been conducted at the facility (2-4). This fact plus the extensive paricle generation and characterization techniques which have become available in recent years [e.g. Liu (5), and Lundgren et al. (6)] made the time auspicious for a light absorption workshop. Existing techniques were used during the workshop to produce a population of aerosol particles with known sizes, numbers, and compositions. The particle population was sampled simultaneously with the variety of light-absorption instrumentation assembled at the workshop.

This paper describes the system developed to generate, transport, and characterize the aerosol particles during the workshop. Conclusions are made concerning the characteristics of the particle mass concentrations and size distributions. Detailed descriptions of the aerosol generated for each experiment are presented in Appendices A and B.

II. LABORATORY SET-UP

The laboratory was set up to meet the particle generation, transport, and characterization needs determined by the workshop organizing committee. The committee specified the following required aerosol particles: carbonaceous (highly absorbing), ammonium sulfate (nonabsorbing in the visible spectrum), a mixture of $(NH_4)_2SO_4$ and methylene blue dye (absorber), methylene blue, ambient (from outside the laboratory), and resuspended soil particles.[3] Further, they specified that two aerosol streams be available to the investigators: a high-concentration stream with particle loading of ~ 1 mg m^{-3}

---

[3] *Arizona road dust (constituents in wt %): SiO_2 = 65 to 76, Al_2O_3 = 11 to 17, Fe_2O_3 = 2.5 to 5, CaO = 3 to 6, Na_2O = 2 to 4, MgO = 0.5 to 1.5, TiO_2 = 0.5 to 1, V_2O_3 = 0.1, ZrO = 0.1, BaO = 0.1, and loss on ignition 2 to 4; size distribution (in number %): 0- to 5-μm = 39 ± 2%, 5- to 10-μm = 18 ± 3%, 10- to 20-μm = 16 ± 3%, 20- to 40-μm = 18 ± 3%, 40- to 80-μm = 9 ± 3%.*

GENERATION AND CHARACTERIZATION OF AEROSOLS 73

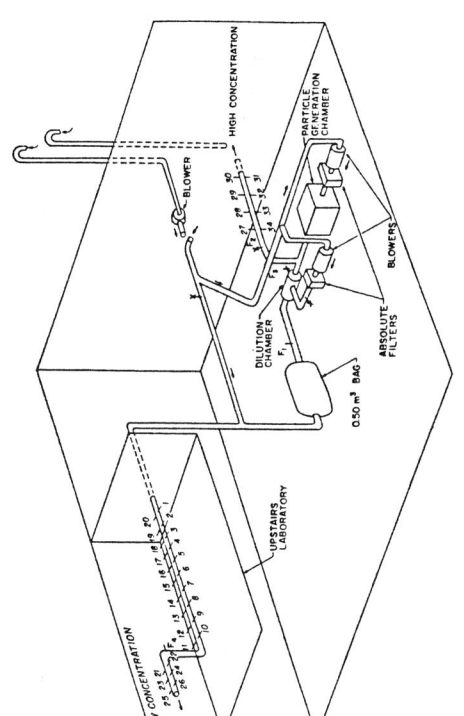

FIGURE 1. Schematic of the laboratory apparatus used to generate and transport aerosol particles used at the workshop (see text for explanation). The investigators sampling stations were as follows: 1. Hindman (TSI 3200, 3030, Royco 220), 2. Hindman (filter), 3. Heintzenberg (filter), 4. Grams/Wyman (polar nephelometer), 5. Patterson/Marshall (filter), 6. blank, 7. Wright/Steffanutti/Stedman (photoacoustic), 8-10. blank, 11. Shaw (filter), 12-13. blank, 14. Bodhaine (4λ nephelometer), 15-16. Waggoner/Clarke (filters), 17. Waggoner/Clarke (1λ nephelometer, Royco 220), 18-19. Rosen (filters), 20. Hindman (AS/SP), 21-23. blank, 24. Gerber/Daniels (flux divergence), 25. Hindman (filter), 26. blank, 27. Hindman (filter), 28. Palmer/McClenny (dichotomous sampler), 29. Patty/Bennett (dichotomous sampler), 30. Voltz/Egan (filters), 31. Patterson/Marshall (filter), 32. Gillespie (filter), 33. Cowen (filter), 34. Gillespie (filter), Box and Shaw set up remote sensors outside but nearby the laboratory.

and a low concentration stream with loading of $\sim 50\ \mu g\ m^{-3}$. Some instruments required bulk particle samples, while others required low volume aerosol samples. Consequently, the two streams were necessary so that both types of instruments could sample simultaneously and for similar periods. Finally, the aerosol streams needed to be steady for up to 2 hours.

The apparatus for generating, transporting, and characterizing the aerosol particles in illustrated in Fig. 1. The particle generation occurred either in the particle generation chamber or with a generator attached to the chamber. The types of particles generated and the generation techniques are listed in Table 1, The formulas for the solutions used to generate particles are listed in Table 2.

The generated particles were transported through the system shown in Fig. 1 in the following manner: a particle-free airstream flowed continuously through the generation chamber at $\sim 500\ l\ min^{-1}$. The aerosol was either generated in the chamber or injected into the chamber. The particle-laden flow split downstream of the chamber. One flow (F_2), the high-concentration aerosol, exhausted directly outside the laboratory at $\sim 475\ l\ min^{-1}$ after passing sampling points which led to investigators' instruments. The other flow (F_3), the approximately $25\ l\ min^{-1}$ remaining of the high concentration flow, was mixed with $\sim 475\ l\ min^{-1}$ of particle-free air in a dynamic dilution chamber. The dilution chamber is shown in Fig. 2. The aerosol which emerged from the chamber was diluted by about 20 times. The diluted aerosol then flowed through a 0.5-m^3 aluminized Mylar bag to dampen any fluctuations in the particle generator. After leaving the bag, the aerosol traveled through the horizontal tube in the upstairs laboratory and exhausted into an exhaust hood. The instruments were connected to ports along the horizontal tube. The residence time between the generation chamber and the exhaust hood was about 2 minutes.

The ambient aerosol particles were brought into the laboratory in two ways. First, the upper blower shown in Fig. 1 was connected to the low concentration line. The appropriate valves (symbols in Fig. 1) were adjusted so the outside air moved quickly to the sampling tube in the upstairs laboratory. The transport blower was a sealed squirrel-cage type and was not a source or sink of particles. The ambient air flow was $500\ l\ min^{-1}$. Second, the high-concentration line was connected to the vertical chimney (outside the building), and outside air was drawn through the high-concentration line by attaching a blower near valve F_2 (Fig. 1).

The flows F_1, F_2, and F_4 in Fig. 1 were measured with sensitive pitot tubes. The blowers upstream of the filters were high-pressure types, and the flow was constricted down to $500\ l\ min^{-1}$ by downstream valves. The reason for the high-pressure blowers was to provide a pressure "head" which would minimize flow irregularities in the sampling lines caused by instruments being turned on and off.

TABLE 1. Aerosol Particle Generation Techniques.

Particle type	Generation technique	Description of generator	Operating conditions
Carbonaceous	Combustion of propane	Ring burner 8 cm diameter with 34 4-mm-dia. holes, located in generating chamber	0.056 m^3/hr (2 ft^3/hr) flow as indicated by a rotometer
Ammonium sulfate	Bubbling salt solution and dessication of solution droplets	Humidified compressed air bubbled through the salt solution using an immersed capillary (3-mm dia. drawn to a 1-mm jet), and the resulting aerosol passed, before dessication, through an impactor consisting of a similar jet placed 2 mm from a flat glass surface	14-psi, particle-free, regulated compressed air flow; 10% salt in distilled water (all percentages by weight)
Ammonium sulfate plus methylene blue	Same technique as for pure ammonium sulfate	Same generator as for pure ammonium sulfate	10% salt in distilled water, 0.3, 1, 4% methylene blue (ultrasonic nebulizer used with 4% solution)
Methylene blue	Nebulize methylene blue solution and dessication of droplets	Monaghan, Co., Littleton, Colorado. Model 670 ultrasonic nebulizer attached to generating chamber	1% methylene blue in distilled water; set two nebulizers at 3.5 and combine effluent
External mixture of carbonaceous and ammonium sulfate	Combustion of propane and bubbling of salt solution	Mix cooled and diluted effluent from bubbler; bubbler attached downstream of diluter	Propane flow: 2 solution CFH, compressed air: 14 psi, 10% salt solution (31 July, 4 Aug); 1% salt solution (7 Aug)
Arizona road dust (AC Spark Plug Div., GMC, Flint, Michigan)	Resuspension	Thermo Systems Inc. St. Paul, Minnesota, model 3400 fluidized-bed particle generator	Chain speed = 60 Bed purge = 40 Bed flow = 40

TABLE 2. Solution Formulas

Ammonium sulfate:
 1000 ml distilled H_2O + 100 g $(NH_4)_2SO_4$
Ammonium sulfate + 1% methylene blue (MB):
 1000 ml distilled H_2O + 100 g $(NH_4)_2SO_4$ + 1 g MB
Ammonium sulfate + 0.3% methylene blue:
 1000 ml distilled H_2O + 100 g $(NH_4)_2SO_4$ + 0.3 g MB
Ammonium sulfate + ~ 4% methylene blue:
 200 ml of the above $(NH_4)_2SO_4$ + 1% MB solution added to 150 ml of the following solution: 1000 ml distilled H_2O + 100 g $(NH_4)_2SO_4$ + 10 g MB
Methylene blue:
 1000 ml distilled H_2O + 10 g MB

The aerosols generated during the workshop were routinely and continuously characterized using particle and state-parameter measuring instruments. The instruments and their performances are listed in Table 3. All of the instruments sampled the aerosol in the upstairs laboratory at the extreme upstream end of the horizontal sample tube. Additionally, filter collections were made in the high-concentration aerosol stream in the downstairs laboratory.

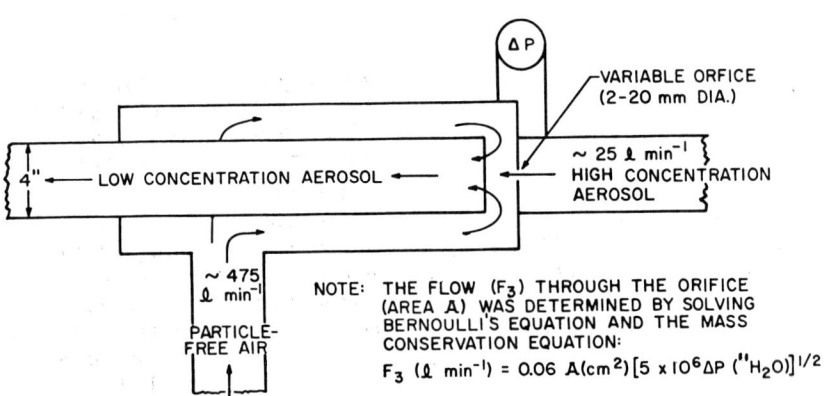

FIGURE 2. Dynamic dilution system. The flow (F_3) through the orifice (area A) was determined by solving Bernoulli's equation and the mass conservation equation: $F_3(l/min^{-1}) = 0.06\ A(cm^2)\ [5 \times 10^6\ \Delta P("H_2O)]^{1/2}$.

GENERATION AND CHARACTERIZATION OF AEROSOLS 77

TABLE 3. Aerosol Particle, State Parameter, and Data Acquisition Instruments.

Instrument type	Measurement technique	Range/resolution	Relative error ± %	Sampling rate
TSI model 3030 electrical aerosol analyzer (EAA)	Aerosol charging, electrometer sensing	$0.006 \leq d \leq 0.56$ μm, 8 channels	10	$4 \, l/min^{-1}$
PMS ASASP-X optical particle counter	He-Ne laser illumination, photo-diode detection	$0.09 \leq d \leq 3.0$ μm, 60 channels	1	$1 \, cm^3 \, s^{-1}$
Royco model 220 optical particle counter (OPC)	Incandscent lamp, photodiode detection	$0.5 \leq d \leq 9.0$ μm, 4 channels	5	$3 \, l/min^{-1}$
TSI model 3200A particle mass monitor (PMM)	Electrostatic precipititator, quartz microbalance	$0.01 \leq d \leq 20$ μm, 1 channel, $2 - 2 \times 10^4$ μg m^{-3}, ± 0.006 μg m^{-3}	10	$1 \, l/min^{-1}$
TSI model 3012 neutralizer	KR-85, 2 m ci, upstream of EAA and PMM	---	---	---
Metricel filters (0.45 μm)	Mechanical beam balance	$m \geq 1 \times 10^{-5}$ g, $\pm 10^{-5}$ g	---	---
Nuclepore filters (47 mm, 0.4-μm pore)	Scanning electron microscope	$d \geq 0.1$ μm	---	---
YSI model 44212 precision temperature thermistor (T)	RTD device, custom interface	$0C \leq T \leq 50C$	0.1 C	---
EG&G model 9022-01 dew point hygrometer (T_d)	Electro-cooled mirror, optical dew detection	$-50C \leq T_d \leq 50C$	0.5 C	---
Belfort recording microbaragraph	Mechanical bellows	$600 \leq p \leq 900$ mb	---	---
PMS-DAS model 2D-32	Digital/analog data acquision system	Particle probes, state parameters	---	---
Pertec 9-track tape drive	Data recorder, computer compatible	---	---	---

III. PROCEDURES

A typical experiment began with particle-free air flowing through the system shown in Fig. 1. After a 5- to 10-minute period of particle-free air, one of the particle generators described in Table 1 was activated. The generator was adjusted until the desired particle mass concentration was detected with the

TSI mass monitor (PPM)[4] and with the University of Washington's integrating nephelometer [see Clarke and Waggoner (7)]. The mass concentration values were monitored for a few minutes to assure a steady-state aerosol. Then, a signal was given to all investigators to begin sampling. The investigators sampled as often as required by their procedures during the typical 1- to 1½-hour particle-generation periods. The periods terminated when all investigators were satisfied that they had collected sufficient samples. Thereafter, the generator was turned off and particle-free air flushed the system. It took about 5 minutes to reestablish particle-free air in the system. The process was repeated for the next aerosol and so on throughout the workshop.

During the particle-generation periods, aerosol particle measurements were continuously collected using the equipment described in Table 3. Discrete samples were obtained with the filters early and late in the periods. The data were collected in the following manner.

A Particle Measuring System's DAS-2D-32 was used for automated data logging of state parameters as well as particle data from various instruments (see Table 3). The data sample rate was 1 Hz for the single particle counters (Royco 220, ASASP-X) and 0.1 Hz for the analog data (temperature, T; dew point, T_d; PMM; EAA). This data-rate selection gave adequate time resolution for the charactrization of the aerosol and kept magnetic tape requirements at a convenient level of 1 per day.

The Royco 220 (OPC) was modified to produce a "one-of-sixteen + strobe" format for each of its four size bins. This allowed it to be recorded identically with the ASASP-X. The TSI-3200A particle mass monitor was modified to output its 0- to 10-kHz square wave directly into the "A" housekeeping section of the DAS probe.

The TSI electrical aerosol analyzer (EAA) was modified such that a pulse applied every 30 sec (for an external circuit) would increment a step voltage change and remain at that step until the next pulse. This resulted in two to three consecutive DAS samples per EAA size step. The resulting format (using only 8 voltage values) yielded an average of one complete cycle in 4 minutes with approximately three samples per step. Range and data values of the EAA were input to the DAS analog data section.

State parameters were measured by a YSI linear thermistor inserted into the sample airstream (0.1 °C accuracy) and an EG&G dew-point hygrometer utilizing a 0.01-μm Nuclepore inline filter to eliminate contamination of the mirror surface by particles.

---

[4]*The crystals in the mass monitor had to be cleaned regularly during the generation period. Further, the instrument was inoperable with soot, because the soot particles built up a feathery-like structure with time and gave unreliable mass concentration values.*

GENERATION AND CHARACTERIZATION OF AEROSOLS

All collected data were recorded on computer compatible 9-track tapes. Tapes from each day's experiments were removed and processed on a facility-owned minicomputer system for "day's end" results. Software was written to validate samples, do 5-minute averaging, and calculate and print tabulated values.

The aerosol particle data were reduced to provide size distributions and number concentrations of the aerosol particles. The data were averaged over 5-minute periods, and the uncertainties of the averages were determined by calculating $\sigma/(n-1)^{1/2}$ where σ is the standard deviation and n is the number of samples in the 5-minute period. The values of current from the EAA (averages and uncertainty of the averages) were multiplied by the manufacturer's calibration factors (8) [$N(cm^{-3})$ proportional to current] to produce particle concentrations in the eight size channels. The 60-particle size ranges for the ASASP-X were established using either a response curve for carbon particles and latex spheres (see Fig. 3) which were derived theoretically by Garvey (9) or the response curve for latex spheres provided by the manufacturer (see Table 4). When carbon particles and methylene blue particles were generated, the data were reduced using the carbon response curve. When ammonium sulfate particles were generated, the data were reduced using the latex sphere response curve. For the mixed ammonium sulfate and methylene blue particles and ammonium sulfate and carbon particles, the data were reduced twice, once using the carbon response curve and once using the latex sphere response curve. The particle concentrations in the 60 size ranges were determined by dividing the average numbers and uncertainties by the volume sampled (1 cm^3 s^{-1} x 300 s). The four particle size ranges for the OPC were established by the manufacturer using latex spheres; no correction for carbon particles was available. These data were reduced in an identical fashion to the ASASP-X data. The OPC sample flow was 2.8 l min^{-1} (indicated).

The mass concentrations of the particles were determined in three ways. First, during the 5-minute averaging periods, the positive frequency shifts from the PMM were summed. The maximum frequency shift for the period was determined and multiplied by the manufacturer's calibration given by mass concentration ($\mu g\ m^{-3}$) = 333 Δf/300s, where Δf is the frequency shift in Hz. Second, particle mass concentrations were calculated from the average particle number csoncentrations for each size interval for the EAA, ASASP-X, and OPC. Particle densities were approximately 2 g cm^{-3}. Particle sizes were approximated by the geometric-mean particle diameter which is defined by $(d_i \times d_{i+1})^{1/2}$ where i is a size channel. Third, the particle collections on the membrane filters were used to determine the particle mass concentrations. The filters were carefully weighed before and after exposure to the aerosol. The mass of collected aerosol divided by the true volume sampled provided a measure of the particle mass concentration.

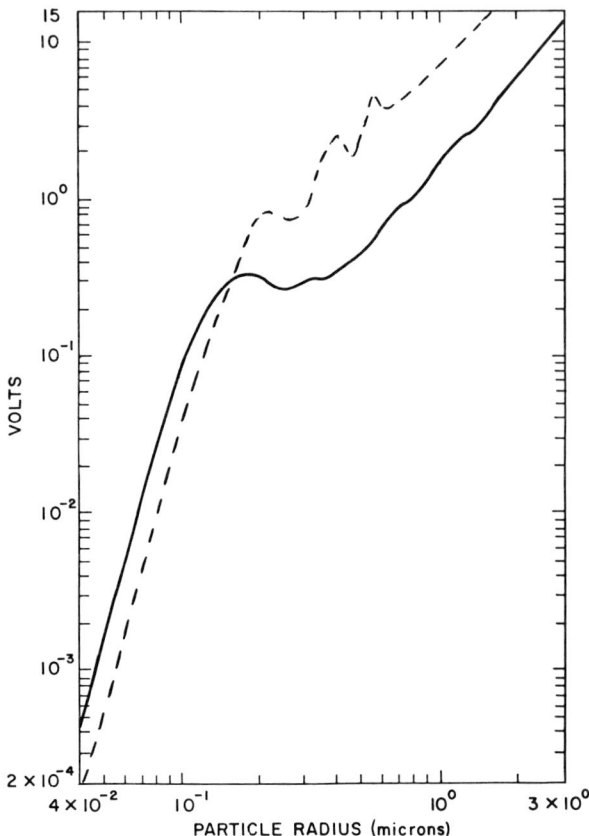

FIGURE 3. *Theoretical response of ASASP-X to carbon particles (solid curve, 1.9500-0.66i) and latex spheres (dashed curve, 1.5920-0.0i) (from Ref. 9).*

The reduced aerosol particle data consisted of 5-minute averages of number and mass concentrations from the EAA (eight diameter ranges bounded by 0.006, 0.01, 0.18, 0.032, 0.056, 0.10, 0.178, 0.316, and 0.562 μm), the ASASP-X (60 diameter ranges (see Table 4), and the OPC (four diameter ranges bounded by 0.5, 3, 5, 7, and 9 μm). The data from these instruments were combined by using five of the eight EAA intervals (0.006 to 0.10 μm), using the size ranges from the ASASP-X as defined in Table 4, and using four of the five OPC intervals (3 to 9 μm). Combining the data in this fashion produced particle size distributions with no overlapping size intervals. The mass concentrations corresponding to the combined size intervals were summed to produce an

GENERATION AND CHARACTERIZATION OF AEROSOLS 81

TABLE 4. Calibration of ASASP-X for Carbon and Latex Particles (Diameter Intervals in μm)[a]

Channel	Volts	Range 0 Carbon	Range 0 Latex	Volts	Range 1 Carbon	Range 1 Latex	Volts	Range 2 Carbon	Range 2 Latex	Volts	Range 3 Carbon	Range 3 Latex
1	0.60	1.20	0.60	0.070	0.20	0.24	0.00608	0.122	0.15	0.000355	0.077	0.090
2	1.42	1.92↑	0.76↑	0.127	0.23↑	0.28↑	0.00865	0.131	0.16	0.000576	0.084	0.097
3	2.11	2.32	0.92	0.194	0.26	0.32	0.0120	0.139	0.17	0.000837	0.088	0.104↑
4	2.75	2.80	1.08	0.270	0.40↓	0.36	0.0165	0.146↑	0.18↑	0.00121	0.094	0.111
5	3.39	3.08↓	1.24↓	0.356	0.80↓	0.40	0.0223	0.156	0.19	0.00172	0.100↑	0.118
6	4.04	3.36	1.40↓	0.459	1.02	0.44	0.0293	0.162	0.20	0.00236	0.106	0.125
7	4.68	3.64	1.56	0.557	1.12	0.48	0.0378	0.170	0.21	0.00318	0.110	0.132
8	5.30	3.84	1.72	0.674	1.21	0.52↑	0.0479	0.178	0.22	0.00419	0.116	0.139
9	5.92	4.08	1.88	0.797	1.31↑	0.56	0.0589	0.184	0.23	0.00542	0.1220	0.146
10	6.54	4.48	2.04↑	0.923	1.54	0.60	0.0709	0.194	0.24	0.00689	0.127	0.153
11	7.16	4.76	2.20	1.05	1.66	0.64	0.0837	0.198	0.25	0.00869	0.131	0.160
12	7.76	4.80	2.36	1.18	1.74	0.68	0.0974	0.200	0.26	0.0109	0.135	0.167
13	8.35	4.92	2.52	1.32	1.86	0.72	0.112	0.208	0.27	0.0134	0.141	0.174
14	8.92	5.04	2.68	1.45	1.92↓	0.76↓	0.127	0.216	0.28↓	0.0164	0.146↓	0.181↓
15	9.47	5.16	2.84	1.58	2.00	0.80	0.143	0.220	0.29	0.0201	0.152	0.188
	10.00	5.40	3.00↓	1.72	2.06	0.84	0.160	0.230↓	0.30	0.0246	0.1560	0.195

[a] Arrows indicate size ranges used to analyze the data.

integrated mass concentration for particles from 0.006 μm to 9.0 μm in diameter. The temperature and dew point values were averaged for the 5-minute periods.

IV. RESULTS

A total of 24 experiments were conducted during the 10-day duration of the workshop (28 July to 8 August). The day, time, and aerosol generated for each experiment are listed in Table 5. A four-wavelength nephelometer was operated continuously for each experiment by Bodhaine (10). His results provide a measure of the light scattering coefficient (σ_s) of the aerosol. They are useful to the discussion here because they demonstrate the steady-state nature of a majority of the aerosols generated.

TABLE 5. Workshop Summary

Date 1980	Time MDT	Aerosol	Remarks
29 July	1114-1330	$(NH_4)_2SO_4$	
29 July	1420-1609	$(NH_4)_2SO_4$	Some large particles in low concentration aerosol
29 July	1613-1734	$(NH_4)_2SO_4$ + 1% MB	(MB = methylene blue)
30 July	1101-1232	$(NH_4)_2SO_4$	Some large particles in low concentration aerosol
30 July	1305-1530	$(NH_4)_2SO_4$ + 0.3% MB	Large flakes in high concentration aerosol
30 July	1624-1800	$(NH_4)_2SO_4$ + ~4% MB	Cubic particles
31 July	0940-1421	Soot	
31 July	1433-1737	Soot + $(NH_4)_2SO_4$	$(NH_4)_2SO_4$ particles volatilized
31 July	1737-1750	Soot	
1 August	1130-1347	MB	
1 August	1405-1600	Soot	
4 August	1005-1129	Soot	
4 August	1129-1412	Soot + $(NH_4)_2SO_4$	~50-50 external mixture
4 August	1418-1445	Soot	
4 August	1454-1558	MB	
5 August	1015-2400	Ambient air	
6 August	0000-0830	Ambient air	
6 August	1017-1336	Arizona road dust	
6 August	1429-1708	Soot + $(NH_4)_2SO_4$	~ 4% soot, external mixture
6 August	1708-1807	Soot	
7 August	1101-1340	Soot + $(NH_4)_2SO_4$	~ 5% soot, external mixture
7 August	1445-1552	Soot + $(NH_4)_2SO_4$ + moisture	RH varied from 50% at intake of low concentration line to ~ 100% at exhaust
8 August	0933-1115	MB	
8 August	1130-1430	$(NH_4)_2SO_4$	

The mass concentrations of the generated aerosols, listed in Table 6, were determined from simultaneous measurements of particle mass (PMM) and particle size distributions and from nearly simultaneous filter collections. The horizontal lines in the table denote that no data were collected during the period with that device. The "no data" entries in the table indicate that the device collected uninterpretable data. The "latex calib." entry in the table indicates that the latex sphere calibration (for nonabsorbing particles) was used to reduce ASASP-X data collected for absorbing particles.

The particle size distributions and mass distributions corresponding to the mass concentrations in Table 6 are plotted and tabulated in Appendix A. The scanning electron microscope photographs of the generated particles are shown in Appendix B. Note, the volume-sampled values in the photograph captions correspond to the volume which flowed through the entire 47-mm diameter filter surface.

V. DISCUSSION

The mass concentrations listed in Table 6 have been categorized into particle types in Figs. 4 through 9. In these figures the mass concentrations determined from the PMM are plotted against simultaneous mass concentrations determined by integrating the size distributions and from the filter collections. The purpose of these figures is to characterize the mass concentration measurements and the particle measurements for the various generated aerosols.

In Fig. 4 it can be seen that the mass concentrations from the three measurement techniques for the $(NH_4)_2SO_4$ particles agree. This result is reasonable because, as can be seen from Appendix B, the $(NH_4)_2SO_4$ particles are primarily spherical particles. The EAA, OPC, and PMM instruments were designed and calibrated for these types of particles. Furthermore, the mass distributions for the $(NH_4)_2SO_4$ particles shown in Appendix A are distributed unimodally with a peak at about 0.4 μm to 0.5 μm diameter with little mass above 1 μm. This result is consistent with the primarily submicron $(NH_4)_2SO_4$ particles shown in Appendix B. Consequently, it is concluded that the $(NH_4)_2SO_4$ particle sizes, numbers, and mass concentrations agree to within ± 20% (as determined from Fig. 4).

The mass concentrations measured for the ammonium sulfate particles which were doped with methylene blue are given in Fig. 5. It can be seen that the measurements from the PMM and particle counters (using the latex-sphere correction instead of the carbon correction) agreed within ± 20%. In contrast, the PMM and particle counter measurements (using the carbon correction for the counter data) did not agree well. Apparently, the percentage of methylene blue was sufficiently low so the ammonium sulfate particles behaved as latex spheres in the particle counters. The PMM and filter measurements were

TABLE 6. Mass Concentration Measurements ($\mu g\ m^{-3}$).

Date	Time (MDT)	Aerosol	PMM	Integrated size distribution	Filters Low concentration aerosol	Filters High concentration aerosol
29 July	1425-1430	$(NH_4)_2SO_4$	293.04	219.9	---	---
29 July	1446-1501	$(NH_4)_2SO_4$	---	---	308	4038
29 July	1540-1545	$(NH_4)_2SO_4$	No data	193.5	---	---
29 July	1643-1648	$(NH_4)_2SO_4$ + 1% MB	224.8	541.8	---	---
29 July	1640-1655	$(NH_4)_2SO_4$ + 1% MB	---	---	120	2918
29 July	1708-1713	$(NH_4)_2SO_4$ + 1% MB	222.0	540.1	---	---
30 July	1129-1144	$(NH_4)_2SO_4$	---	---	No data	3188
30 July	1205-1210	$(NH_4)_2SO_4$	136.2	143.3	---	---
30 July	1250-1255	Filtered air	4.9	0	--	---
30 July	1405-1410	$(NH_4)_2SO_4$ + 0.3% MB	57.6	70.0 (latex calib.)	---	---
30 July	1715-1720	$(NH_4)_2SO_4$ + 4% MB	68.9	70.36 (latex calib.)	---	---
30 July	1651-1731	$(NH_4)_2SO_4$ + 4% MB	---	---	138	---
30 July	1656-1716	$(NH_4)_2SO_4$ + 4% MB	---	---	---	2506
31 July	1151-1206	Carbonaceous	---	---	No data	---
31 July	1153-1158	Carbonaceous	---	---	---	1687
31 July	1230-1235	Carbonaceous	No data	94.9	---	---
31 July	1628-1713	Soot + salt	---	---	154	---
31 July	1653-1708	Soot + salt	---	---	---	4844
31 July	1705-1710	Soot + salt	No data	145.0 (latex calib.)	---	---
31 July	1745-1750	Carbonaceous	No data	86.3	---	---
1 August	1215-1301	Methylene blue	---	---	140	---
1 August	1235-1251	Methylene blue	---	---	---	6438
1 August	1330-1335	Methylene blue	171.4	405.1	---	---
1 August	1330-1335	Methylene blue	171.4	36.34 (latex calib.)	---	---
1 August	1500-1505	Carbonaceous	No data	60.7	---	---
1 August	1521-1551	Carbonaceous	---	---	81.7	---
1 August	1526-1536	Carbonaceous	---	---	---	2174
4 August	1055-1100	Carbonaceous	No data	69.3	---	---
4 August	1211-1241	Soot + salt	---	---	468.3	---
4 August	1214-1223	Soot + salt	---	---	---	6699
4 August	1340-1345	Soot + salt	No data	241.6 (latex calib.)	---	---
4 August	1342-1357	Soot + salt	---	---	591	---
4 August	1344-1349	Soot + salt	---	---	---	6858
4 August	1430-1435	Carbonaceous	16.0	64.1	---	---
4 August	1420-1445	Carbonaceous	---	---	195	---

GENERATION AND CHARACTERIZATION OF AEROSOLS

TABLE 6. Continued

Date	Time (MDT)	Aerosol	PMM	Integrated size distribution	Filters Low concentration aerosol	Filters High concentration aerosol
4 August	1425-1440	Carbonaceous	---	---	---	1931
4 August	1521-1526	Methylene blue	186	1035	---	---
4 August	1521-1526	Methylene blue	186	84.1 (latex calib.)	---	---
4 August	1523-1553	Methylene blue	---	---	1009	---
4 August	1525-1536	Methylene blue	---	---	---	10779
5 August	1105-1110	Ambient	12.5	15.1	---	---
5 August	1210-1711	Ambient	---	---	17.9	19.7
5 August	1615-1620	Ambient	23.8	34.5	---	---
5-6 August	1736-0826	Ambient	---	---	17.0	13.6
5 August	2346-2351	Ambient	8.2	6.0	---	---
6 August	1201-1206	Road dust	104	993.20	---	---
6 August	1201-1254	Road dust	---	---	132.7	644.8
6 August	1358-1403	Carbonaceous	No data	34.1	---	---
6 August	1459-1536	Soot + salt	---	---	731	---
6 August	1546-1551	Soot + salt	No data	585 (latex calib.)	---	---
6 August	1552-1622	Soot + salt	---	---	516	---
6 August	1740-1745	Carbonaceous	40	84.8	---	---
7 August	1030-1035	Carbonaceous	10.7	16.5	---	---
7 August	1129-1213	Soot + salt	---	---	355	---
7 August	1135-1140	Soot + salt	377	168.1 (latex calib.)	---	---
7 August	1228-1301	Soot + salt	---	---	487	---
7 August	1355-1400	Carbonaceous	30	11.7	---	---
7 August	1525-1530	Soot + salt + moisture	901.4	165	---	---
7 August	1555-1600	Carbonaceous	55.5	21.1	---	---
7 August	1615-1620	Filtered air	13.1	0.06	---	---
8 August	1020-1025	Methylene blue	432.9	1343.9	---	---
8 August	1025-1030	Methylene blue	419.0	116.8 (latex calib.)	---	---
8 August	1056-1112	Methylene blue	---	---	802.6	12146
8 August	1235-1240	$(NH_4)_2SO_4$	31	28.5	---	---
8 August	1320-1419	$(NH_4)_2SO_4$	---	---	No data	---
8 August	1323-1340	$(NH_4)_2SO_4$	---	---	---	489
8 August	1435-1440	$(NH_4)_2SO_4$	39	37	---	---

within ± 90% of each other. Consequently, the representative particle mass measurements for the ammonium sulfate particles doped with methylene blue were obtained with the PMM and particle counters (latex calibration).

The ammonium sulfate particles doped with 0.3% and 1% methylene blue were spheres and the 4% particles (produced by ultrasonic nebulization instead

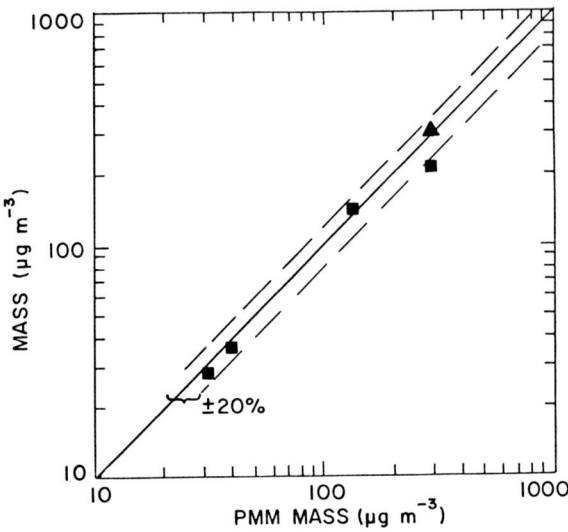

FIGURE 4. Mass concentrations for ammonium sulfate particles from simultaneous measurements with particle mass monitor (PMM), particle counter (■), and filters (▲).

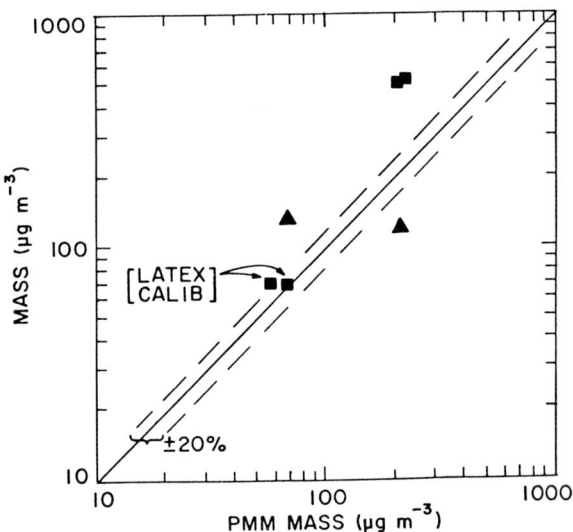

FIGURE 5. Mass concentrations for ammonium sulfate particles doped with methylene blue from simultaneous measurements with particle mass monitor (PMM), particle counters (■), and filters (▲).

GENERATION AND CHARACTERIZATION OF AEROSOLS

of bubbling the solution) were cubic, as seen in Appendix B. Nevertheless, the particle counters apparently responded to the cubes as spheres. Inspecting the size distributions (Appendix A) and particle photographs (Appendix B) for the 0.3% and 4% doped particles, it can be seen that they are consistent. The 0.3% particles are primarily submicron with a few supermicron particles as seen in the photographs, and the mass distribution has a peak in the sub- and super-micron sizes as seen in the size distributions. The 4% particles are primarily submicron as seen in the photographs and the mass distribution is peaked in the submicron as seen in the size distributions. This agreement supports the conclusion that the PMM and particle-counter-derived mass concentrations (latex calibration) agree to within ± 20% for the ammonium sulfate particles doped with methylene blue.

The mass concentrations for the carbonaceous particles are shown in Fig. 6. It is clear from the figure that there is little correlation between the simultaneous PMM and particle-counter-derived mass concentrations. This was due to the fact that the PMM would drift systematically with time. Starting with clean crystals, the PMM values would be the highest. Then, the values would continuously decrease. It was found that the aggregated nature of the particles (see photographs in Appendix B) prevented a good attachment of the particles to the crystals. Consequently, as more mass deposited, the particles

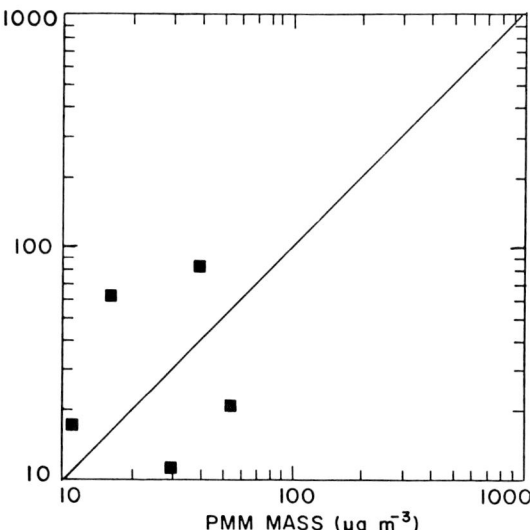

FIGURE 6. Mass concentrations for carbonaceous particles from simultaneous measurements with particle mass monitor (PMM), particle counter (■).

vibrated out of phase with the substrate and caused an apparent loss of mass. Inspection of the particle size distributions for carbonaceous particles (Appendix A) shows consistent bimodal mass distributions and reproducible mass concentrations: 77 ± 6 μg m^{-1} (for those experiments with similar generation and transport configurations). Consequently, it is concluded that the mass concentrations of the carbonaceous particles determined from the particle size and number measurements are more representative than those measured with the PMM.

An external mixture of carbonaceous particles and ammonium sulfate particles was produced by first generating the soot particles and then adding the salt particles. The mass concentrations measured for these mixtures are given in Fig. 7. It can be seen that the mass concentrations determined from simultaneous filter collections and PMM measurements were within ± 20% of each other. This result is reasonable, because most of the particle mass was from the salt particles as seen in the particle photographs (Appendix B, 6 and 7 August 1980, photographs). Further, it can be seen from the photographs that the carbonaceous and salt particles were aggregates. Three of the five comparisons of masses derived from the particle counter measurements and the filter collections were within ± 20% of each other. However, uncertainties with the calibration of the particle counters to such mixed particles (absorbers and nonabsorbers) reduces the meaning of these results. Consequently, it is concluded that the most representative mass concentration measurements for the soot-salt mixture were obtained with the PMM and the filters.

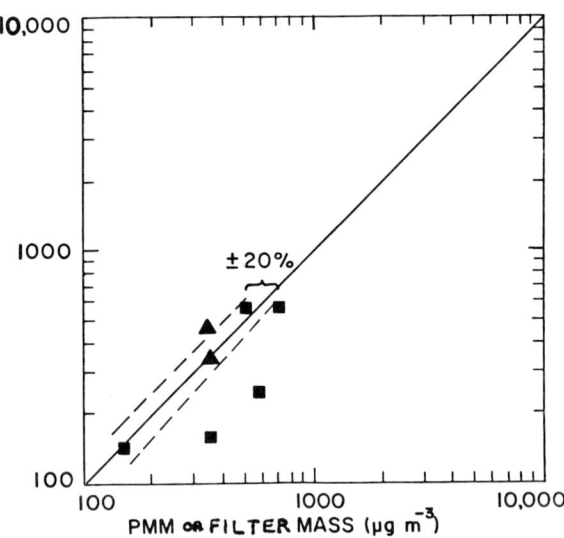

FIGURE 7. Mass concentrations for externally mixed carbonaceous and ammonium sulfate particles. Particle counter derived masses vs. filter collections (■). Filter collections vs. particle mass monitor measurements (▲).

GENERATION AND CHARACTERIZATION OF AEROSOLS 89

The mass concentrations of the methylene blue particles are given in Fig. 8. It can be seen that the mass concentrations determined from the particle measurements were highly dependent on the calibration used. The latex sphere calibration produced masses significantly lower ($\pm$ 100% lower) than the PMM masses, and the carbon calibration produced masses significantly higher than the PMM masses. However, two of the three simultaneous masses from the filters and PMM were within $\pm$ 80% of each other. Therefore, it is concluded that the masses from the particle counters are unreliable due to the uncertainties with the calibration. Thus, the most reliable mass concentration measurements for the methylene blue particles are from the PMM and the filters. The particles are spherical, as shown in Appendix B. This fact supports the conclusion the PMM mass data are reliable.

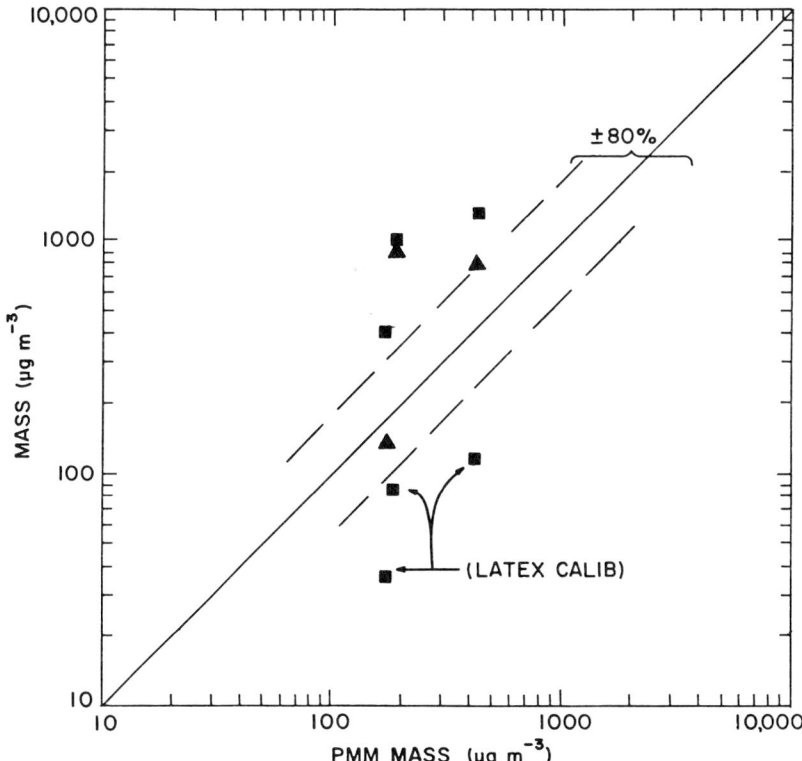

FIGURE 8. Mass concentrations for methylene blue particles from simultaneous measurements with particle mass monitor (PMM), particle counters (■), and filters (▲).

The mass concentrations for the ambient particle brought into the laboratory from outside are shown in Fig. 9. It can be seen from the figure that the mass concentrations determined from the PMM, the particle measurements, and the filters were within ± 30% of each other. This result indicates the mass concentrations from each instrument agree to ± 30%.

The Arizona road dust aerosol mass concentrations are listed in Table 6. It can be seen that the filter-derived and PMM mass concentrations are in good agreement (133 vs 104 μg m^{-3}). However, the mass concentrations from the particle measurements are almost a factor of 10 or greater. It is clear from the size distributions shown in Appendix A and the photographs in Appendix B that most of the mass was in supermicron particles. Apparently, the highly nonspherical particles were oversized by the OPC, and hence the mass concentrations were overestimated. It is concluded that the most representative road dust mass concentrations were obtained using the PMM and filters.

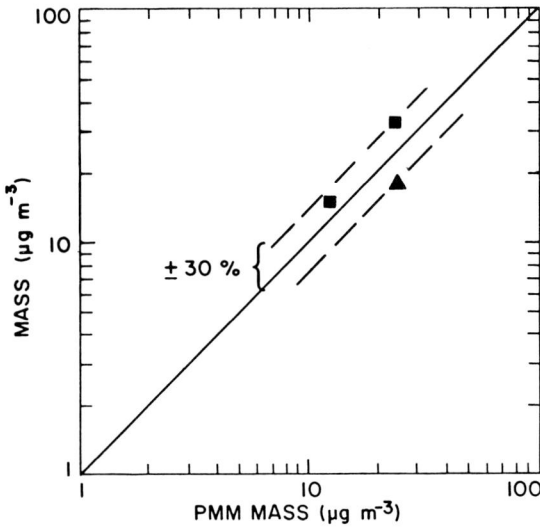

FIGURE 9. *Mass concentrations of ambient particles for simultaneous measurements with particle mass monitor (PMM), particle counters (■), and filters (▲).*

VI. CONCLUSIONS

Large quantities of stable aerosols were generated and transported to investigators using the generation and transport system described in this report. Briefly, submicron-sized, primarily spherical particles of ammonium sulfate and ammonium sulfate doped with methylene blue were generated by bubbling solutions. Pure spherical methylene blue particles were generated by ultrasonic nebulization. Chain-aggregate carbonaceous particles were produced by burning propane. External mixtures of carbonaceous and ammonium sulfate particles were produced; the resulting particles appeared to be aggregates. An Arizona road dust aerosol (highly nonspherical particles) was produced by resuspending the dust. Ambient particles were transported from outdoors for investigation.

The sizes (0.01 μm to 10 μm diameter) and number concentrations (10^{-7} cm^{-3} to 10^{-5} cm^{-3}) of the generated particles were determined using an electrical aerosol analyzer, an active scattering spectrometer probe, and an optical particle counter. The mass concentrations of the particles were determined in three ways: from measurements with a particle mass monitor (PMM), by integrating the particle size measurements, and by weighing particle collections. A combined analysis of the particle size distributions and mass concentrations revealed the following:

1. The size distributions and mass concentrations for the ammonium sulfate, ammonium-sulfate/methylene-blue, and ambient particles could be determined to between ± 20% to ± 30% agreement using the PMM, integrated particle size measurements, and filter collections.

2. The size distributions and mass concentrations of the carbonaceous particles were determined from the integrated particle size measurements. The PMM could not properly collect these soot particles.

3. The mass concentrations of the methylene blue particles and the externally mixed carbonaceous and ammonium sulfate particles were determined satisfactorily from the PMM and filter measurements. The size distributions of these particles are questionable due to uncertain calibrations of the optical particle counters.

4. The size distributions and mass concentrations of the ambient particles were determined to ± 30% agreement using the PMM, particle counter, and filter measurements.

5. The mass concentrations of the Arizona road dust were determined satisfactorily using the PMM and filter measurements. The optical particle counter was not calibrated for the highly nonspherical particles.

The authors recommend that future efforts in the field include:

1. Developing techniques for using the particle mass monitor to accurately determine mass concentrations of carbonaceous aerosols.

2. Determining proper calibrations for the ASASP-X and Royco instruments for methylene blue particles.

ACKNOWLEDGMENTS

The authors are grateful to Mr. C. Swain for developing most of the particle transport and generation system, Mr. D. Clair for performing the mass concentration measurements from the filter collections and arranging for the SEM photographs, Mr. S. Laxmann for developing the computer software to reduce the data tapes, and Mrs. L. McCall for plotting all of the particle size distributions and mass distributions.

REFERENCES

1. Gerber, H. E., Optical techniques for the measurement of light absorption by particulates, In *Particulate Carbon: Atmospheric Life Cycle* (G. T. Wolff and R. L. Klimisch, eds.), Plenum Press, New York (1981).
2. Grant, L. O., *The Second International Workshop on Condensation and Ice Nuclei,* Department Atmospheric Science, Colorado State University, Fort Collins, Colorado (1971).
3. Garvey, D. M., Measurement of high concentrations of ice nuclei by nine instruments, *Preprints-Conference on Cloud Physics,* pages 357-360, American Meteorological Society, Boston (1974).
4. Hindman, E. E., II, Garvey, D. M., Langer, G., Odencrantz, F. K., and Gregory, G. L., Laboratory investigations of cloud nuclei from combustion of space shuttle propellant, *J. Appl. Meteor.* 19, 175-184 (1980).
5. Liu, B. Y. H., *Fine Particles: Aerosol Generation, Measurement, Sampling and Analysis,* Academic Press, New York (1976).
6. Lundgren, D. A., Harris, F. S., Jr., Marlow, W. H., Lippmann, M., Clark, W. E., and Durham, M. D., *Aerosol Measurement,* University Presses of Florida, Gainesville (1979).
7. Clarke, A. D., and Waggoner, A. P., Results from University of Washington participation in First International Workshop on Light Absorption by Aerosol Particles, In *Light Absorption by Aerosol Particles* (H. E. Gerber and E. E. Hindman, eds.), Spectrum Press, Hampton, Virginia (1982).
8. Liu, B. Y. H., and Pui, D. Y. H., On the performance of the electrical aerosol analyzer, *Aerosol Sci.* 6, 249 (1981)
9. Garvey, D. M., Pinnick, R. G., and Bruce, C. W., Calibration and application of two Knollenberg optical particle counters, In *Proceedings Eighth International Conference on Cloud Physics,* pages 677-679 (1980).
10. Bodhaine, B. A., The GMCC four-wavelength nephelometer, In *Light Absorption by Aerosol Particles* (H. E. Gerber and E. E. Hindman, eds.), Spectrum Press, Hampton, Virginia (1982).

APPENDIX A

Tables and Graphs for Particle-Size and Particle-Mass Distributions

Symbols and definitions: d_g = *geometric mean diameter of channel size interval;* $\Delta N/\Delta d$ = *number of particles per size interval per aerosol volume;* $\Delta M_v = (\pi/6)d_g^3 \varrho \Delta N$ = *mass of particles per aerosol volume in each size interval.*

Graphs are shown in Figs. A-1—A-35 and the corresponding tabulations are shown in Tables A-1—A-9.

TABLE A-1. Size and Mass Distribution with Latex and Carbon Calibrations Corresponding to Figs. A-1–A-4.

Size interval (μm)	d_g (μm)	Latex (Fig. A-1) $\Delta N/\Delta d$ ($cm^{-3}\,\mu m^{-1}$)	Latex (Fig. A-1) ΔM_v ($\mu g\,m^{-3}$)	Latex (Fig. A-2) $\Delta N/\Delta d$ ($cm^{-3}\,\mu m^{-1}$)	Latex (Fig. A-2) ΔM_v ($\mu g\,m^{-3}$)	Carbon (Fig. A-3) $\Delta N/\Delta d$ ($cm^{-3}\,\mu m^{-1}$)	Carbon (Fig. A-3) ΔM_v ($\mu g\,m^{-3}$)	Carbon (Fig. A-4) $\Delta N/\Delta d$ ($cm^{-3}\,\mu m^{-1}$)	Carbon (Fig. A-4) ΔM_v ($\mu g\,m^{-3}$)
0.006–0.01	0.0077	---	---	---	---	---	---	---	---
0.01–0.018	0.013	---	---	8.9×10^6	1.57×10^{-1}	---	---	3.8×10^6	7.61×10^{-2}
0.018–0.032	0.024	4.5×10^6	7.1×10^{-4}	6.8×10^6	1.18×10^0	3.0×10^6	5.6×10^{-1}	3.5×10^6	6.75×10^{-1}
0.032–0.056	0.042	1.0×10^6	1.9×10^{-3}	7.6×10^5	1.27×10^0	1.2×10^6	2.2×10^0	9.8×10^5	1.85×10^0
0.056–0.100	0.074	8.5×10^5	1.8×10^{-2}	6.7×10^5	1.15×10^1	5.9×10^5	1.2×10^1	7.3×10^5	1.41×10^1
0.1–0.146	0.12	---	---	---	---	1.1×10^5	9.97×10^0	1.0×10^5	9.038×10^0
0.100–0.181	0.140	1.2×10^5	2.5×10^{-2}	9.4×10^4	1.876×10^1	---	---	---	---
0.146–0.230	0.18	---	---	---	---	5.4×10^4	2.831×10^1	4.9×10^4	2.582×10^1
0.181–0.280	0.22	5.5×10^4	7.2×10^{-2}	4.6×10^4	4.683×10^1	---	---	---	---
0.230–0.4	0.30	---	---	---	---	5.9×10^3	2.38×10^1	5.5×10^3	2.101×10^1
0.280–0.520	0.38	1.2×10^4	1.2×10^{-1}	9.0×10^3	1.006×10^2	---	---	---	---
0.4–0.8	0.57	---	---	---	---	9.5×10^2	7.18×10^1	8.5×10^2	6.47×10^1
0.520–0.76	0.63	1.3×10^2	3.2×10^{-2}	2.1×10^2	8.98×10^0	---	---	---	---
0.76–1.24	0.97	1.5×10^0	4×10^{-4}	3.6×10^0	1.063×10^0	---	---	---	---
0.8–1.31	1.0	---	---	---	---	7.8×10^2	3.723×10^2	7.9×10^2	3.77×10^2
1.24–2.04	1.6	2.3×10^{-2}	1.2×10^{-4}	3.1×10^{-2}	1.237×10^{-1}	---	---	---	---
1.31–1.92	1.6	---	---	---	---	6.6×10^0	1.622×10^1	7.0×10^0	1.788×10^1
1.92–3.0	2.4	---	---	---	---	2.2×10^{-1}	2.68×10^0	2.1×10^{-1}	2.704×10^0
2.04–3.0	2.5	---	---	---	---	---	---	---	---
3.0–5.0	3.9	3.0×10^{-3}	3.2×10^{-4}	1.1×10^{-2}	1.16×10^0	7.5×10^{-4}	9.12×10^{-2}	3.2×10^{-4}	3.91×10^{-2}
5.0–7.0	5.9	3.4×10^{-4}	1.2×10^{-4}	7.2×10^{-5}	2.74×10^{-2}	2.5×10^{-4}	1.08×10^{-1}	1.4×10^{-4}	6.19×10^{-2}
7.0–9.0	7.9	---	---	---	---	---	---	3.6×10^{-5}	3.74×10^{-2}

GENERATION AND CHARACTERIZATION OF AEROSOLS 95

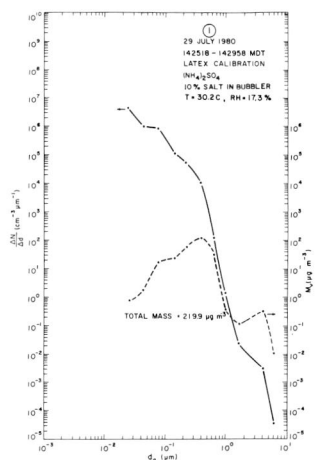
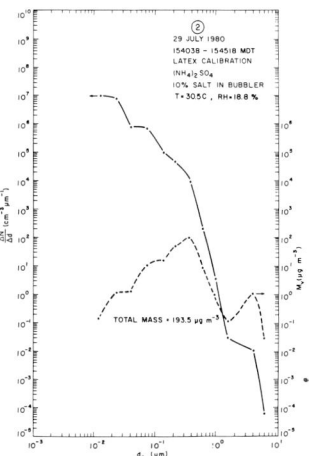
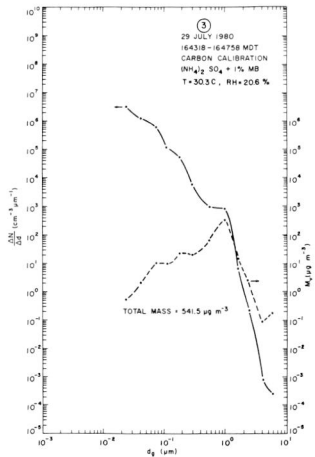
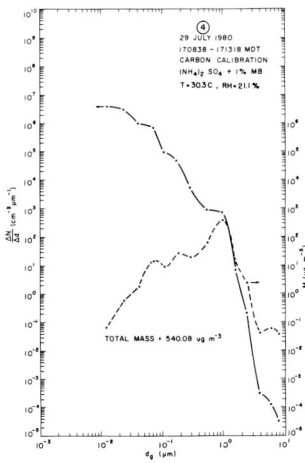

TABLE A-2. Size and Mass Distribution with Latex and Carbon Calibrations Corresponding to Figs. A-5—A-8.

Size interval (μm)	d_g (μm)	Latex (Fig. A-5) $\Delta N/\Delta d$ ($cm^{-3}\,\mu m^{-1}$)	ΔM_v ($\mu g\,m^{-3}$)	Latex (Fig. A-6) $\Delta N/\Delta d$ ($cm^{-3}\,\mu m^{-1}$)	ΔM_v ($\mu g\,m^{-3}$)	Latex (Fig. A-7) $\Delta N/\Delta d$ ($cm^{-3}\,\mu m^{-1}$)	ΔM_v ($\mu g\,m^{-3}$)	Carbon (Fig. A-8) $\Delta N/\Delta d$ ($cm^{-3}\,\mu m^{-1}$)	ΔM_v ($\mu g\,m^{-3}$)
0.006-0.01	0.0077	---	---	---	---	---	---	---	---
0.01-0.018	0.013	5.6×10^5	9.5×10^{-3}	---	---	9.4×10^5	1.6×10^{-2}	---	---
0.018-0.032	0.024	1.3×10^6	2.2×10^{-1}	4.4×10^5	5.0×10^{-4}	1.0×10^6	1.8×10^{-1}	2.0×10^6	3.8×10^{-1}
0.032-0.056	0.042	3.0×10^6	5.3×10^{-1}	2.5×10^5	4.2×10^{-1}	2.6×10^5	4.6×10^{-1}	9.6×10^5	1.8×10^0
0.056-0.10	0.074	2.3×10^5	3.9×10^0	1.1×10^5	1.8×10^0	2.0×10^5	3.6×10^0	6.4×10^5	1.2×10^1
0.10-0.146	0.12							2.4×10^5	1.9×10^1
0.10-0.181	0.14	3.9×10^4	7.2×10^0	1.4×10^4	2.8×10^0	3.5×10^4	6.7×10^0		
0.146-0.230	0.18							3.8×10^4	1.4×10^1
0.181-0.280	0.22	1.6×10^4	1.7×10^1	5.2×10^3	5.8×10^0	1.4×10^4	1.4×10^1		
0.230-0.40	0.30							4.0×10^2	1.3×10^0
0.280-0.520	0.38	3.3×10^3	4.4×10^1	1.2×10^3	1.5×10^1	3.2×10^3	3.7×10^1		
0.40-0.80	0.57							1.6×10^1	1.2×10^0
0.520-0.76	0.63	4.0×10^2	1.9×10^1	1.4×10^2	7.8×10^0	1.1×10^2	5.6×10^0		
0.76-1.24	0.97	5.6×10^1	1.8×10^1	4.0×10^1	1.4×10^1	7.1×10^0	2.2×10^0		
0.80-1.31	1.0							1.2×10^1	7.6×10^0
1.24-2.04	1.6	1.5×10^0	3.5×10^0	2.1×10^0	4.8×10^0	1.1×10^{-1}	2.4×10^{-1}		
1.31-1.92	1.6							2.0×10^{-1}	5.4×10^0
1.92-3.0	2.4							5.8×10^{-1}	1.1×10^1
2.04-3.0	2.5	1.8×10^{-2}	1.1×10^{-1}	5.2×10^{-2}	4.6×10^{-1}	---	---		
3.0-5.0	3.9	2.5×10^{-1}	2.6×10^1	1.5×10^{-1}	1.5×10^1	1.1×10^{-3}	1.8×10^{-1}	4.1×10^{-2}	5.1×10^0
5.0-7.0	5.9	7.0×10^{-4}	3.2×10^{-1}	5.5×10^{-4}	2.2×10^{-1}	---	---	7.0×10^{-3}	3.0×10^0
7.0-9.0	7.9	---	---	7.0×10^{-5}	7.0×10^{-2}	---	---	1.5×10^{-3}	1.5×10^0

GENERATION AND CHARACTERIZATION OF AEROSOLS 97

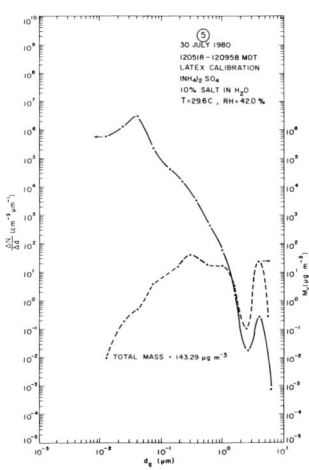

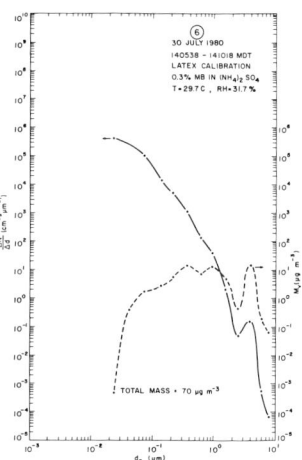

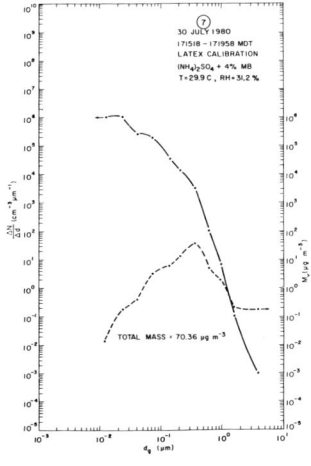

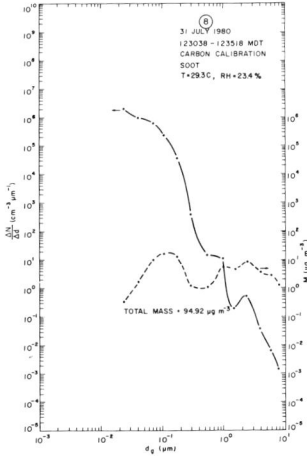

TABLE A-3. *Size and Mass Distribution with Latex and Carbon Calibrations Corresponding to Figs. A-9—A-12.*

Size interval (μm)	d_g (μm)	Latex (Fig. A-9) $\Delta N/\Delta d$ ($cm^{-3} \mu m^{-1}$)	Latex (Fig. A-9) ΔM_v ($\mu g\, m^{-3}$)	Carbon (Fig. A-10) $\Delta N/\Delta d$ ($cm^{-3} \mu m^{-1}$)	Carbon (Fig. A-10) ΔM_v ($\mu g\, m^{-3}$)	Carbon (Fig. A-11) $\Delta N/\Delta d$ ($cm^{-3} \mu m^{-1}$)	Carbon (Fig. A-11) ΔM_v ($\mu g\, m^{-3}$)	Latex (Fig. A-12) $\Delta N/\Delta d$ ($cm^{-3} \mu m^{-1}$)	Latex (Fig. A-12) ΔM_v ($\mu g\, m^{-3}$)
0.006-0.01	0.0077	2.2×10^9	4.4×10^0	---	---	---	---	---	---
0.01-0.018	0.013	---	---	---	---	1.6×10^5	3.1×10^{-3}	1.6×10^5	2.75×10^{-3}
0.018-0.032	0.024	4.6×10^7	7.9×10^0	---	---	5.9×10^4	1.2×10^{-2}	6.0×10^4	1.03×10^{-2}
0.032-0.056	0.042	4.6×10^6	7.6×10^0	6.3×10^5	1.2×10^0	7.1×10^3	1.3×10^{-2}	7.3×10^3	1.2×10^{-2}
0.056-0.10	0.074	1.8×10^6	3.1×10^1	6.4×10^5	1.2×10^1	2.0×10^3	3.8×10^{-2}	2.0×10^3	3.5×10^{-2}
0.10-0.146	0.12	---	---	2.0×10^5	1.6×10^1	2.8×10^2	2.7×10^{-2}	---	---
0.10-0.181	0.14	1.1×10^5	2.3×10^1	---	---	---	---	2.0×10^2	4.6×10^{-1}
0.146-0.230	0.18	3.4×10^4	3.0×10^1	3.0×10^4	1.2×10^1	3.3×10^2	2.2×10^{-1}	---	---
0.181-0.280	0.22	---	---	---	---	---	---	2.3×10^2	2.8×10^{-1}
0.230-0.40	0.30	---	---	3.3×10^2	1.1×10^0	4.0×10^2	1.1×10^0	---	---
0.280-0.520	0.38	2.8×10^3	3.5×10^1	---	---	---	---	1.2×10^3	1.8×10^1
0.40-0.80	0.57	---	---	1.3×10^1	1.0×10^0	1.5×10^2	2.1×10^1	---	---
0.520-0.76	0.63	1.4×10^2	3.2×10^0	---	---	---	---	1.7×10^2	8.1×10^0
0.76-1.24	0.97	1.8×10^0	5.5×10^{-1}	---	---	---	---	1.2×10^1	3.8×10^{-1}
0.80-1.31	1.0	---	---	1.0×10^1	6.5×10^{-1}	2.9×10^2	1.6×10^2	---	---
1.24-2.04	1.6	8.4×10^{-2}	1.9×10^{-2}	---	---	---	---	2.1×10^{-1}	4.4×10^{-1}
1.31-1.92	1.6	---	---	1.5×10^0	4.5×10^{-1}	6.9×10^1	1.4×10^2	---	---
1.92-3.0	2.4	---	---	5.3×10^{-1}	8.5×10^0	4.7×10^0	7.1×10^1	---	---
2.04-3.0	2.5	---	---	---	---	---	---	---	---
3.0-5.0	3.9	8.0×10^{-3}	8.5×10^{-1}	4.3×10^{-2}	5.2×10^0	4.7×10^{-2}	5.6×10^0	4.7×10^{-2}	5.0×10^0
5.0-7.0	5.9	8.5×10^{-4}	3.5×10^{-1}	7.0×10^{-3}	3.3×10^0	9.0×10^{-4}	4.0×10^{-1}	9.3×10^{-4}	3.6×10^{-1}
7.0-9.0	7.9	1.8×10^{-4}	1.7×10^{-1}	2.2×10^{-3}	2.3×10^0	1.4×10^{-4}	1.5×10^{-1}	1.4×10^{-4}	1.3×10^{-1}

GENERATION AND CHARACTERIZATION OF AEROSOLS 99

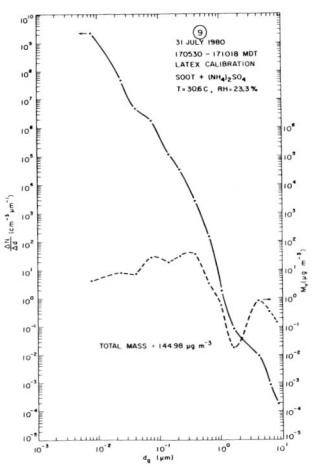
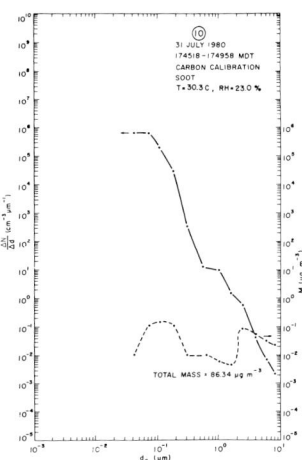
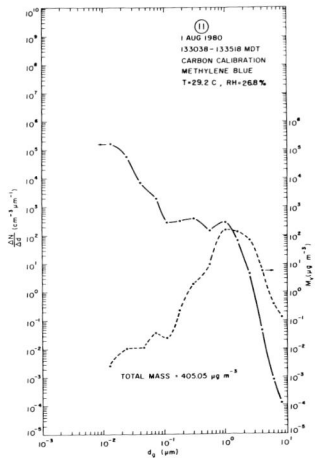
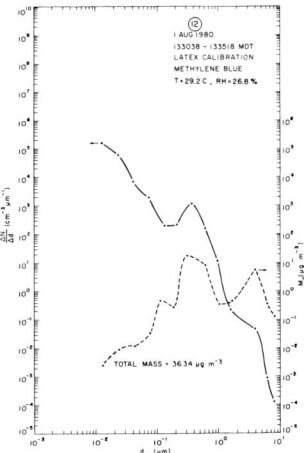

TABLE A-4. Size and Mass Distribution with Latex and Carbon Calibrations Corresponding to Figs. A-13–A-16.

Size interval (μm)	d_g (μm)	Carbon (Fig. A-13) $\Delta N/\Delta d$ ($cm^{-3} \mu m^{-1}$)	ΔM_v ($\mu g\ m^{-3}$)	Carbon (Fig. A-14) $\Delta N/\Delta d$ ($cm^{-3} \mu m^{-1}$)	ΔM_v ($\mu g\ m^{-3}$)	Latex (Fig. A-15) $\Delta N/\Delta d$ ($cm^{-3} \mu m^{-1}$)	ΔM_v ($\mu g\ m^{-3}$)	Carbon (Fig. A-16) $\Delta N/\Delta d$ ($cm^{-3} \mu m^{-1}$)	ΔM_v ($\mu g\ m^{-3}$)
0.006-0.01	0.0077	---	---	---	---	---	---	---	---
0.01-0.018	0.013	---	---	---	---	---	---	---	---
0.018-0.032	0.024	9.3×10^5	1.7×10^{-1}	4.1×10^6	8.2×10^{-2}	---	---	---	---
0.032-0.056	0.042	6.7×10^5	1.1×10^0	1.1×10^6	2.2×10^{-1}	3.3×10^6	5.6×10^{-1}	2.1×10^6	4.2×10^{-1}
0.056-0.10	0.074	3.2×10^5	6.2×10^0	4.0×10^5	7.5×10^{-1}	4.6×10^5	7.7×10^{-1}	9.2×10^5	1.7×10^0
0.10-0.146	0.12	1.8×10^5	1.5×10^1	1.8×10^5	3.4×10^0	8.0×10^5	1.4×10^1	3.1×10^5	6.1×10^0
0.10-0.181	0.14			1.8×10^5	1.6×10^1			1.7×10^5	1.4×10^1
0.146-0.230	0.18	2.5×10^4	9.5×10^0	2.8×10^4	1.1×10^1	1.5×10^5	2.9×10^1	2.2×10^4	9.0×10^0
0.181-0.280	0.22								
0.230-0.40	0.30	2.6×10^2	8.5×10^{-1}	3.1×10^2	1.0×10^{-1}	4.5×10^4	4.2×10^1	2.4×10^2	8.1×10^{-1}
0.280-0.520	0.38								
0.40-0.80	0.57	1.0×10^2	2.9×10^{-1}	1.2×10^1	9.0×10^{-1}	6.6×10^3	7.9×10^1	1.1×10^1	8.3×10^{-1}
0.52-0.76	0.63					5.8×10^2	2.8×10^1		
0.76-1.24	0.97					3.5×10^1	1.1×10^1		
0.80-1.31	1.0	7.5×10^0	4.1×10^0	1.1×10^1	6.3×10^0			8.0×10^0	4.8×10^0
1.24-2.04	1.6					6.9×10^{-1}	1.4×10^0		
1.31-1.92	1.6	2.1×10^0	4.8×10^0	1.4×10^0	3.7×10^0			1.1×10^0	3.6×10^0
1.92-3.0	2.4	2.2×10^{-1}	3.6×10^0	5.4×10^{-1}	9.0×10^0			4.7×10^{-1}	8.1×10^0
2.04-3.0	2.5					8.7×10^{-3}	1.4×10^{-1}		
3.0-5.0	3.9	1.7×10^{-2}	2.2×10^0	3.3×10^{-2}	4.1×10^0	2.8×10^{-1}	3.0×10^1	3.1×10^{-2}	3.7×10^0
5.0-7.0	5.9	3.3×10^{-3}	1.4×10^0	4.8×10^{-3}	2.1×10^0	7.0×10^{-3}	2.7×10^0	3.9×10^{-3}	1.7×10^0
7.0-9.0	7.9	1.2×10^{-3}	1.2×10^0	1.9×10^{-3}	2.0×10^0	2.0×10^{-3}	1.9×10^0	6.4×10^{-4}	6.7×10^{-1}

GENERATION AND CHARACTERIZATION OF AEROSOLS

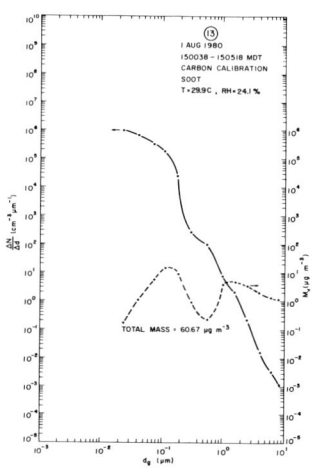

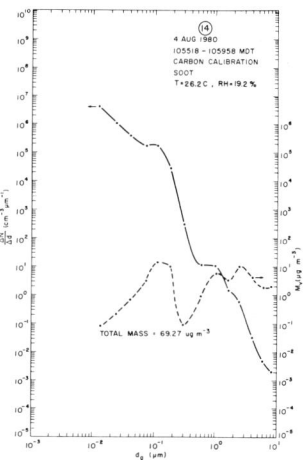

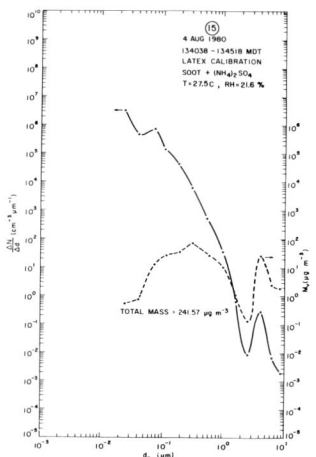

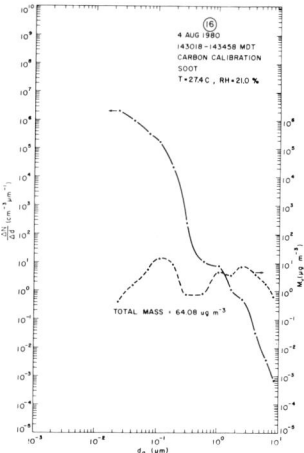

TABLE A-5. Size and Mass Distribution with Latex and Carbon Calibrations Corresponding to Figs. A-17—A-20.

Size interval (μm)	d_g (μm)	Carbon (Fig. A-17) ΔN/Δd (cm^{-3} μm^{-1})	Carbon (Fig. A-17) ΔM$_v$ (μg m^{-3})	Latex (Fig. A-18) ΔN/Δd (cm^{-3} μm^{-1})	Latex (Fig. A-18) ΔM$_v$ (μg m^{-3})	Latex (Fig. A-19) ΔN/Δd (cm^{-3} μm^{-1})	Latex (Fig. A-19) ΔM$_v$ (μg m^{-3})	Latex (Fig. A-20) ΔN/Δd (cm^{-3} μm^{-1})	Latex (Fig. A-20) ΔM$_v$ (μg m^{-3})
0.006-0.01	0.0077	---	---	---	---	---	---	---	---
0.01-0.018	0.013	4.8x10^4	9.5x10^{-4}	4.8x10^4	8.4x10^{-4}	8.9x10^5	1.6x10^{-2}	2.1x10^6	3.6x10^{-2}
0.018-0.032	0.024	1.0x10^5	2.0x10^{-2}	1.0x10^5	1.8x10^{-2}	4.9x10^5	8.5x10^{-2}	5.8x10^5	1.0x10^{-1}
0.032-0.056	0.042	5.7x10^4	1.1x10^{-1}	5.7x10^4	9.5x10^{-2}	8.1x10^4	1.4x10^{-1}	1.5x10^5	2.6x10^{-1}
0.056-0.10	0.074	1.3x10^4	2.5x10^{-1}	1.3x10^4	2.2x10^{-1}	4.7x10^4	8.1x10^{-1}	7.1x10^4	1.2x10^0
0.10-0.146	0.12	7.9x10^2	7.5x10^{-2}	5.7x10^2	1.1x10^{-1}	6.4x10^3	1.2x10^0	1.6x10^4	3.0x10^0
0.10-0.181	0.14	---	---	---	---	---	---	---	---
0.146-0.230	0.18	1.0x10^3	6.3x10^{-1}	6.9x10^2	8.0x10^{-1}	1.3x10^3	1.2x10^0	3.8x10^3	3.6x10^0
0.181-0.280	0.22	---	---	---	---	---	---	---	---
0.230-0.40	0.30	7.9x10^2	3.9x10^0	2.5x10^3	3.9x10^1	1.0x10^2	1.1x10^0	3.6x10^2	3.9x10^0
0.280-0.520	0.38	3.2x10^2	2.4x10^1	---	---	---	---	---	---
0.40-0.80	0.57	---	---	4.8x10^2	2.3x10^1	6.0x10^0	3.2x10^{-1}	1.2x10^1	6.2x10^{-1}
0.52-0.76	0.63	---	---	3.7x10^1	1.2x10^1	1.5x10^0	5.7x10^{-1}	3.2x10^0	1.2x10^0
0.76-1.24	0.97	---	---	---	---	---	---	---	---
0.80-1.31	1.0	7.3x10^2	4.7x10^2	3.4x10^{-1}	6.7x10^{-1}	2.1x10^{-1}	6.2x10^{-1}	5.5x10^{-1}	1.5x10^0
1.24-2.04	1.6	---	---	---	---	---	---	---	---
1.31-1.92	1.6	1.1x10^2	3.0x10^1	---	---	---	---	---	---
1.92-3.0	2.4	1.5x10^1	9.3x10^0	---	---	6.9x10^{-2}	8.8x10^{-1}	1.0x10^{-1}	1.2x10^0
2.04-3.0	2.5	---	---	---	---	---	---	---	---
3.0-5.0	3.9	7.6x10^{-2}	9.3x10^0	7.6x10^{-2}	8.2x10^0	4.8x10^{-2}	5.1x10^0	9.5x10^{-2}	1.0x10^1
5.0-7.0	5.9	3.9x10^{-4}	1.7x10^{-1}	3.9x10^{-4}	1.5x10^{-1}	5.4x10^{-3}	2.1x10^0	1.2x10^{-2}	4.6x10^0
7.0-9.0	7.9	---	---	---	---	7.9x10^{-4}	7.3x10^{-1}	2.8x10^{-3}	2.6x10^0

GENERATION AND CHARACTERIZATION OF AEROSOLS 103

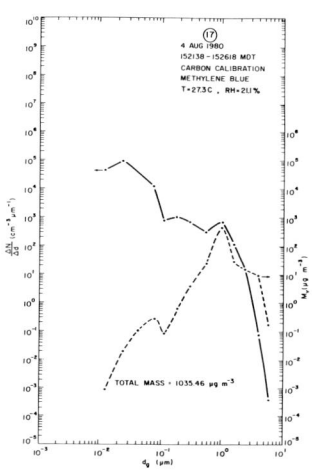
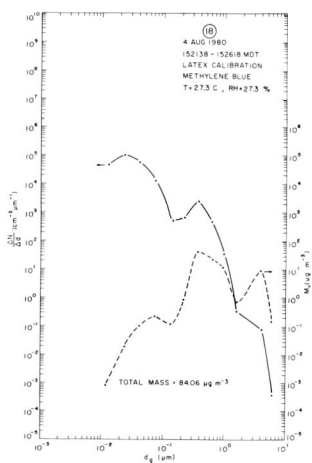
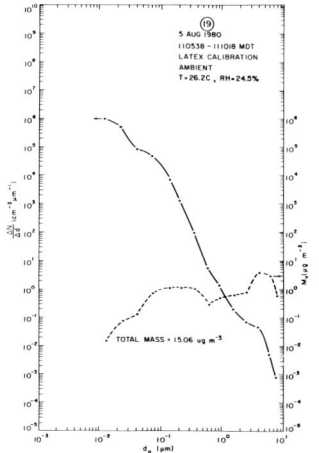
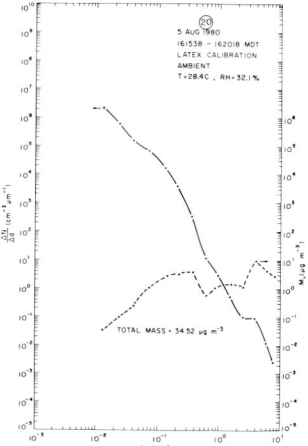

TABLE A-6. Size and Mass Distribution with Latex and Carbon Calibrations Corresponding to Figs. A-21—A-24.

Size interval (μm)	d_g (μm)	Latex (Fig. A-21) $\Delta N/\Delta d$ ($cm^{-3} \mu m^{-1}$)	ΔM_v ($\mu g\ m^{-3}$)	Latex (Fig. A-22) $\Delta N/\Delta d$ ($cm^{-3} \mu m^{-1}$)	ΔM_v ($\mu g\ m^{-3}$)	Carbon (Fig. A-23) $\Delta N/\Delta d$ ($cm^{-3} \mu m^{-1}$)	ΔM_v ($\mu g\ m^{-3}$)	Latex (Fig. A-24) $\Delta N/\Delta d$ ($cm^{-3} \mu m^{-1}$)	ΔM_v ($\mu g\ m^{-3}$)
0.006-0.01	0.0077	---	---	---	---	---	---	---	---
0.01-0.018	0.013	$1.8x10^{-3}$	$1.0x10^5$	$5.2x10^4$	$9.2x10^{-4}$	---	---	$1.7x10^6$	$3.0x10^{-2}$
0.018-0.032	0.024	$1.2x10^{-2}$	$7.1x10^4$	$2.5x10^5$	$4.4x10^{-2}$	---	---	$1.5x10^7$	$2.6x10^0$
0.032-0.056	0.042	$3.1x10^{-2}$	$1.9x10^5$	$8.5x10^3$	$1.4x10^{-2}$	$3.9x10^5$	$7.3x10^{-1}$	---	---
0.056-0.10	0.074	$2.2x10^{-1}$	$1.3x10^4$	$1.6x10^4$	$2.7x10^{-1}$	$4.9x10^5$	$9.5x10^0$	$2.3x10^6$	$3.9x10^1$
0.10-0.146	0.12					$9.2x10^4$	$7.5x10^0$		
0.10-0.181	0.14	$7.2x10^{-1}$	$3.8x10^3$	$4.6x10^3$	$9.1x10^{-1}$			$1.3x10^5$	$2.6x10^1$
0.146-0.230	0.18					$9.6x10^3$	$3.8x10^0$		
0.181-0.280	0.22	$9.4x10^{-1}$	$1.0x10^2$	$1.9x10^3$	$1.9x10^0$	$8.5x10^1$	$2.8x10^{-1}$	$6.8x10^4$	$6.9x10^1$
0.230-0.40	0.30								
0.280-0.520	0.38	$6.3x10^0$	$6.7x10^1$	$5.0x10^2$	$6.4x10^0$	$3.7x10^0$	$2.8x10^{-1}$	$1.5x10^4$	$1.8x10^2$
0.40-0.80	0.57								
0.520-0.76	0.63	$6.6x10^{-2}$	$1.3x10^0$	$1.2x10^2$	$6.8x10^0$			$1.8x10^3$	$8.8x10^1$
0.76-1.24	0.97	$1.2x10^{-1}$	$3.1x10^{-1}$	$3.8x10^1$	$1.5x10^1$			$2.6x10^2$	$8.7x10^1$
0.80-1.31	1.0					$2.7x10^0$	$1.6x10^0$		
1.24-2.04	1.6	$2.0x10^{-1}$	$7.4x10^{-2}$	$1.0x10^2$	$2.9x10^1$			$2.4x10^0$	$4.5x10^0$
1.31-1.92	1.6					$3.8x10^{-1}$	$1.1x10^0$		
1.92-3.0	2.4					$1.3x10^{-1}$	$2.4x10^0$		
2.04-3.0	2.5	$9.1x10^{-3}$	$9.5x10^{-3}$	$3.2x10^0$	$4.3x10^1$			$8.7x10^{-3}$	$9.1x10^{-2}$
3.0-5.0	3.9	$1.3x10^0$	$1.2x10^{-2}$	$3.5x10^0$	$3.7x10^2$	$8.4x10^{-3}$	$1.0x10^0$	$1.4x10^1$	$1.6x10^3$
5.0-7.0	5.9	$7.6x10^{-1}$	$2.0x10^{-3}$	$8.0x10^{-1}$	$3.1x10^2$	$1.7x10^{-3}$	$7.3x10^{-1}$	$6.5x10^{-2}$	$2.5x10^1$
7.0-9.0	7.9	$7.9x10^{-1}$	$8.5x10^{-4}$	$2.3x10^{-1}$	$2.1x10^2$	$5.0x10^{-4}$	$5.2x10^{-1}$	$1.6x10^{-3}$	$1.5x10^0$

GENERATION AND CHARACTERIZATION OF AEROSOLS 105

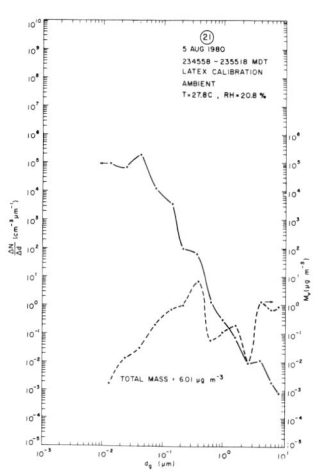

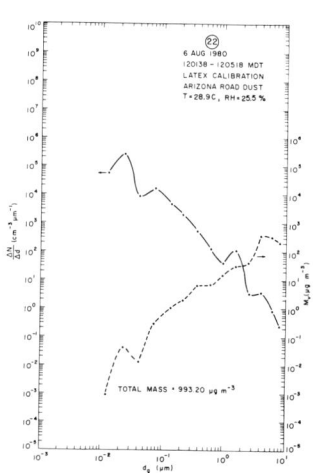

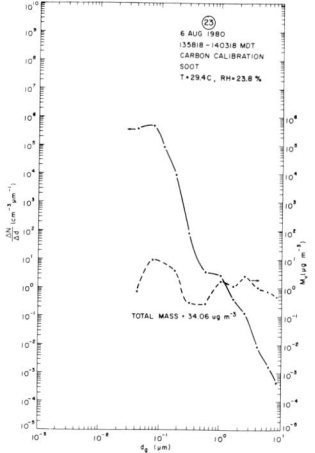

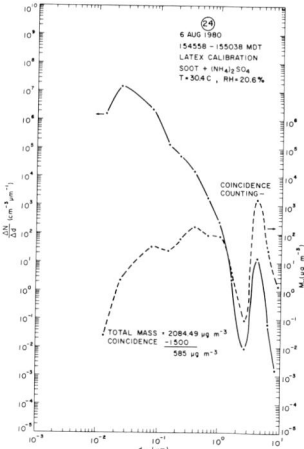

TABLE A-7. Size and Mass Distribution with Latex and Carbon Calibrations Corresponding to Figs. A-25—A-28.

Size interval (μm)	d_g (μm)	Carbon (Fig. A-25) $\Delta N/\Delta d$ ($cm^{-3}\,\mu m^{-1}$)	ΔM_v ($\mu g\,m^{-3}$)	Carbon (Fig. A-26) $\Delta N/\Delta d$ ($cm^{-3}\,\mu m^{-1}$)	ΔM_v ($\mu g\,m^{-3}$)	Latex (Fig. A-27) $\Delta N/\Delta d$ ($cm^{-3}\,\mu m^{-1}$)	ΔM_v ($\mu g\,m^{-3}$)	Carbon (Fig. A-28) $\Delta N/\Delta d$ ($cm^{-3}\,\mu m^{-1}$)	ΔM_v ($\mu g\,m^{-3}$)
0.006-0.01	0.0077	---	---	---	---	$2.7x10^9$	$4.2x10^0$	---	---
0.01-0.018	0.013	$1.5x10^7$	$2.9x10^{-1}$	$1.4x10^6$	$2.8x10^{-2}$	$4.2x10^4$	$7.3x10^{-4}$	$1.1x10^6$	$2.3x10^{-2}$
0.018-0.032	0.024	$2.6x10^6$	$5.1x10^{-1}$	$3.5x10^5$	$6.8x10^{-2}$	---	---	$2.0x10^5$	$3.9x10^{-2}$
0.032-0.056	0.042	$3.2x10^5$	$6.0x10^{-1}$	---	---	$1.5x10^3$	$2.4x10^{-3}$	$3.6x10^4$	$6.8x10^{-2}$
0.056-0.10	0.074	$8.4x10^5$	$1.6x10^1$	$1.7x10^5$	$3.4x10^0$	$2.8x10^6$	$4.8x10^1$	$9.8x10^4$	$1.9x10^0$
0.10-0.146	0.12	$2.0x10^6$	$1.7x10^1$	$4.3x10^4$	$2.9x10^0$			$3.2x10^4$	$2.7x10^0$
0.10-0.181	0.14					$1.9x10^5$	$3.7x10^1$		
0.146-0.230	0.18	$2.8x10^4$	$1.1x10^1$	$4.9x10^3$	$1.9x10^0$			$3.7x10^3$	$1.5x10^0$
0.181-0.280	0.22					$6.7x10^4$	$5.8x10^1$		
0.230-0.40	0.30	$2.9x10^2$	$9.5x10^{-1}$	$4.9x10^2$	$1.6x10^{-1}$			$4.3x10^1$	$1.5x10^{-2}$
0.280-0.520	0.38					$1.7x10^3$	$1.4x10^1$		
0.40-0.80	0.57	$1.0x10^1$	$7.8x10^{-1}$	$1.9x10^0$	$1.5x10^{-1}$			$1.4x10^0$	$1.0x10^{-1}$
0.52-0.76	0.63					$1.7x10^0$	$6.9x10^{-2}$		
0.76-1.24	0.97					$2.4x10^{-1}$	$1.1x10^{-1}$		
0.80-1.31	1.0	$9.2x10^0$	$5.7x10^0$	$1.4x10^0$	$8.5x10^{-1}$			$1.5x10^0$	$9.3x10^{-1}$
1.24-2.04	1.6					$5.2x10^{-2}$	$1.0x10^{-1}$		
1.31-1.92	1.6	$1.4x10^0$	$4.4x10^0$	$2.7x10^{-1}$	$6.4x10^{-1}$			$3.0x10^{-1}$	$9.7x10^{-1}$
1.92-3.0	2.4	$4.7x10^{-1}$	$8.0x10^0$	$1.1x10^{-1}$	$1.6x10^0$			$4.3x10^{-2}$	$8.3x10^{-1}$
2.04-3.0	2.5					---	---		
3.0-5.0	3.9	$4.5x10^{-2}$	$5.4x10^0$	$7.8x10^{-3}$	$9.4x10^1$	$8.9x10^{-3}$	$9.5x10^{-1}$	$6.6x10^{-3}$	$8.0x10^{-1}$
5.0-7.0	5.9	$6.9x10^{-3}$	$3.0x10^0$	$1.8x10^{-3}$	$7.6x10^1$	$1.1x10^{-3}$	$4.2x10^{-1}$	$1.0x10^{-3}$	$4.5x10^{-1}$
7.0-9.0	7.9	$1.4x10^{-3}$	$1.4x10^0$	$1.8x10^{-4}$	$1.9x10^1$	$6.8x10^{-4}$	$6.3x10^{-1}$	$1.4x10^{-4}$	$1.5x10^{-1}$

GENERATION AND CHARACTERIZATION OF AEROSOLS

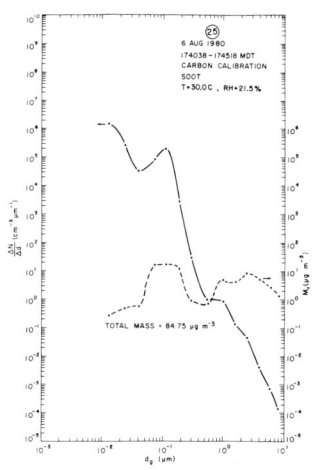

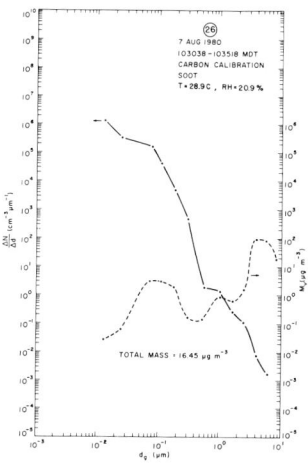

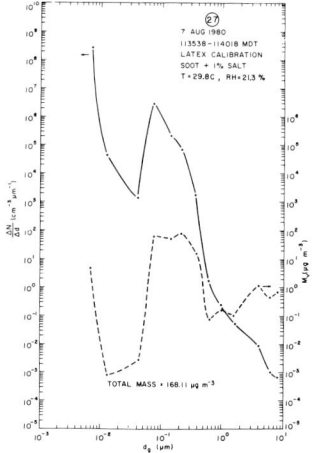

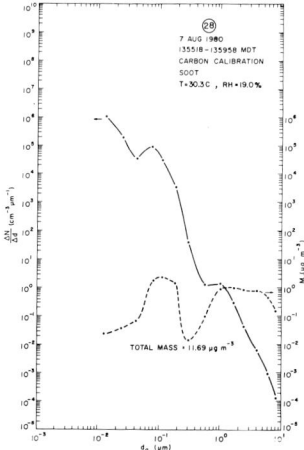

TABLE A-8. Size and Mass Distribution with Latex and Carbon Calibrations Corresponding to Figs. A-29—A-32.

Size interval (μm)	d_g (μm)	Latex (Fig. A-29) $\Delta N/\Delta d$ ($cm^{-3} \mu m^{-1}$)	ΔM_v ($\mu g\ m^{-3}$)	Carbon (Fig. A-30) $\Delta N/\Delta d$ ($cm^{-3} \mu m^{-1}$)	ΔM_v ($\mu g\ m^{-3}$)	Latex (Fig. A-31) $\Delta N/\Delta d$ ($cm^{-3} \mu m^{-1}$)	ΔM_v ($\mu g\ m^{-3}$)	Carbon (Fig. A-32) $\Delta N/\Delta d$ ($cm^{-3} \mu m^{-1}$)	ΔM_v ($\mu g\ m^{-3}$)
0.006-0.01	0.0077	---	---	---	---	---	---	$7.9x10^6$	$1.39x10^{-2}$
0.01-0.018	0.013	---	---	$5.2x10^7$	$1.0x10^0$	---	---	---	---
0.018-0.032	0.024	---	---	$4.6x10^6$	$8.9x10^{-1}$	---	---	$3.6x10^4$	$7.0x10^{-3}$
0.032-0.056	0.042	$9.7x10^5$	$1.61x10^0$	$6.5x10^5$	$1.2x10^0$	---	---	$7.3x10^3$	$1.4x10^{-2}$
0.056-0.10	0.074	$7.3x10^5$	$1.26x10^1$	$1.3x10^5$	$2.6x10^0$	---	---	$2.7x10^3$	$5.2x10^{-2}$
0.10-0.146	0.12	---	---	$5.3x10^4$	$4.4x10^0$	---	---	$9.1x10^2$	$8.5x10^{-2}$
0.10-0.181	0.14	$1.5x10^5$	$3.1x10^1$	---	---	$1.1x10^2$	$2.3x10^{-2}$	---	---
0.146-0.230	0.18	---	---	$5.9x10^3$	$2.4x10^0$	---	---	$1.0x10^3$	$6.4x10^{-1}$
0.181-0.280	0.22	$8.3x10^4$	$8.0x10^1$	---	---	$2.7x10^1$	$2.5x10^{-2}$	---	---
0.230-0.40	0.30	---	---	$9.2x10^1$	$1.1x10^0$	---	---	$4.0x10^2$	$1.8x10^0$
0.280-0.520	0.38	$4.4x10^3$	$3.6x10^1$	---	---	$1.1x10^0$	$1.0x10^{-2}$	---	---
0.40-0.80	0.57	---	---	$5.5x10^0$	$4.1x10^{-1}$	---	---	$2.7x10^2$	$2.1x10^1$
0.52-0.76	0.63	$2.6x10^0$	$1.0x10^{-1}$	---	---	---	---	---	---
0.76-1.24	0.97	$1.2x10^{-1}$	$3.9x10^{-2}$	---	---	---	---	---	---
0.80-1.31	1.0	---	---	$4.9x10^0$	$2.6x10^0$	---	---	$1.1x10^3$	$7.5x10^3$
1.24-2.04	1.6	$3.1x10^{-2}$	$9.3x10^{-2}$	---	---	---	---	---	---
1.31-1.92	1.6	---	---	$3.3x10^{-1}$	$9.4x10^{-1}$	---	---	$1.5x10^2$	$3.9x10^2$
1.92-3.0	2.4	---	---	$5.0x10^{-2}$	$9.0x10^{-1}$	---	---	$1.2x10^1$	$1.6x10^2$
2.04-3.0	2.5	---	---	---	---	---	---	---	---
3.0-5.0	3.9	$8.9x10^{-3}$	$9.5x10^{-1}$	$4.6x10^{-3}$	$5.6x10^{-1}$	$3.6x10^{-5}$	$3.8x10^{-3}$	$1.6x10^{-1}$	$1.9x10^1$
5.0-7.0	5.9	$7.2x10^{-4}$	$2.7x10^{-1}$	$7.4x10^{-4}$	$3.2x10^{-1}$	---	---	$1.1x10^{-4}$	$4.6x10^{-2}$
7.0-9.0	7.9	$1.1x10^{-4}$	$9.9x10^{-2}$	$2.5x10^{-4}$	$2.6x10^{-1}$	---	---	---	---

GENERATION AND CHARACTERIZATION OF AEROSOLS 109

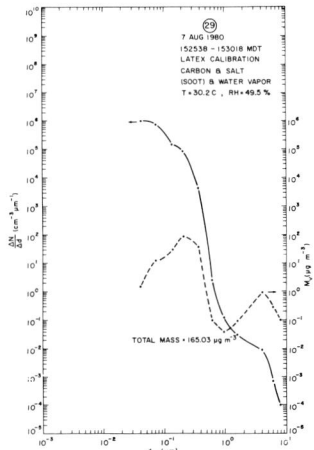

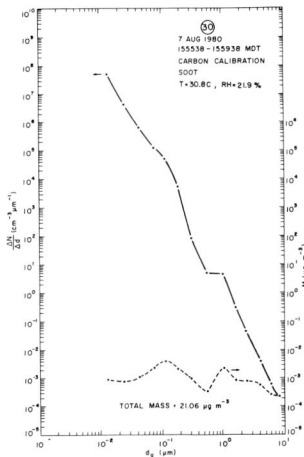

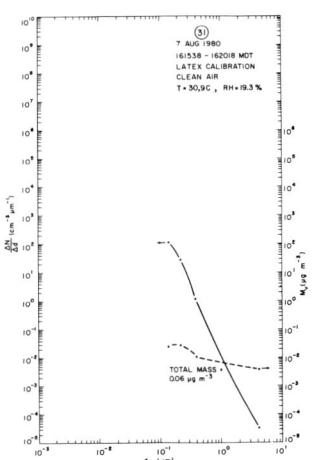

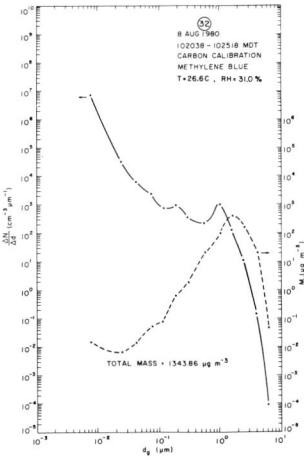

TABLE A-9. Size and Mass Distribution with Latex and Carbon Calibrations Corresponding to Figs. A-33—A-35.

Size interval (µm)	d_g (µm)	Latex (Fig. A-33) $\Delta N/\Delta d$ ($cm^{-3}\,\mu m^{-1}$)	ΔM_v (µg m^{-3})	Latex (Fig. A-34) $\Delta N/\Delta d$ ($cm^{-3}\,\mu m^{-1}$)	ΔM_v (µg m^{-3})	Latex (Fig. A-35) $\Delta N/\Delta d$ ($cm^{-3}\,\mu m^{-1}$)	ΔM_v (µg m^{-3})	$\Delta N/\Delta d$	ΔM_v
0.006-0.01	0.0077	$4.8x10^6$	$7.4x10^{-3}$	---	---	---	---		
0.01-0.018	0.013	---	---	$5.3x10^5$	$9.3x10^{-3}$	$1.0x10^5$	$1.8x10^{-3}$		
0.018-0.032	0.024	$4.4x10^4$	$7.6x10^{-3}$	$2.2x10^5$	$3.9x10^{-2}$	$1.9x10^5$	$3.3x10^{-2}$		
0.032-0.056	0.042	$1.1x10^4$	$1.8x10^{-2}$	$7.0x10^4$	$1.2x10^{-1}$	$7.6x10^4$	$1.3x10^{-1}$		
0.056-0.10	0.074	$2.7x10^3$	$4.6x10^{-2}$	$6.5x10^4$	$1.1x10^0$	$4.4x10^4$	$7.5x10^{-1}$		
0.10-0.146	0.12								
0.10-0.181	0.14	$6.6x10^2$	$2.9x10^{-2}$	$8.9x10^3$	$1.7x10^0$	$8.8x10^3$	$1.7x10^0$		
0.146-0.230	0.18								
0.181-0.280	0.22	$7.6x10^2$	$9.0x10^{-1}$	$3.3x10^3$	$3.4x10^0$	$3.3x10^2$	$3.3x10^0$		
0.230-0.40	0.30								
0.280-0.520	0.38	$2.7x10^4$	$5.1x10^1$	$7.2x10^2$	$8.3x10^1$	$7.0x10^2$	$8.1x10^0$		
0.40-0.80	0.57								
0.52-0.76	0.63	$7.8x10^2$	$3.5x10^1$	$6.9x10^1$	$3.9x10^0$	$6.8x10^1$	$3.8x10^0$		
0.76-1.24	0.97	$3.2x10^1$	$9.6x10^1$	$2.0x10^1$	$7.4x10^0$	$2.0x10^1$	$7.2x10^0$		
0.80-1.31	1.0								
1.24-2.04	1.6	$1.7x10^{-1}$	$2.9x10^{-1}$	$7.0x10^{-1}$	$1.3x10^0$	$2.0x10^9$	$4.7x10^0$		
1.31-1.92	1.6								
1.92-3.0	2.4								
2.04-3.0	2.5	---	---	---	---	$2.6x10^{-2}$	$3.5x10^{-1}$		
3.0-5.0	3.9	$1.8x10^{-1}$	$2.0x10^1$	$9.4x10^{-3}$	$1.0x10^0$	$6.2x10^{-2}$	$6.6x10^0$		
5.0-7.0	5.9	$1.1x10^{-4}$	$4.1x10^{-2}$	$3.6x10^{-5}$	$1.4x10^{-2}$	$2.1x10^{-4}$	$8.2x10^{-2}$		
7.0-9.0	7.9	---	---	---	---	---	---		

GENERATION AND CHARACTERIZATION OF AEROSOLS 111

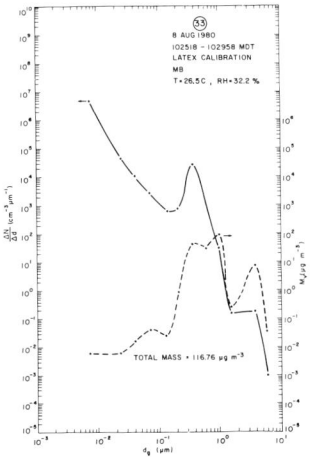

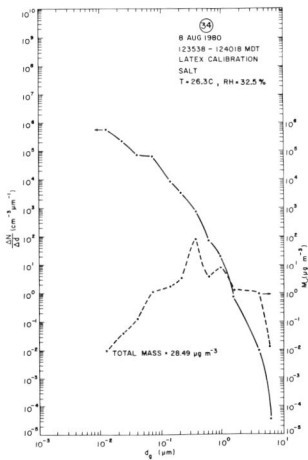

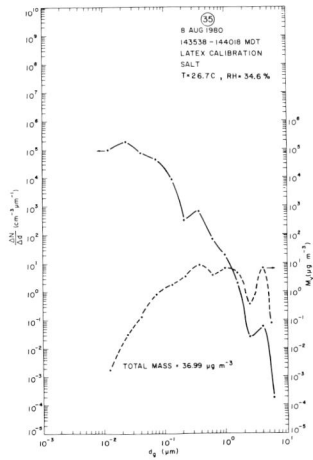

APPENDIX B

Scanning electron microscope photographs of workshop aerosol particles. The volume-sampled values in the photograph captions correspond to the volume which flowed through the entire 47-mm-diameter filter surface.

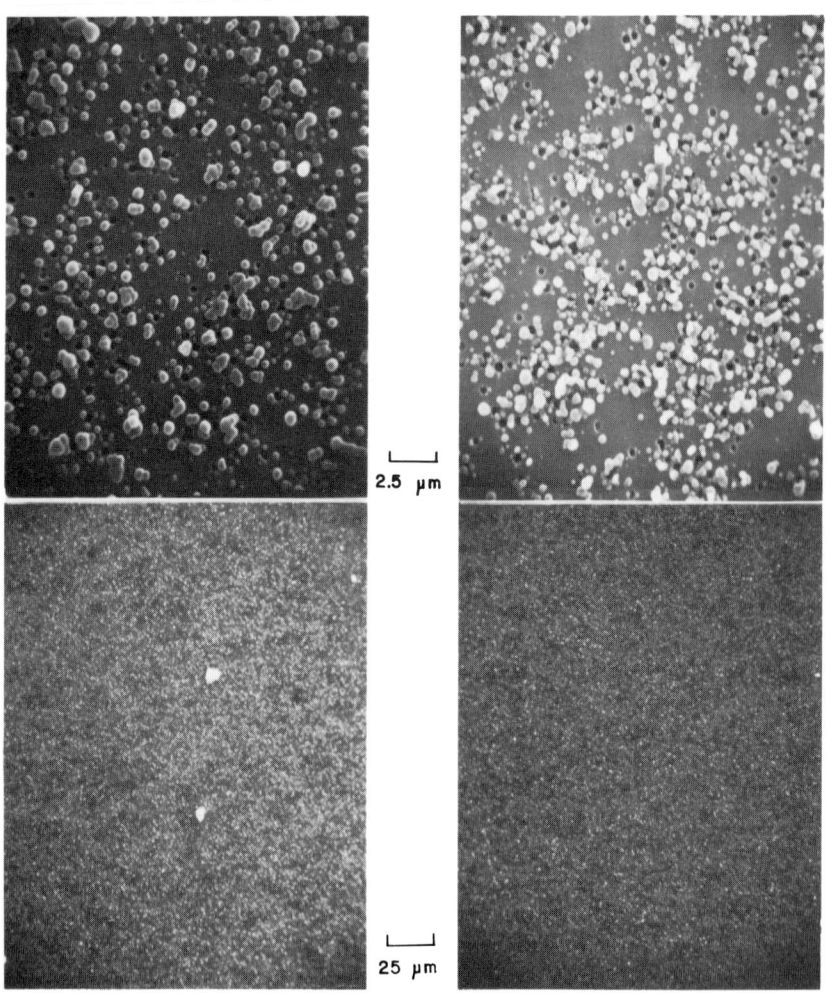

FIGURE B-1. (a) *29 July 1980, 1512-1527 MDT; $(NH_4)_2SO_4$; volume sampled = 1784 l; low concentration aerosol.* (b) *29 July 1980, 1704-1719 MDT; $(NH_4)_2SO_4$ + 1% methylene blue; volume sampled = 1798 l; low concentration aerosol.*

GENERATION AND CHARACTERIZATION OF AEROSOLS

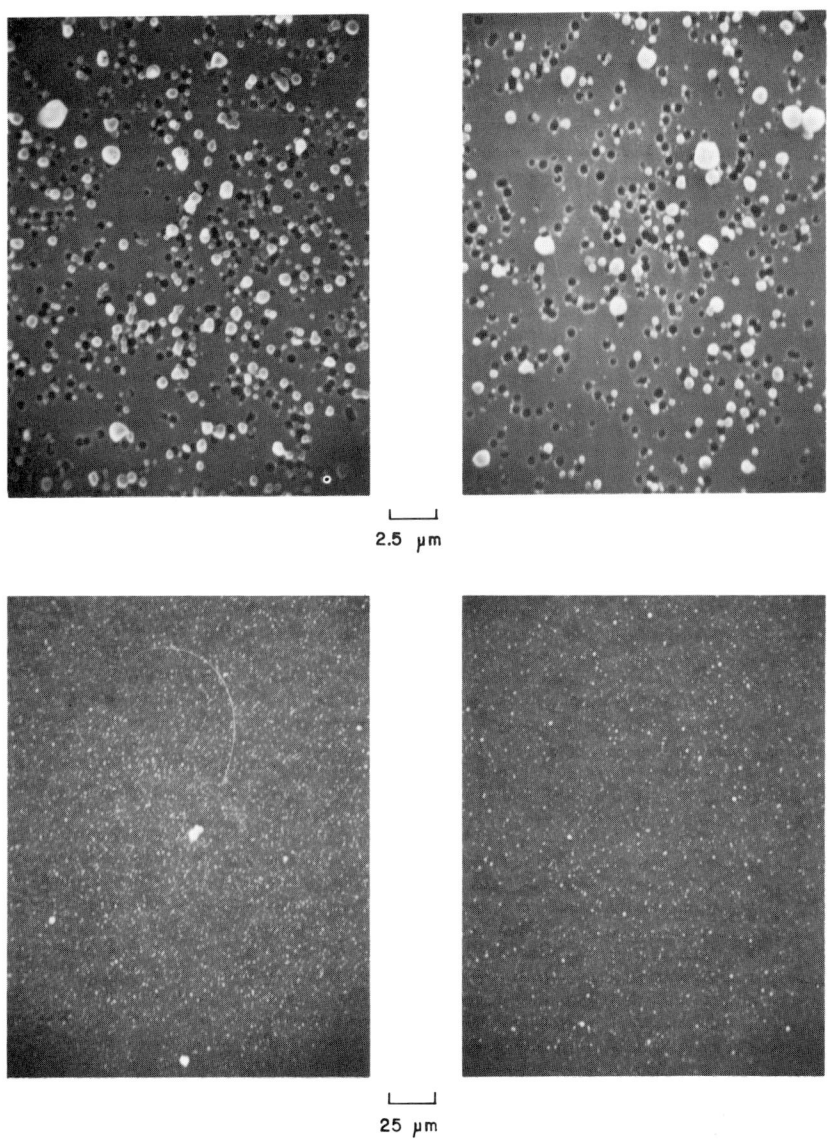

FIGURE B-2 (a) 30 July 1980, 1154-1210 MDT; $(NH_4)_2SO_4$; volume sampled = 1991 l; low concentration aerosol. (b) 30 July 1980, 1426-1456 MDT; $(NH_4)_2SO_4$ + 0.3% methylene blue; volume sampled = 3508 l; low concentration aerosol.

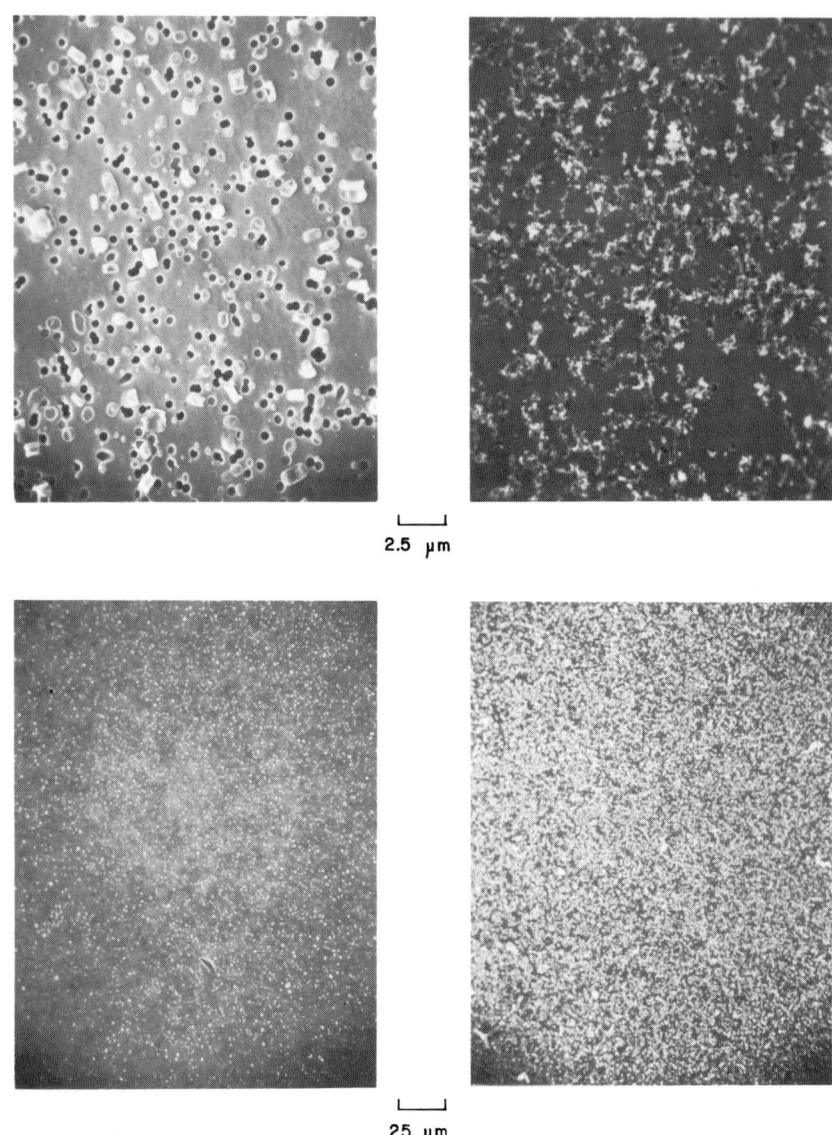

FIGURE B-3 (a) 30 July 1980, 1734-1755 MDT; $(NH_4)_2SO_4$ + 4% methylene blue; volume sampled = 2596 l; low concentration aerosol. (b) 31 July 1980, 1127-1137 MDT; carbonaceous; volume sampled = 1221 l; low concentration aerosol.

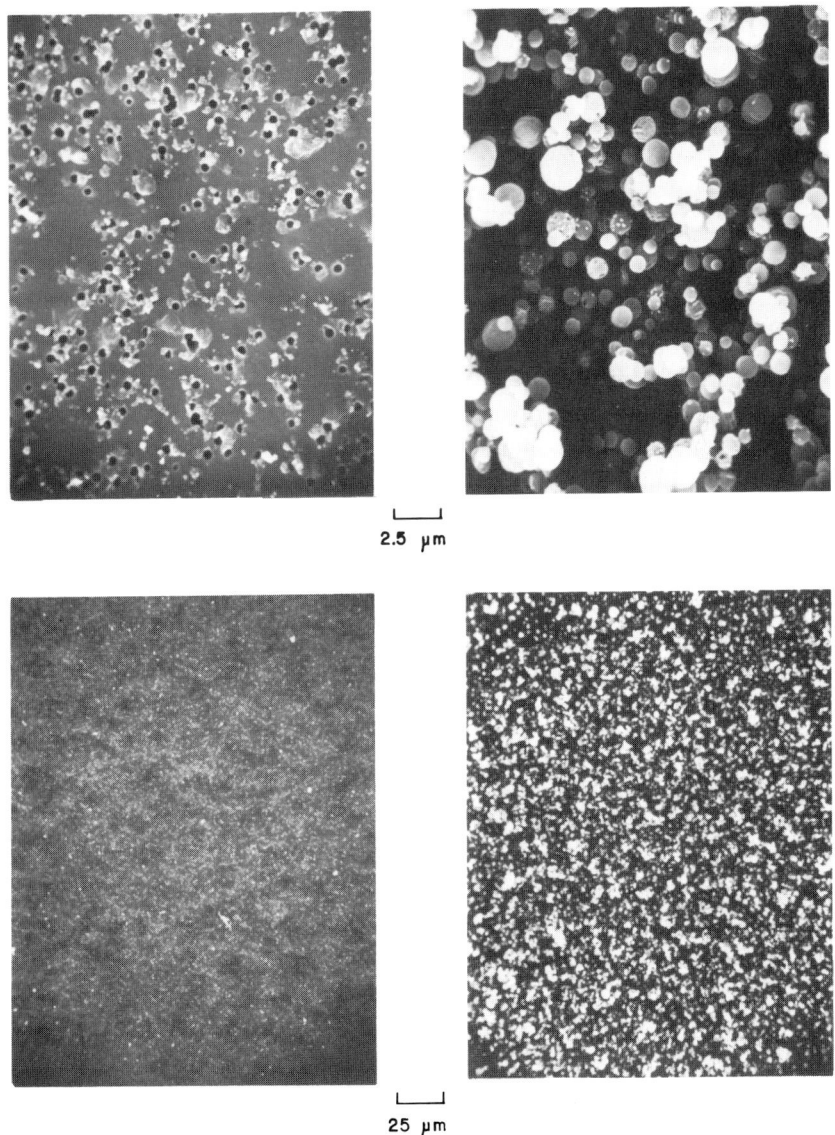

FIGURE B-4 (a) 31 July 1980, 1532-1547 MDT; soot + $(NH_4)_2SO_4$; volume sampled = 1802 l; low concentration aerosol. (b) 1 August 1980, 1311-1316 MDT; methylene blue; volume sampled = 523 l; high concentration aerosol.

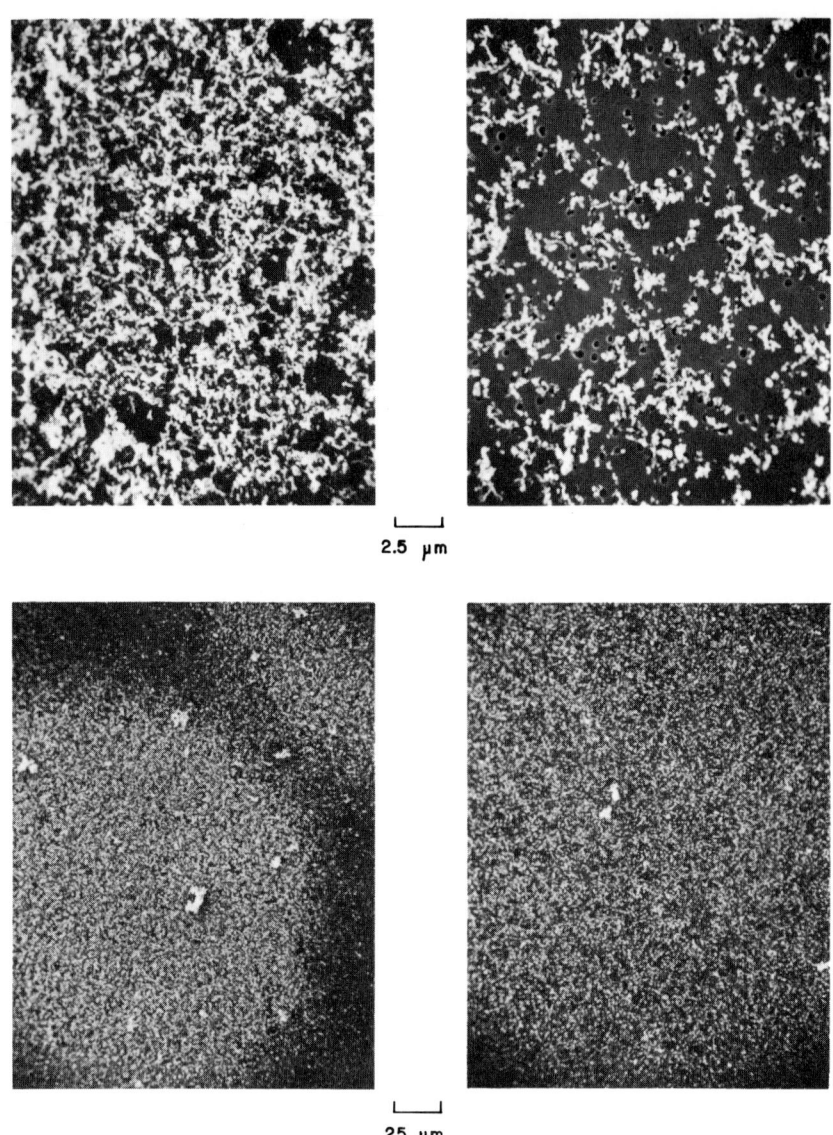

FIGURE B-5. (a) 1 August 1980, 1504-1505 MDT; carbonaceous; volume sampled = 92 l; high concentration aerosol. (b) 4 August 1980, 1118-1125 MDT; carbonaceous; volume sampled = 798 l; low concentration aerosol.

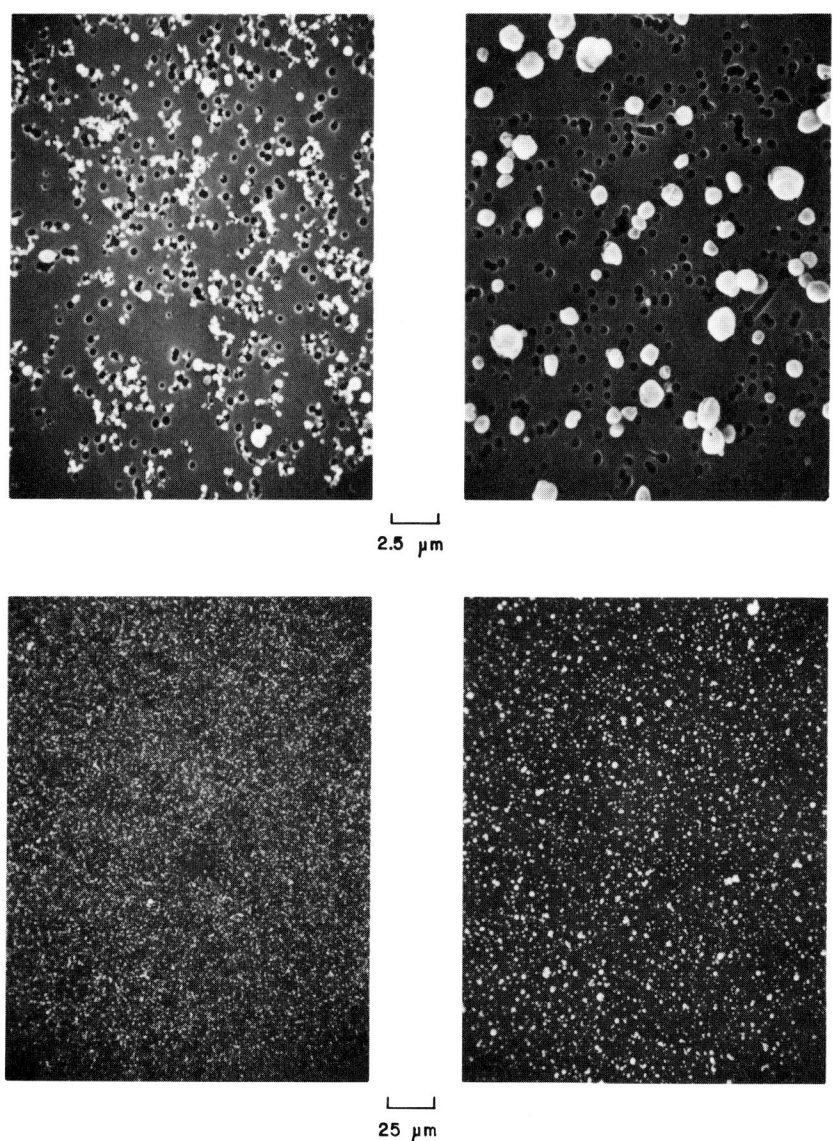

FIGURE B-6. (a) 4 August 1980, 1154-1156 MDT; soot + $(NH_4)_2SO_4$ (~50-50 mix); volume sampled = 243 l; low concentration aerosol. (b) 4 August 1980, 1510-1517 MDT; methylene blue; volume sampled = 869 l; low concentration aerosol.

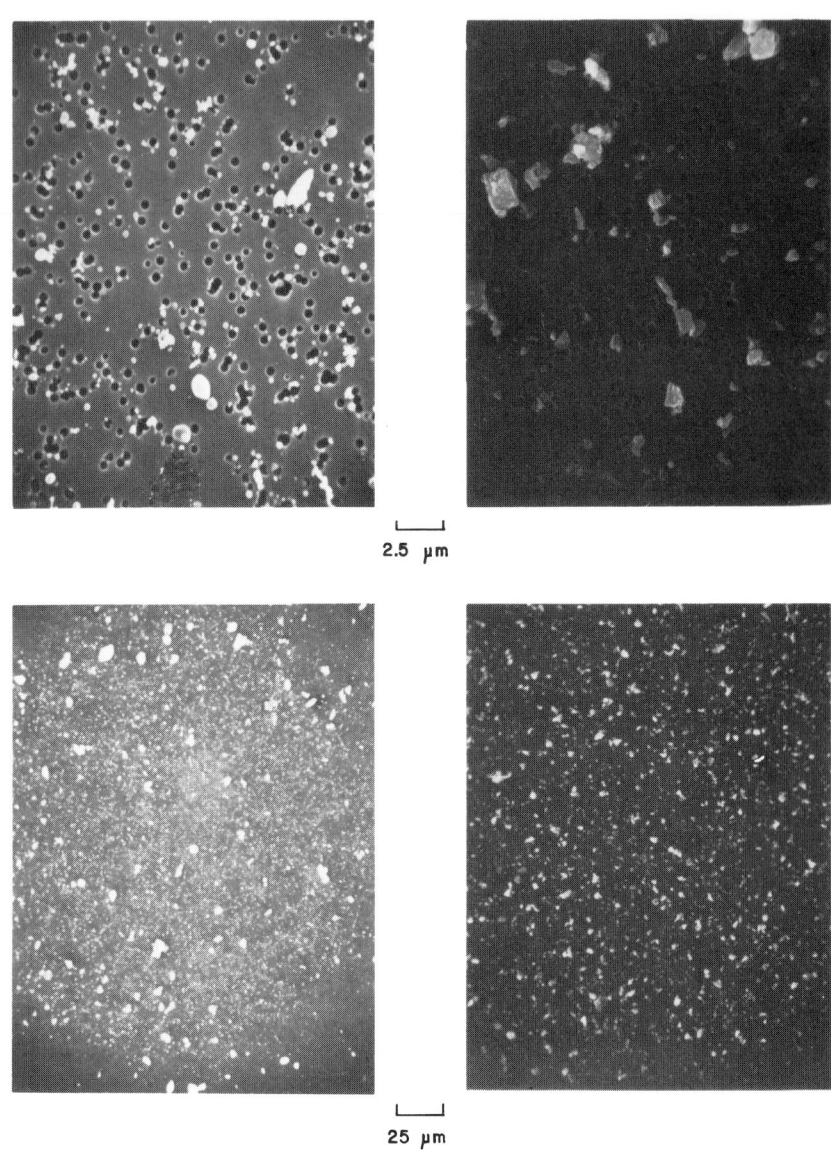

FIGURE B-7. (a) 5 August 1980, 1042-1213 MDT; ambient air; volume sampled = 11,384 l; upstairs sampling line. (b) 6 August 1980, 1056-1109 MDT; Arizona road dust; volume sampled = 1629 l; low concentration aerosol.

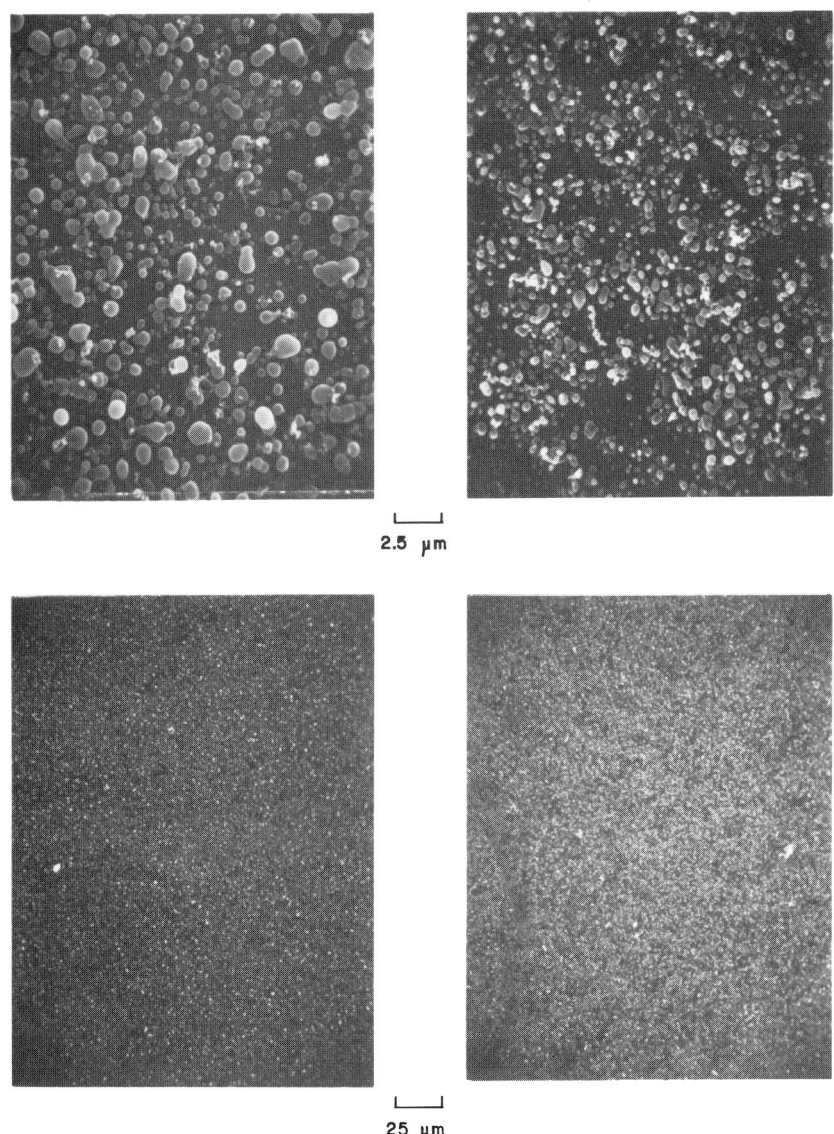

FIGURE B-8. (a) 6 August 1980, 1539-1550 MDT; soot + $(NH_4)_2SO_4$ ($\sim$ 4% soot); volume sampled = 1343 l; low concentration aerosol. (b) 7 August 1980, 1215-1225 MDT; soot + $(NH_4)_2SO_4$ ($\sim$ 5% soot); volume sampled = 1220 l; low concentration aerosol.

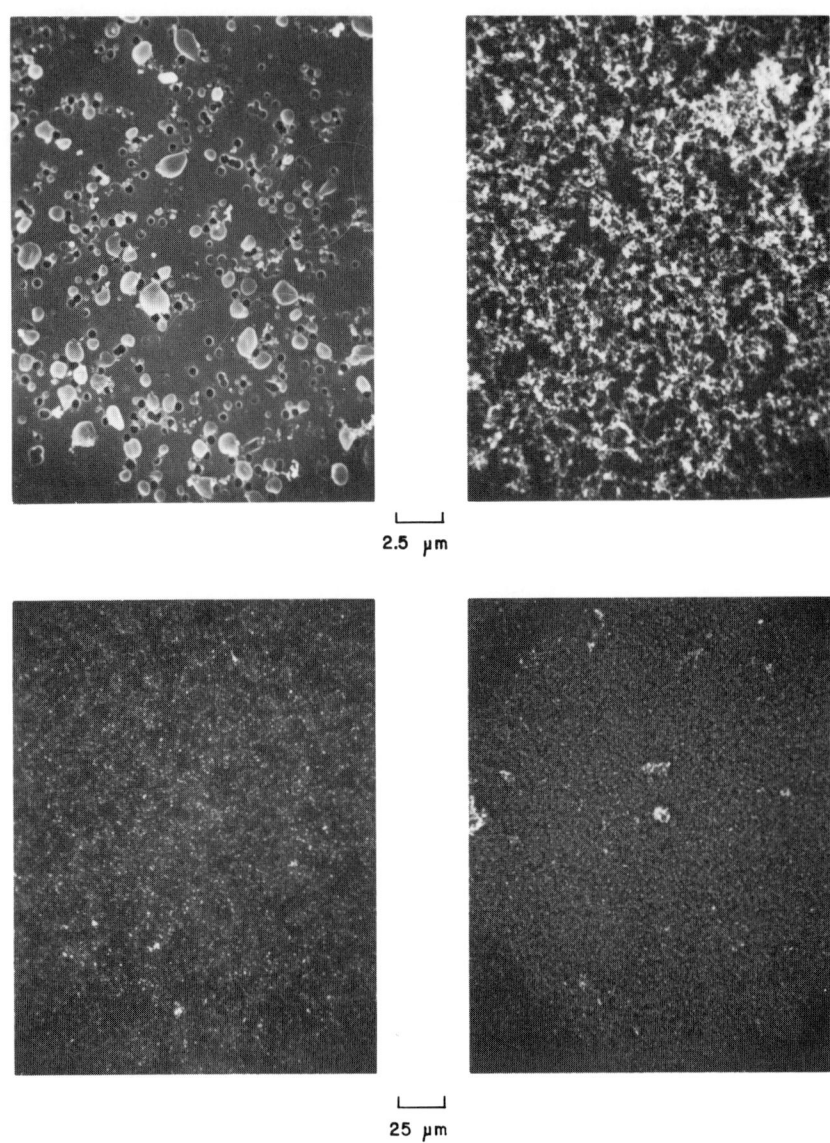

FIGURE B-9. 7 August 1980, 1527-1536 MDT; soot + $(NH_4)_2SO_4$ + moisture; volume sampled = 1042 l; low concentration aerosol. (b) 7 August 1980, 153100-153130 MDT; soot; volume sampled = 54 l; high concentration aerosol.

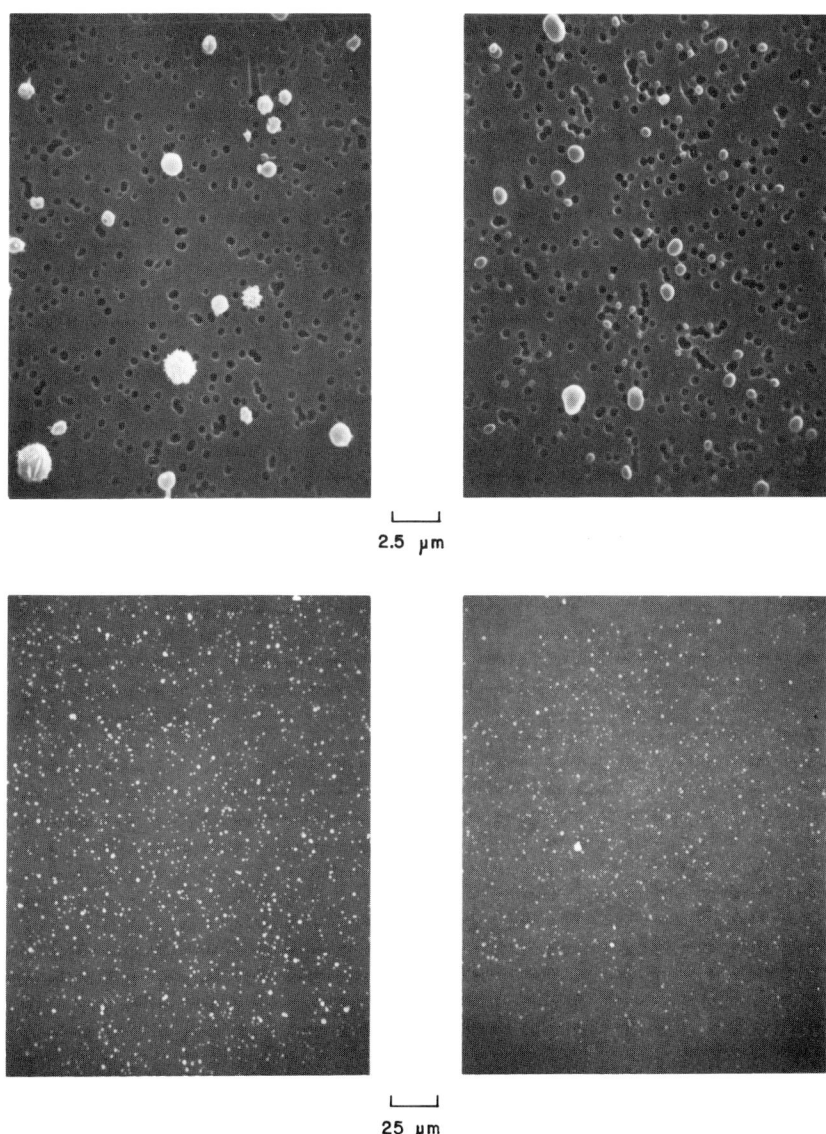

FIGURE B-10. (a) 8 August 1980, 1035-1037 MDT; methylene blue; volume sampled = 249 l; low concentration aerosol. (b) 8 August 1980, 124000-124030 MDT; $(NH_4)_2SO_4$ volume sampled = 61 l; high concentration aerosol.

INSTRUMENTATION SUMMARY

compiled by
H. Gerber and E. Hindman

The workshop participants (both those who brought instruments to the workshop and those who submitted contributions to the workshop proceedings) were asked to complete a form which summarized the important aspects of their instrumentation. Table 1 gives the summary for the light-absorption instrumentation, and Table 2 describes the instruments which were brought to the workshop to support the absorption measurements.

The following instructions were given to the participants for completing Table 1. In the column marked "Instrument Type" one of the following identifications was to be used:

particle sample—particles of high or low concentration are collected in or on a substrate; includes integrating plate, transmission, reflectance, and integrating sphere methods;

optical measurement and inversion—includes point and remote-sensing techniques;

direct measurement—measurement of extinction and scattering through a layer of aerosol;

spectrophone—includes *in situ* and filter methods.

In the column labeled "Parameter Measured" the choice consisted of the following:

σ_a—aerosol absorption coefficient;

n_2—imaginary part of the refractive index;

$\tilde{\omega}$— albedo of single scattering.

The "Relative Error" was defined as the range in the value of the parameter (listed under "Parameter Measured") corresponding to the 90% confidence interval for the parameter, divided by the value of the parameter, and expressed in ± %. The confidence interval included estimates of both systematic and random errors. In order to make the comparison of the relative errors listed by the participants meaningful, all errors were to be estimated for hypothetical measurements with the instruments in an atmosphere with the following properties:

$n_2 = 0.02$
$\sigma_a = 0.8 \times 10^{-5}$ m^{-1}
$\tilde{\omega} = 0.85$
$N(r) = 0.75\ r^{-4}$
$\int N(r) \approx 2000$ cm^{-3}, (0.05 μm - 10 μm)
$M_v \approx 25$ μg m^{-3}

If the preceding values were too small to estimate errors, the following were to be used: $n_2 = 0.1$, $\sigma_a = 7.08 \times 10^{-5}$ m^{-1}, $\tilde{\omega} = 0.57$, $N(r) = 2.25\ r^{-4}$, $N \approx 6000$, and $M_v \approx 75$. The use of these larger values was to be noted by the participants. It was assumed that all the values held for all wavelengths of light, and that there was a homogeneous distribution of the particles in the lower 2 km of the atmosphere.

INSTRUMENTATION SUMMARY

TABLE 1. Light Absorption Instrumentation.

Investigator	Instrument type	Parameter measured	Wavelength of light (μm)	Relative error (±%)	Sample volume (liter)	Sampling time (min)	Cost of field equipment ($)	Required auxiliary equipment	Labor needed for one data point (min)
Bennett, C. A., Jr. Patty, R. R.	spectrophone (filter deposit)	σ_a	0.6328	30[a]	variable	0–10	} 75,000	microbalance	30–60
	particle sample (integrating plate)	σ_a	0.6328	20[a]	variable	0–10			
Box, M. A. Copp, J. D.	optical measurement and inversion (diffuse and global fluxes)	n_2	channels from 0.4 to 1.0	50	whole atmosphere	several hours for complete data set	prototype	minicomputer	a few man-hours
Clarke, A. D. Waggoner, A. P.	particle sample (integrating plate)	σ_a	0.55	5	200–10,000	10–1,000	250	absorption photometer, microbalance (for B_a meas.)	25
Cowen, S.	particle sample (integrating plate)	σ_a, n_2	0.19–0.87	25	141	10	8,000	Cahn microbalance, vacuum pump, filter holder	45
Egan, W. G.	particle sample (diffuse reflectance and transmission)	n_1, n_2	0.185–1.105[b]	<10	20–14,000	2–1,300	400	modified Perkin Elmer 13U spectrometer, $BaSO_4$ integrating sphere, IBM 370/168 computer, Mettler microbalance	60 per wavelength
Gerber, H. E.	direct measurement (transmission and scattering in cell)	σ_g $\tilde\omega$	0.6328 0.6328	75[c] 25[c]	} 10	} 3	} 10,000	none	} 5
		n_2	0.6328	75[c]				none	
Gillespie, J. B.	particle sample (diffuse reflectance)	n_2	0.3–1.7	15			100	minicomputer, spectrometer, integrating sphere, microbalance, mixer mill	90

TABLE 1. Continued.

Investigator	Instrument type	Parameter measured	Wavelength of light (μm)	Relative error (±%)	Sample volume (liter)	Sampling time (min)	Cost of field equipment ($)	Required auxiliary equipment	Labor needed for one data point (min)
Grams, G. W. Coletti, A.	optical measurement and inversion (polar nephelometer)	$P_1(\theta),^d$ $P_2(\theta)$ in 5° steps from $\theta = 10°$ to 170°	0.6328	5 in $P(\theta)$	5-20	15-30	approx. 50,000-75,000	computer	10-20 per $P(\theta)$ scan
Heintzenberg, J.	particle sample (integrating plate, hydrosol in integrating sphere)	σ_a	0.55	e	20-1,200	2-120	200	integrating sphere photometer	60
Hänel, G. Hillenbrand, Ch.	spectrophone (in situ)	σ_a^f	0.29-3.5	10-30	0.5	10	25,000	none	60 (at present)
Hänel, G. Schloss, R.	particle sample (integrating plate)	σ_a	0.29-3.5	8-20	1,000-5,000	200-500	15,000	none	120 (at present)
Japar, S. M., et al.	spectrophone (in situ)	σ_a	0.5145 ^{g}	25	4 x 10^5	0.25^g	20,000	none	10
Patterson, E. M.	particle sample (diffuse reflectance)	n_2	0.3-0.7	40	4 x 10^5	200	400	spectrophotometer	60
Marshall, B. T.	particle sample (integrating plate)	σ_a	0.633	20-50	10^3	60	400	laser, transmission monitor	75
Roessler, D. M. Faxvog, F. R.	spectrophone (in situ)	σ_a $\tilde{\omega}$	0.5145 10.6 0.5145 10.6	15 30 11 11	10^{-2} 8 x 10^{-3}	10^{-2} 10^{-2}	75,000 30,000	microbalance none	60 20
Röhl, R., et al.	spectrophone (filter deposit)	n_2^h	0.2-1.8	49^i	17-170	1-10	6,000	Cahn microbalance, photoacoustic spectrometer, desk calculator, dichotomous sampler	60
Rosen, H.	particle sample (transmission)	σ_a	0.6328	20	200	>10	200	none	5

126 H. GERBER & E. HINDMAN

INSTRUMENTATION SUMMARY

TABLE 1. Continued.

Investigator	Instrument type	Parameter measured	Wavelength of light (μm)	Relative error (±%)	Sample volume (liter)	Sampling time (min)	Cost of field equipment ($)	Required auxiliary equipment	Labor needed for one data point (min)
Shaw, G. E.	optical measurement and inversion (diffuse and global fluxes)	$\tilde{\omega}$	0.4–1.0	15[j]	whole atmosphere		5,000		10
Volz, F. E.	particle sample[k] (transmission through KBr pellet)	n_2	0.38–0.94	50				pellet press, microbalance, computer photometer[l]	10
		n_2	2.5–30	30				pellet press, microbalance, computer spectrometer[m]	10
Wright, W. M., et al.	spectrophone (in situ)	σ_a	0.488 0.514	20[n]	0.26	~3	10,000	none	10

[a] Perturbation to the absorptivity measurements by the integrating plate method and the spectrophone is due to scattering from the particles.
[b] This wavelength range reflects the workshop measurements, a range of 0.185 μm to 45 μm is possible.
[c] These estimated errors reflect the workshop performance of the system, part of which was malfunctioning. Errors smaller by a factor of about 10 are possible.
[d] $P_1(\theta)$, $P_2(\theta)$ are the two components of polarized light for the scattering phase function. Given uniform and spherical aerosol particles, and an accurate determination of their size distribution, it is possible to invert the phase function to give n_2 within a factor of ± 2 for $n_2 \gtrsim 10^{-3}$.
[e] $\pm 50\%$ for the worst case (see workshop data for 7/29/1); normally $\pm 10\%$ for hydrosol measurements. For the integrating-plate results no error limits can be given, because only average blank values could be used. Single cases with high individual blank values and/or low filter loadings yield very unrealistic results. A $\pm 20\%$ uncertainty from the aerosol sampling itself must be added.
[f] σ_a is the effective absorption coefficient for the given wavelength range which is from a light source that simulates the solar spectrum.
[g] The error corresponds to $\sigma_a = 7.08 \times 10^{-5}$ m^{-1} and the sampling time of 0.25 min. At the $\sigma_a = 0.8 \times 10^{-5}$ m^{-1} value, time averaging over ≈ 30 min. is needed for the same accuracy.
[h] Assuming a particle density of 2.0 gm cm^{-3} and using the relation $n_2 = B_a \rho \lambda / 4\pi$, where B_a denotes absorptivity (specific absorption coefficient), ρ is particle density, and λ is the wavelength of light.
[i] $\pm 49\%$ at 0.6 μm; $\pm 36\%$ at 0.78 μm; $\pm 2\%$ at 1.2 μm; $\pm 2\%$ at 1.6 μm.
[j] Mainly due to systematic errors in determining optical depth and incoming sky flux.
[k] The format of the summary is not entirely applicable to these techniques, since they require > 3 mg of aerosol-particle matter.
[l] Extinction meter with wide angle receiver.
[m] Infrared spectrograph.
[n] For $\sigma_a = 7.08 \times 10^{-5}$ m^{-1}.

TABLE 2. Supporting Instrumentation

Investigator	Instrument type	Parameter measured	Relative error (± %)	Wavelength of light (μm)	Sampling rate	Particle size range (μm)
Bennett, C. A., Jr. Patty, R. R.	dichotomous sampler	filter loading	5	-	16.7 l/min	<3.5
Box, M. A. Copp, J. D.	spectral radiometer to measure wavelength dependent aerosol optical depth	spectral optical thickness	2	10 channels, 0.4-1.0	once every min. or longer as desired	0.1-2.0
Clarke, A. D. Waggoner, A. P.	modified Royco optical particle counter and Apple computer	particle size distribution	15	0.55	0.1 l/min.	0.3-10
Clarke, A. D. Waggoner, A. P.	integrating nephelometer	σ_s	10	0.53	25 l/min	0.1-1.0
Hänel, G. Busen, R.	integrating radiometer	circum-global radiation	6-9	0.29-3.5	-	-
Shaw, G. E.	impactor and x-ray spectrometer	particle composition[a]	-	-	17.5 l/min	0.1-10

[a]To derive the index of refraction for Mie calculations, i.e., to determine if sulfate or crustal aerosol particles.

AN EVALUATION OF PHOTOACOUSTIC AND TRANSMISSION TECHNIQUES FOR MONITORING PARTICULATE CARBON COLLECTED ON TEFLON FILTERS[1]

C. A. Bennett, Jr.
R. R. Patty

Physics Department
North Carolina State University
Raleigh, North Carolina

The First International Workshop on Light Absorption by Aerosol Particles (Fort Collins, Colorado, August 1980) provided an excellent opportunity to obtain particulate samples of laboratory-generated carbon, ammonium sulfate, and carbon/ammonium-sulfate mixtures collected on filters, and to investigate the problems which can be encountered when one tries to ascertain carbon loadings from ambient samples. These results indicate that the photoacoustic technique is preferable to the transmission technique (integrating plate method) for ambient samples with low filter loadings since the presence of a nonabsorbing, scattering aerosol (ammonium sulfate) only slightly perturbs the photoacoustic signal and significantly affects the transmitted signal. Measurements indicate that the photoacoustic signal depends not only on the energy absorbed from the incident beam but also on the existence of thermal wave interference effects and, especially for heavily loaded filters, on the presence of a nonabsorbing, scattering aerosol. In addition to evaluating these techniques it was possible to determine the specific absorptivity from a plot of optical thickness versus loading for pure carbon samples and thus determine the imaginary index of refraction.

[1]*This work was supported in part by the Environmental Protection Agency under Grant R80533202 and by National Science Foundation Grant ATM 8005356 for participation in the workshop.*

I. INTRODUCTION

The generation of various particulate samples at the First International Workshop on Light Absorption by Aerosol Particles provided an opportunity to evaluate techniques for ambient monitoring. It is widely accepted that the absorbing component of essentially all ambient atmospheric particulate samples is due largely to elemental carbon (1). For the past few years this laboratory has been investigating the feasibility of detecting and monitoring ambient levels of atmospheric elemental carbon via two absorptive optical techniques, (a) the integrating plate method (IPM) (2) and (b) photoacoustic detection (PAD). For both methods the sampling technique involves the deposition of particulate carbon onto filter substrates as a preconcentration step prior to analysis [this distinguishes this method of photoacoustic analysis from that of real-time photoacoustic measurements made on suspended carbon particles (see Ref. 3)]. Since both techniques involve the absorptivity of this material, it is possible to estimate this absorptivity and hence to infer an imaginary index of refraction.

Both the IPM and PAD techniques are discussed in light of the limitations associated with the complicated nature of the interactions taking place within the sample during analysis. Simplifying assumptions must be applied with caution. For example, although the assumption may be valid that IPM measurements on pure soot are not affected by multiple scattering effects between the soot and the filter medium for certain filter materials, it has been found that the presence of scattering particles such as $(NH_4)_2SO_4$ can significantly affect absorptivity measurements. Also, for ambient samples it has been found that assumptions regarding the thermal properties and corresponding "saturation levels" of the photoacoustic signal associated with the PAD technique should be carefully applied, since there is strong dependence upon chopping frequency and upon the relative amount of scattering to absorbing particles in the sample. Results indicate that the PAD technique applied to samples collected on filters may be one of the best techniques for ambient monitoring and, in some cases, one of the worst for absorptivity measurements.

II. EXPERIMENTAL TECHNIQUE

The detector used in these measurements consisted of a small cylindrical cell 1.3 cm in diameter and 0.3 cm long equipped with a Knowles model BT 1759 microphone. The entrance window was glass and the output window was opal glass. An RCA 1P39 photodiode, placed immediately behind the opal glass, was used to measure the fraction of light transmitted by each filter. Light from a HeNe laser ($\lambda = 632.8$ nm) was modulated by a PAR model 192 variable speed chopper prior to impinging on the particulate samples contained on filter substrates mounted in the photoacoustic cell. The output of the microphone was measured with a PAR model 128A lock-in amplifier, and the output of the photodiode was measured with a PAR model HR8 lock-in amplifier. The particles were deposited on the filter substrates at a flow rate of 16.7 l/min. The Teflon filters (Giha Corp., 1 μm) were stored in a dessicator for at least 48 hours prior to weighing on a Mettler balance capable of yielding filter weights before and after loading to within ± 60 μg.

III. TRANSMISSION MEASUREMENTS (IPM) FOR PURE CARBON

This relatively simple and straightforward method assumes single scattering by the particles making up the sample which is deposited on the filter substrate (2); the method is most accurate for highly absorbing particles such as soot. Plots of optical thickness, the product of absorption coefficient σ_a (m^{-1}) and thickness l (m), versus loading (μg/cm^2) for pure carbon samples generated at Colorado State University (CSU) and North Carolina State University (NCSU), are shown in Fig. 1. Difficulties with the balance at CSU resulted in significant scatter and in a nonzero y-intercept; however, the slopes are comparable, and it is clear that loadings can be obtained from measurements of optical thickness. The slope of the line through the CSU data points is $B_a = \sigma_a/M_v = 5.6$ m^2/g where M_v is the density of soot in g/m^3 (this compares with a slope of 5.1 m^2/g for the NCSU data points). This slope will be used later to calculate an imaginary index of refraction of soot at 632.8 nm.

It has been reported that the slope in Fig. 1 depends on the type of filter material being used; in particular, absorptivities varying up to a factor of three were determined from measurements made with Nuclepore filters and those made with "fibrous" filters such as Millipore and Quartz (4). Since the Teflon filters could be described as fibrous, a limited series of experiments were performed for pure soot samples. It was found, within the limits of experimental error (about ± 5%), that the results obtained with Teflon filters agreed with those obtained with Nuclepore filters and that Millipore filters gave unusual results (actually, some apparent nonlinearity in the range $0 \leq \sigma_a l < 0.5$). Although these results need to be checked more carefully, it appears that the dependence of absorptivity on filter material is not due solely to the fibrous nature of the filter.

IV. PHOTOACOUSTIC MEASUREMENTS FOR PURE CARBON

Early in the course of this photoacoustic study of ambient elemental carbon, a lack of saturation was observed in the photoacoustic signal as the optical thickness of the pure carbon deposit exceeded 3.0. This failure of the photoacoustic signal to saturate as the sample became opaque seemed surprising, since at the low chopping frequency of 100 Hz originally used for this

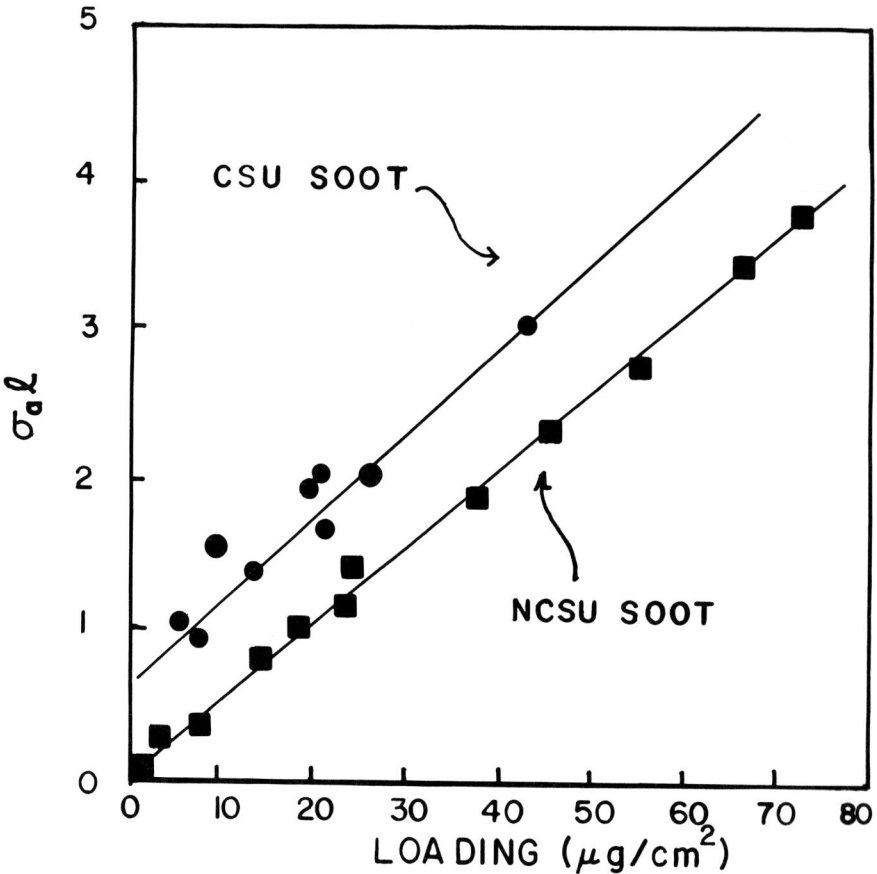

FIGURE 1. Plot of optical thickness $\sigma_a l$ (measured by IPM) versus loading for pure soot samples. CSU samples (circles) collected 31 July, 0940-1421 MDT are compared with samples collected at NCSU (squares).

study one would expect that the thermal diffusion length of the pure carbon sample would far exceed all sample thicknesses under consideration. This is apparently the case, and here this lack of saturation in the photoacoustic signal is related to interference effects from the thermal waves generated within the very thin carbon samples. That thermal-wave interference will influence the photoacoustic response is well understood (5,6); some interesting experimental results are presented here involving the nature of these effects for soot produced at NCSU, soot produced at CSU, and soot plus $(NH_4)_2SO_4$ mixtures produced at CSU. Not only are these results important to present objectives concerning the calibration of an ambient soot monitor, they are of importance to anyone wishing to make absorptivity measurements using PAD with filter samples.

Since elemental carbon is so highly absorbing (absorption coefficient $\sigma_a > 10^5$ cm^{-1}), the theory of Rosencwaig and Gersho (7) is appropriate for the description of the photoacoustic response. The photoacoustic signal arises from the energy absorbed from an amplitude-modulated beam of light which impinges upon the sample. This absorbed energy initiates rapidly damped traveling thermal waves within the sample, and the resulting thermal wave which is transmitted to the gas within the photoacoustic cell produces the acoustic signal. If the sample is thin enough, thermal-wave interference will affect the amplitude of this transmitted thermal wave and hence will affect the photoacoustic response.

The thermal waves in the filter backing, sample, and gas regions of the photoacoustic cell are illustrated in Fig. 2. In each region, $\sigma_i = (1 + j)a_i$ where $j = \sqrt{-1}$ and where a_i is the wavenumber of the thermal wave in the region i.

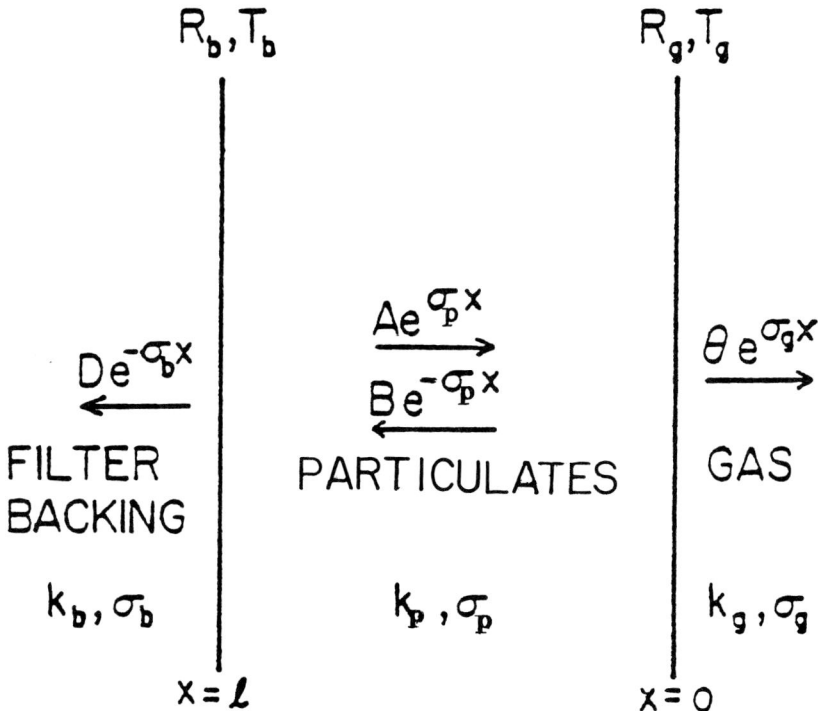

FIGURE 2. *Thermal waves in each of the three regions of the photoacoustic cell. Note that the direction of positive x is to the left.*

As well as being dependent upon the thermal properties of the material in which the thermal wave is propagating, a_i is dependent upon the angular frequency ω at which the incident light beam is modulated; in particular, $a_i \propto \sqrt{\omega}$, or equivalently, $\lambda_i \propto 1/\sqrt{\omega}$, where λ_i is the wavelength of the thermal wave in the region i. Therefore, the wavelength of the thermal waves may be varied in each of the three regions by varying the rate at which the incident light beam is modulated. The thermal waves initiated within the sample travel to the sample-backing and sample-gas boundaries. By applying the boundary conditions of temperature and flux continuity, the thermal wave reflection coefficient R_i and the thermal wave transmission coefficient T_i may be computed at each of these boundaries. These coefficients have the form (6)

$$R_b = \frac{1-b}{1+b}$$

$$T_b = \frac{2}{1+b}$$

$$R_g = \frac{1-g}{1+g}$$

$$T_g = \frac{2}{1+g} \tag{1}$$

where

$$b = \frac{k_b \sigma_b}{k_p \sigma_p}$$

$$g = \frac{k_g \sigma_g}{k_p \sigma_p}$$

and where k_i is the thermal conductivity (W/cm K) in the medium i, and σ_i is defined above. It is important to note that thermal waves are rapidly damped; in one period the wave will be attenuated by a factor of $e^{-2\pi}$. The backing and gas regions of the cell are assumed to be thicker than one thermal wavelength in that region so that there is only a transmitted right traveling wave in the gas and a transmitted left traveling wave in the backing (see Fig. 2). However, if the sample is less than a thermal wavelength in thickness, the thermal waves within the sample will reflect multiple times between the two boundaries, and the resulting thermal wave interference will affect the amplitudes of the transmitted waves.

To see how this occurs, let the incident light beam have intensity $\frac{1}{2}I_0(1 + e^{j\omega t})$. Since the photoacoustic signal arises from periodic temperature fluctuations, the nonperiodic portion of the incident flux will be ignored so that at any point x within the sample (see Fig. 2) the light absorbed between x and x + dx is $\frac{1}{2}I_0\sigma_a e^{-\sigma_a x}dx$ in steady state, where σ_a is the absorption coefficient of the sample in cm^{-1}. It is assumed that this absorbed radiation is converted nonradiatively into heat with a conversion efficiency of unity; this heat flux will propagate through the sample as thermal waves. In the one-dimensional case, half of the energy will diffuse toward the gas and half will diffuse toward the backing. Hence the thermal waves initiated by light absorbed between x and x + dx will have amplitude $(\sigma_a I_0 e^{-\sigma_a x}/4k_p\sigma_p)dx$. These thermal waves travel through the sample and reflect multiple times between the sample-backing and sample-gas boundaries. Consider the thermal wave which originally travels toward the gas; it will contribute a series of transmitted terms at x = 0 equal to

$$\frac{\sigma_a I_0 e^{-\sigma_a x}}{4k_p\sigma_p} T_g \{e^{-\sigma_p x} + R_b R_g e^{-\sigma_p(2l+x)} + \ldots$$
$$+ (R_b R_g)^n e^{-\sigma_p(2nl+x)} + \ldots \} dx \quad (2)$$

where n = 0, 1, 2, ... Similarly, the thermal wave which originally travels toward the backing will contribute a series of transmitted terms at x = 0 equal to

$$\frac{\sigma_a I_0 e^{-\sigma_a x}}{4k_p\sigma_p} T_g\{R_b e^{-\sigma_p(2l-x)} + (R_b R_g)R_b e^{-\sigma_p(4l-x)} + \ldots$$
$$+ (R_b R_g)^n R_b e^{-\sigma_p[2(n+1)l-x]} + \ldots\}dx \quad (3)$$

Both of these series are geometric, with ratio $R_b R_g e^{-2\sigma_p l}$. Summing both series and adding gives the amplitude of the thermal wave transmitted to the gas due to light absorbed between x and x + dx. Integrating from 0 to l, the amplitude θ of the thermal wave transmitted to the gas due to radiation absorbed from all points within the sample is computed

$$\theta = \frac{I_0\sigma_a T_g}{4k_p\sigma_p} \left\{ \frac{1}{\sigma_a + \sigma_p} (1-e^{-(\sigma_a+\sigma_p)l}) \right.$$

$$\left. + R_b e^{-2\sigma_p l} \frac{1}{\sigma_a - \sigma_p} (1-e^{-(\sigma_a-\sigma_p)l}) \right\} /$$

$$\{1 - R_b R_g e^{-2\sigma_p l}\} \quad (4)$$

A modest amount of algebra will convert this expression into the result obtained by Rosencwaig and Gersho (7). The first term in the braces represents the contribution from the thermal waves within the sample which originally travel toward the gas, and the second term represents the contribution from the thermal waves within the sample which initially travel toward the backing, hence the factor R_b. Whether or not there is a 180° phase shift for thermal waves reflecting from the sample-backing boundary depends upon the sign of R_b (5,6); this will be an important consideration in interpreting the photoacoustic results.

A simple thermodynamic argument can be used to compute the measured photoacoustic response from this transmitted thermal wave (7); this discussion will be omitted since it will not affect the quantities in Eq. (4) which are essential to the discussion. Hence, the photoacoustic response emerges as a complex function of the absorptive and thermal properties of the sample. Although it has been customary to make certain simplifying assumptions regarding the thermal properties of the sample in the hope that the absorptive properties of the sample could be obtained in a straightforward manner, it will be shown that the existence of thermal wave interference effects in thin samples makes this a very difficult thing to do in practice.

A common assumption involves the quantity $\sigma_p = (1 + j)a_p$; if one assumes that $\sigma_a \gg a_p$ to the extent that $a_p = (2\pi)/(\lambda_p) \cong 0$, then Eq. (4) begins to look very appealing. This is equivalent to stating that the thermal wavelength in the sample is much longer than the sample thickness or that the sample is "thermally thin." However, it is for thermally thin samples that thermal wave interference effects manifest themselves most predominantly. Consider, for example, a thermally thin sample deposited on a substrate for which the reflection coefficient R_b is 1.0; the thermal waves at the sample-gas boundary (see Fig. 2) interfere constructively (very little thermal energy is lost to the backing) and the photoacoustic signal is then very large. However, if the same sample is deposited on a substrate for which R_b is -1.0 then thermal waves reflecting from the sample-backing boundary suffer a 180° phase shift so that the resulting destructive interference at the sample-gas boundary causes the photoacoustic signal to be reduced (hence separate calibrations must be performed on each type of filter material used). Although thermal wave interference does exist, one might assume that, over the range of sample thicknesses from zero to optically opaque, the ratio of sample thickness to thermal wavelength does not change enough for thermal wave interference effects to alter the photoacoustic signal [since $\lambda_p \propto 1/\sqrt{\omega}$, this approximation improves with lower chopping frequencies]. In this case, one would expect a photoacoustic response which is approximately proportional to the energy absorbed by the sample. However, since most filter materials act as diffuse reflectors, the optical reflectivity R_f must be included; this results in an expression for the fraction of the incident energy deposited within the sample which is given by $(1 + R_f e^{-\sigma_a l})(1 - e^{-\sigma_a l})$. In the past, investigators have ignored the reflectivity of the filter and have normalized the photoacoustic signal by assuming a saturation level (8). If saturation exists, a plot of photoacoustic signal versus loading would be that shown in Fig. 3 for the Teflon filters ($R_f \sim 0.7$ which includes the light backscattered by the opal glass). From such data the optical thickness $\sigma_a l$ could be obtained, and, if accurate loadings were known, a plot of optical thickness versus loading would yield the absorptivity. (If the filter material used has an R_f close to 1.0, such as Millipore, this would be easy since the fraction of the energy absorbed is close to $1 - e^{-2\sigma_a l}$.) However, the following data indicate that this approach is not always appropriate.

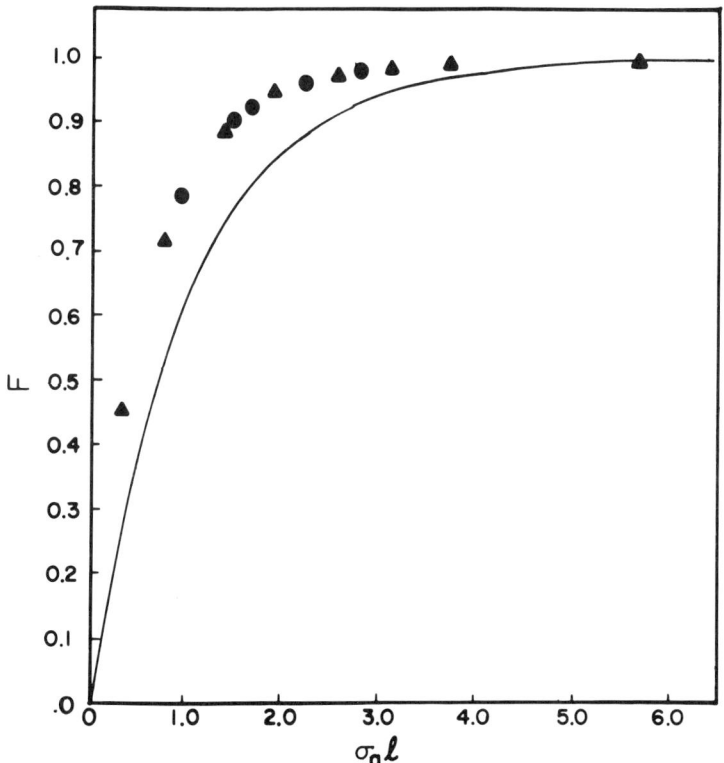

FIGURE 3. *Fraction of the incident energy absorbed by the sample F plotted versus optical thickness $\sigma_a l$ for pure soot samples. Values of F for soot samples collected at CSU were computed from $(1 + Re^{-\sigma_a l})(1 - e^{-\sigma_a l})$ to illustrate the approximate appearance of the photoacoustic signal expected for thermally thin samples in the absence of thermal wave interference. CSU samples taken 7 August, 1455-1552 MDT (circles) and NCSU samples (triangles) were used. For comparison $(1 - e^{-\sigma_a l})$, which ignores the reflectivity R of the filter, is plotted as a solid line.*

Figure 4 compares a plot of measured photoacoustic signal versus $\sigma_a l$ at a 100-Hz chopping frequency for pure soot samples taken at CSU with that for pure soot samples generated at NCSU subsequent to the workshop. It is clear that the CSU data do not saturate, although the nonsaturation is not too severe. The fact that data for the NCSU-generated soot more dramatically

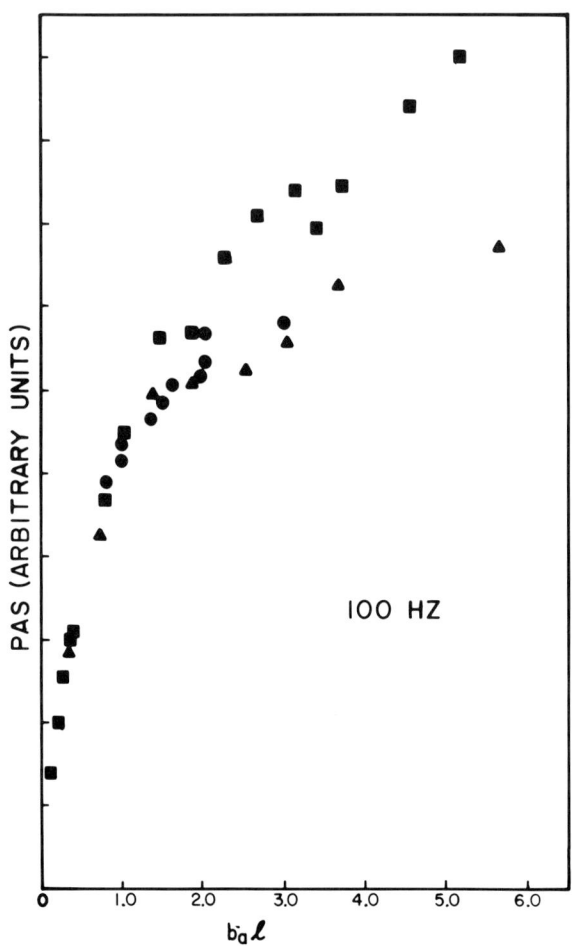

FIGURE 4. Photoacoustic signal plotted versus optical thickness for pure soot samples using a 100-Hz chopping frequency. CSU samples collected 31 July, 0940-1420 MDT (circles); CSU samples collected 7 August, 1455-1552 MDT (triangles); NCSU samples (squares). The "expected" appearance of Fig. 3 is not observed, and a difference, attributed to different thermal properties in CSU and NCSU samples, is noted.

PHOTOACOUSTIC AND TRANSMISSION TECHNIQUES

exhibit nonsaturation is believed to be related to differences in particle size. NCSU samples were produced in a very oxygen-rich blue flame, whereas the CSU samples were generated with a yellow diffusion flame. It appears that CSU samples are "fluffier," resulting in a sample which is less thermally thin than the diffusion flame data. Notice the rather large amount of scatter in both sets of photoacoustic data. It appears that this is due to nonthermally thick Teflon filter substrates; thermal energy transmitted to the filter from the sample is then transmitted to the gas before damping to zero. While it is true that the nonsaturation decreases as the chopping frequency tends toward zero, the magnitude of this scatter increases with decreasing chopping frequency (this is a filter-dependent phenomenon; relatively little scatter is observed when Nuclepore substrates are used). It should be pointed out that if the same filter is repetitively inserted and sampled, the photoacoustic signal for each measurement agrees to within 3%, and the optical thicknesses are determined to within 2%. It appears that the source of this scatter is not associated with the sample or the precision of the measurements, but instead results from the variability in filter substrates. The fact that the photoacoustic signal does not saturate but continues to rise, although the heavily loaded filters are very opaque, indicates that the thermal waves reflecting from the sample-backing boundary are shifted in phase by 180°. The thermal waves at the sample-gas boundary interfere destructively for the thinner samples and interfere more and more constructively as the sample thickens, causing the photoacoustic signal to continue to rise rather than saturating.

In Fig. 5 plots are shown of PAS versus $\sigma_a l$ for the same samples used for Fig. 4 analyzed at a chopping frequency of 3000 Hz. By increasing the chopping frequency, the wavelength of the thermal wave within the samples has been decreased, thus increasing the thermal thickness. The data for the NCSU-generated soot illustrate destructive interference for the thinner samples, constructive interference at a $\sigma_a l$ of about 2.0 and then destructive interference for the thicker samples. These thick samples exhibit a signal which is very close to a final saturation value, attained when the sample thermal thickness becomes so great that thermal waves reflecting from the sample-backing boundary damp to zero before reaching the sample-gas boundary. The CSU data are approaching (but apparently have not reached) a thickness for which constructive interference is attained; this again illustrates the difference in thermal properties of the soot produced under different flame conditions. Notice that the scatter in the photoacoustic data has largely disappeared at the higher chopping frequency, which is in agreement with previous remarks concerning the thermal thickness of the substrate. However, as will be shown later, one should avoid high chopping frequencies when analyzing ambient samples, since the perturbation to the photoacoustic signal associated with the presence of nonabsorbing scattering particles is much more significant for those frequencies, indicating that compounds such as $(NH_4)_2SO_4$ seriously affect the "thermal thinness" of soot samples.

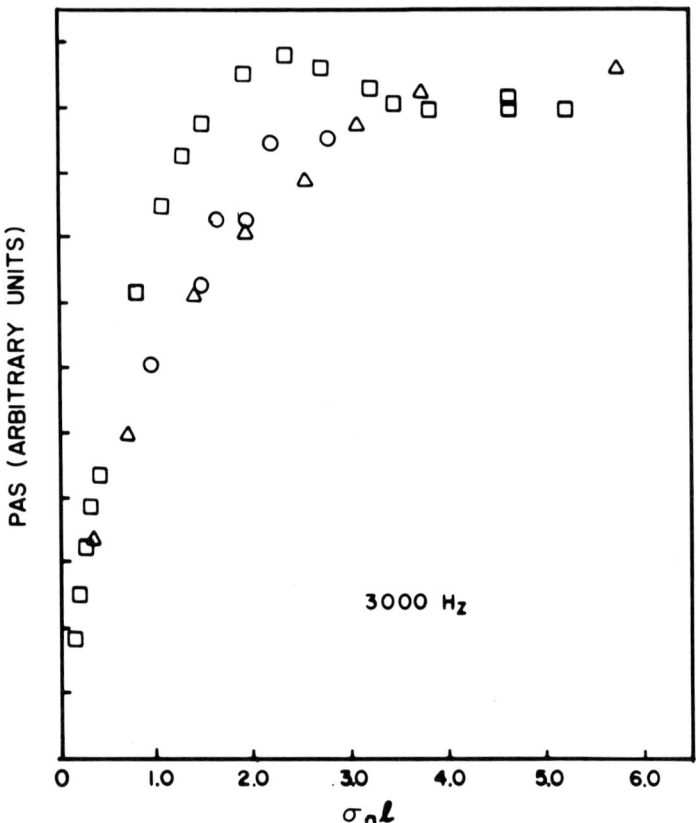

FIGURE 5. Plot of photoacoustic signal versus optical thickness for pure soot samples illustrating the effect of increasing the chopping frequency to 3000 Hz which decreases the wavelength of the thermal waves and thus increases the thermal thickness of the sample. The samples were those used for Fig. 4.

V. INFLUENCE OF AMMONIUM SULFATE ON TRANSMISSION AND PHOTOACOUSTIC MEASUREMENTS

It has been traditional for those involved with both the transmission (IPM) and PAD techniques to assert that the presence of $(NH_4)_2SO_4$ in the sample does not seriously affect their measured valve of absorptivity. The generation of $(NH_4)_2SO_4$ and carbon samples (separate and mixed) at the CSU workshop provided an opportunity to evaluate this assumption. The validity of the assumption is significant, since ambient samples contain $(NH_4)_2SO_4$ and other nonabsorbing scattering particles.

PHOTOACOUSTIC AND TRANSMISSION TECHNIQUES   141

Figure 6 shows a plot of -ln T versus loading ($\mu g/cm^2$) for pure $(NH_4)_2SO_4$ deposited on Teflon filters, where T is the fraction of incident light transmitted by the nonabsorbing salt particles as determined by the integrating plate method (IPM). Note that most of the loadings are very large, much larger than those encountered in practice for ambient samples. The essential feature here is that for loadings of less than 25 $\mu g/cm^2$, the extraneous extinction due to pure salt is probably less than 5%. Pure $(NH_4)_2SO_4$ gives no photoacoustic signal with visible light.

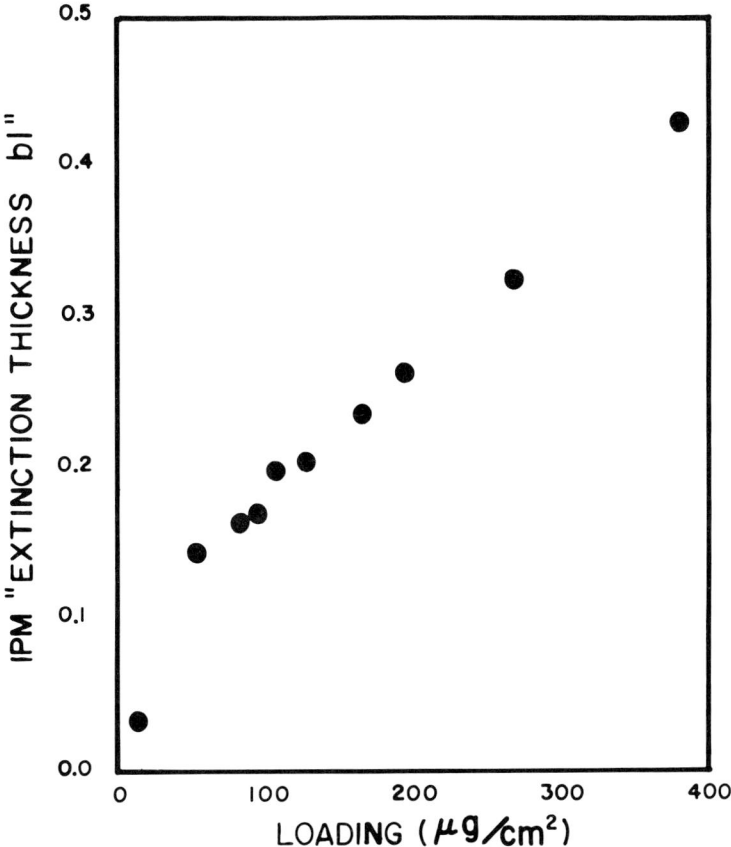

FIGURE 6. An "extinction thickness" for pure $(NH_4)_2SO_4$ particles on Teflon filters (determined by IPM) is plotted versus loading for samples taken 29 July, 1114-1330 MDT. Note that most loadings are very large, and some nonlinearily for $0 <$ "bl" < 0.1 is indicated.

Before proceeding to real mixtures, results are presented of experiments that could be done at CSU which were more accurate than those which could be done for the known mixtures, since in practice preparing known mixtures is difficult. After a pure carbon deposit is analyzed by IPM and PAD, an overload of $(NH_4)_2SO_4$ is placed on top of the carbon and IPM and PAD measurements are then repeated. Although these are not actual mixtures, the results are surprising and are in qualitative agreement with data taken on real mixtures. Figure 7 shows the results of such an experiment for the IPM and PAD techniques for a chopping frequency of 100 Hz. The percent perturbation to the pure soot signal is plotted versus the optical thickness of the pure soot for a 13-μg/cm^2 overlay of salt (triangles), a 6-μg overlay of salt (circles), and a 25-μg/cm^2 overlay of salt (squares). Total loading of salt is estimated using reported concentrations ($\sim$ 500 μg/m^3) and sampling times (12 min., 6 min., and 24 min.). Actual mixtures being simulated could be calculated using

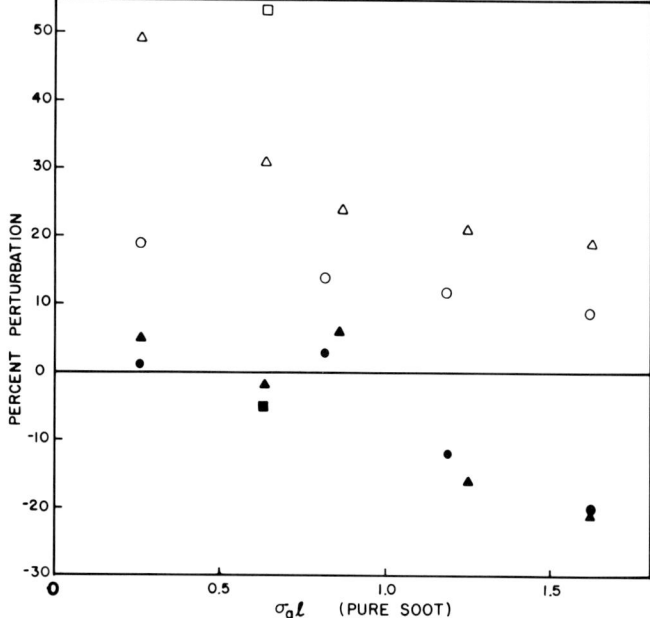

FIGURE 7. The percentage by which the IPM and PAD results for pure soot are perturbed due to overlaying $(NH_4)_2SO_4$ on pure soot is plotted versus optical thickness for the pure soot samples for three different salt overlays: 25 μg/cm^2—PAD (closed squares), IPM (open squares); 12μg/cm^2—PAD (closed triangles), IPM (open triangles); 6 μg/cm^2—PAD (closed circles), IPM (open circles). For low carbon loadings, typical of ambient samples, note that the PAD measurements are much less perturbed.

carbon loading $L = (\sigma_a l)/(5.6 \times 10^{-2} \text{ cm}^2/\mu g)$. Note that the perturbation to the IPM measurements is much larger (about 5 to 10 times) than would be expected on the basis of Fig. 6, indicating that multiple scattering is apparently taking place within the samples (if the salt particles form a true overlay, the effect is probably similar to that of sandwiching an absorbing sample between the filter and the opal glass during IPM measurements). Note also that in the range $0 < \sigma_a l < 1.0$ (a good range for ambient measurements), that the photoacoustic signal is perturbed much less than the IPM measurements. Just why this is so is not clear; multiple scattering leading to increased absorption will increase the photoacoustic signal (PAS), alteration of the thermal properties of pure soot by the salt could decrease or increase the PAS (depending upon the particle size distribution of the salt particles, etc.), and backscatter of radiation away from the sample (small, by Fig. 6) would decrease the PAS. It appears that the PAD technique is much more accurate than IPM for ambient monitoring of particulate carbon on lightly loaded filters.

Figure 8 shows a plot of percent perturbation to the pure soot signal versus salt loading ($\mu g/cm^2$) for a set of five replicate soot samples which all had $\sigma_a l = 1.3 \pm 0.06$ before the salt was overlayed. The results are similar to, but not in quantitative agreement with, the results shown in Fig. 7.

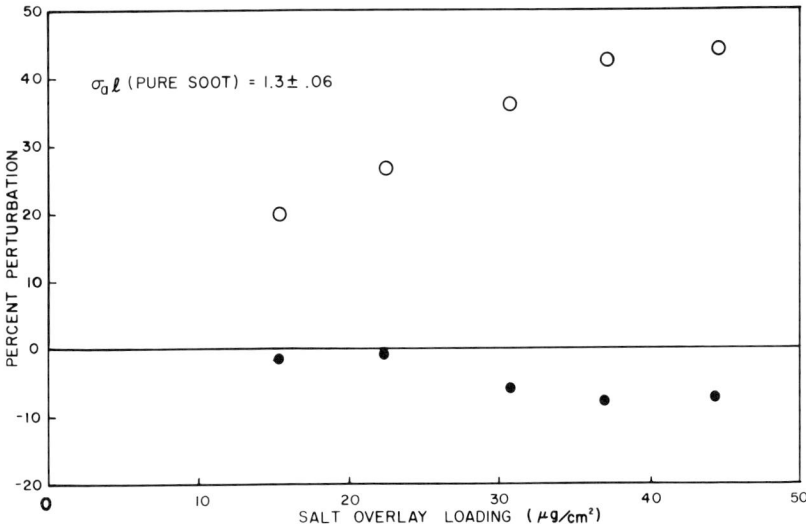

FIGURE 8. The percentage by which the IPM (open circles) and PAD (closed circles) results for pure soot are perturbed due to overlaying varying amounts of $(NH_4)_2SO_4$ on replicate soot samples is plotted versus the $(NH_4)_2SO_4$ loading.

With these qualifications in mind, Fig. 9 shows a plot of PAS at 100 Hz versus $\sigma_a'l$, where $\sigma_a'l$ is the perturbed optical thickness of a mixture sample (determined by IPM) for a 50%-soot/50%-salt mixture and for a 50-50 mix in which the salt particles were introduced in the flame of the soot generator and thus "volitalized." The gross difference in the behavior of the two curves is probably due to different thermal properties of the two mixtures; SEM photographs of each mixture provided at the workshop indicated that the volitalized salt particles were much larger than the nonvolitalized particles. The data are shown in order to emphasize the way in which different particles of the same nonabsorbing material can affect the photoacoustic response. Whether or not thermal wave interference is responsible for the decrease in photoacoustic signal for increasing optical thickness is uncertain at this time.

The advantages of PAD over IPM on mixture (ambient) samples are illustrated by their relative perturbations to a 4% soot mixture taken on 6 August 1980. After sampling the mixture, the salt was turned off so that an accurate measure of $\sigma_a l$ (for pure soot) of the mixture could be deduced. The perturbed $\sigma_a'l$ (from 1.8 m³ sampled) was found to be 1.0, and the unperturbed $\sigma_a l$ was

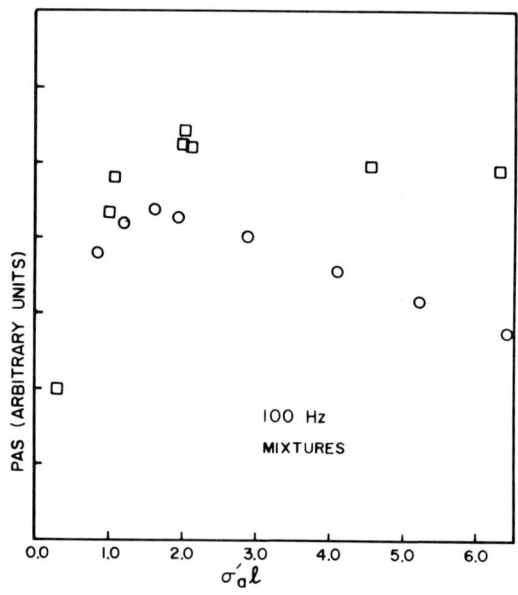

FIGURE 9. Photoacoustic signal plotted versus $\sigma_a'l$ (perturbed optical thickness for salt-soot mixtures): 50-50 mix, salt introduced downstream of flame (squares) 4 August, 1129-1412 MDT; 50-50 mix, salt introduced through flame (circles) 31 July, 1433-1737 MDT. Note the decrease in signal for high loadings.

PHOTOACOUSTIC AND TRANSMISSION TECHNIQUES 145

about 0.45; thus, the IPM technique gave a result that was 120% too high. The photoacoustic signal was about 35% too low. The PAD perturbation might be reduced at lower chopping frequencies.

To summarize these comments, Fig. 10 shows a composite of photoacoustic signals from all soot and soot mixture samples taken at CSU (note that the mixtures are plotted against $\sigma_a'l$). Although there is good agreement at low loadings, this agreement diminishes as the sample thickness increases. The choice of an appropriate "saturation level" must be done with care and is left as an exercise for the individual investigator.

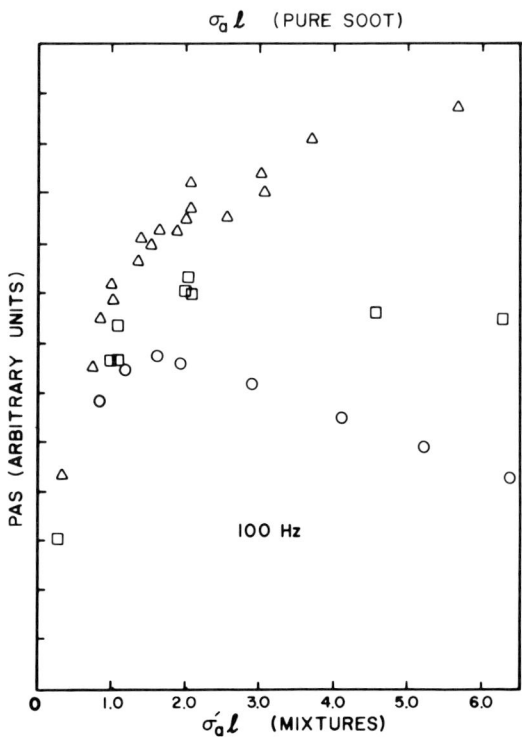

FIGURE 10. Composite plot of photoacoustic signal versus optical thickness for samples collected at CSU illustrating the difficulty of ascertaining a saturation level. Pure soot samples are plotted versus $\sigma_a l$ (triangles); the samples of Fig. 9 are plotted versus $\sigma_a' l$. Apparently the addition of $(NH_4)_2SO_4$ alters the thermal properties of the sample. The chopping frequency was 100 Hz.

VI. THE COMPLEX INDEX OF REFRACTION OF PURE SOOT AT 632.8 NM

In view of the observations concerning IPM and PAD, the authors have reasonable confidence in only one of these methods for determining the optical properties of soot; this is IPM measurements on pure soot samples. The specific absorption coefficient determined from the $\sigma_a l$ versus loading plot is $B_a = 5.6$ m^2/g. It is assumed that the integrating plate (opal glass) has "integrated out" all effects due to scattering. A density for soot of 2.0 g/cm^3 is also assumed (9). The complex index of refraction at 632.8 nm is then $n_2 = (B_a \varrho \lambda)/(4\pi) = 0.56 \pm 0.1$ where the large error bars reflect the rather limited set of loading values.

VII. CONCLUSIONS

The imaginary part of the index of refraction of pure soot at 632.8 nm has been estimated using the IPM with pure soot samples. The difficulty of performing such a measurement with PAD has been discussed.

Also discussed were the effects of thermal wave interference and the relative perturbations to both the IPM and PAD techniques due to the presence of nonabsorbing scattering particles such as $(NH_4)_2SO_4$ in the sample. It appears that the magnitude of these perturbations is much less for PAD than for IPM for lightly loaded samples. When accurate values of the absorbing properties of soot are determined, the PAD technique applied to samples collected on filters may become a powerful tool for determining actual ambient levels of atmospheric absorption due to this type of particle.

REFERENCES

1. Rosen, H., Hansen, A. D. A., Gundel, L., and Novakov, T., Identification of the optically absorbing component in urban aerosols, *Appl. Opt. 17*, 3859-3861 (1978).
2. Lin, C.-I., Baker, M., and Charlson, R. J., Absorption coefficient of atmospheric aerosol: A method for measurement, *Appl. Opt. 12*, 1356-1363 (1973).
3. Wright, W. M., Stedman, D. H., Stefanutti, L, and Terhune, R. W., Measurement of light absorption by aerosols with an optoacoustic detector, *Light Absorption by Aerosol Particles* (H. E. Gerber and E. E. Hindman, eds.), Spectrum Press, Hampton, Virginia (1982).
4. Cadle, S. H., and Groblicki, P. J., An evaluation of methods for the determination of organic and elemental carbon in particulate samples, *Particulate Carbon: Atmospheric Life Cycle* (G. T. Wolff and R. L. Klimisch, eds.), Plenum Press, New York (1981).

5. McDonald, F. A., and Wetsel, G. C., Jr., Generalized theory of the photoacoustic effect, *J. Appl. Phys. 49,* 2313-2322 (1978).
6. McDonald, F. A., Photoacoustic effect and the physics of waves, *Am. J. Phys. 48,* 41-47 (1980).
7. Rosencwaig, A., and Gersho, A., Theory of the photoacoustic effect with solids, *J. Appl. Phys. 47,* 64-69 (1976).
8. Yasa, Z., Amer, N. M., Rosen, H., Hansen, A. D. A., and Novakov, T., Photoacoustic investigation of urban aerosol particles, *Appl. Opt. 18,* 2528-2530 (1979).
9. Roessler, D. M., and Faxvog, F. R., Optical properties of agglomerated acetylene smoke particles at 0.5145-μm and 10.6-μm wavelengths, *J. Opt. Soc. Am. 70,* 230-235 (1980).

THE GMCC FOUR-WAVELENGTH NEPHELOMETER

Barry A. Bodhaine
Environmental Research Laboratories
National Oceanic and Atmospheric Administration
Boulder, Colorado

A four-wavelength nephelometer continuously monitored the test aerosols generated during the 2-week period of the First International Workshop on Light Absorption by Aerosol Particles (Fort Collins, Colorado, August 1980). Data of the aerosol-scattering coefficient at the wavelengths of 450-, 550-, 700-, and 850-nm are tabulated for selected periods, and all the data recorded on a strip chart are shown.

I. INTRODUCTION

The Geophysical Monitoring for Climatic Change (GMCC) program, under the National Oceanic and Atmospheric Administration (NOAA), operates background measurement stations at Barrow, Alaska; Mauna Loa, Hawaii; American Samoa; and the South Pole (1). A variety of atmospheric variables, including carbon dioxide, ozone, aerosols, and meteorological parameters, are measured continuously with a view toward understanding possible anthropogenic effects on climate.

Routine aerosol monitoring at these stations consists of the measurement of condensation nuclei concentration and volumetric light scattering. Continuous hourly data are available for Barrow since May 1976, Mauna Loa since January 1974, Samoa since July 1977, and the South Pole since January 1974. In addition, there are many cooperative programs with other institutions to measure aerosol chemistry, carbonaceous aerosol, aerosol size distribution, and aerosol absorption coefficient.

Aerosol light scattering is measured at all stations with four-wavelength nephelometers of the Ahlquist and Charlson design (2). The GMCC instruments have been described by Bodhaine and Mendonca (3) and Bodhaine (4).

A four-wavelength nephelometer was included in the First International Workshop on Light Absorption by Aerosol Particles in order to study the effects of an absorbing aerosol on the measured light-scattering coefficient. It is obvious that a highly absorbing aerosol could cause an error in the measurement which could be resolved only with additional measurements of the aerosol absorption coefficient. It was also desired to compare the Angstrom exponent of the aerosol (measured with the nephelometer) with the aerosol size distribution as measured by a high-resolution optical counter. Finally, the nephelometer was used to monitor the laboratory aerosol continuously to provide a reference for the other participants in the workshop.

II. INSTRUMENTATION AND CALIBRATION

Four-wavelength nephelometer SN104, manufactured by Meteorology Research, Inc., was installed at the workshop on 28 July 1980. The geometry of this instrument is the same as the standard broad-band nephelometer described by Charlson et al. (5). A continuous flow of air through the sample volume is illuminated by a projection lamp, and light scattered over the angles 7° to 170° is detected by a photomultiplier tube (PMT) using photon counting techniques. A rotating filter wheel in front of the PMT allows the instrument to measure volumetric light scattering at the nominal wavelengths of 450, 550, 700, and 850 nm. In practice, the effective wavelengths of the instrument may be somewhat different because of the spectral characteristics of the lamp, narrow band filters, and PMT. The effective wavelengths may be determined when the instrument is calibrated.

The GMCC nephelometer has an automatic valve on the sample inlet line to allow alternate sampling of ambient and clean air. The instrument steps through three modes to complete a measurement cycle. In the background (BG) mode, filtered air flows through the instrument and the scattering signal is stored in memory. In the normalization mode (CAL), clean air remains in the instrument and a white object is rotated into the field of view of the PMT. This signal, consisting of white object plus background, is stored in memory. Finally, ambient (AMB) air is allowed to flow through the instrument and the scattering signal is stored in memory. The instrument then performs the calculation

$$\log (AMB - BG) - \log (CAL - BG)$$

for all four wavelengths and the output of the instrument is the volume scattering coefficient of the aerosol particles in units of m^{-1}. One entire cycle is completed in approximately 6 minutes. Sensitivity and noise considerations for this instrument have been discussed by Waggoner et al. (6).

Calibration of the nephelometer is accomplished by filling it with a gas of known scattering coefficient. The use of air, argon, carbon dioxide, and Freon-12 (F12) for calibrating nephelometers has been studied in detail by Bodhaine (4). The GMCC program uses carbon dioxide as its calibration standard; however, Freon-12 was used at this workshop.

All nephelometer data were recorded on a Leeds & Northrup Speedomax 250 multipoint chart recorder. This recorder has 0- to 10-mV span and a chart speed of 1 inch per hour.

III. WORKSHOP RESULTS,

The GMCC nephelometer was calibrated at the workshop by filling it with F12 gas. The scattering coefficient for F12 is: σ_{Rg}(F12-air, 550 nm, 840 mb, 20°C) = 1.354×10^{-4} m^{-1}. The calibration gave a strip chart ordinate value of V = 64.5 on a scale of 0 to 100. Since the strip chart data must be corrected for zero and calibration offsets, in general, this gives

$$\sigma_S = 10^{\left(\frac{V + \Delta V \text{ offset}}{20} - 7\right)} m^{-1} \qquad (1)$$

Substituting σ_s = 1.354×10^{-4} m^{-1} at V = 64.5 gives ΔV = -1.8676 or

$$\sigma_S = 0.80653 \times 10^{-7} \times 10^{\frac{V}{20}} m^{-1} \qquad (2)$$

Equation (2) may be used to convert chart recorder ordinate values to volumetric light scattering σ_s.

Assuming that channel 2 of the instrument is centered at 550 nm wavelength, it is possible to calculate the effective wavelengths of the other three channels because the Angstrom exponent for Rayleigh scattering is 4.09 (dispersion effects cause a slight deviation from the exact value of 4).

Using the formula for the Angstrom exponent

$$\alpha_{ij} = - \frac{\log b_i - \log b_j}{\log \lambda_i - \log \lambda_j} \qquad (3)$$

the effective wavelengths of the instrument were calculated using data from the F12 calibration and α = 4.09. These results are tabulated in Table 1. With

TABLE 1. Effective Wavelengths of Nephelometer SN104.

Channel no.	Chart ordinate value	Effective wavelength
1	71.8	450 nm
2	64.5	550 nm
3	58.4	655 nm
4	50.4	820 nm

these known values of effective wavelengths, it is now possible to calculate the aerosol Angstrom exponent for the light-scattering data obtained at the aerosol absorption workshop.

All continuous data for the entire workshop are presented in Fig. 1. Note that all times are given in MDT and that time runs from right to left on the chart, or from the lower edge of the chart portion as placed on this page. In general, the four channels run parallel to each other and the channel numbers

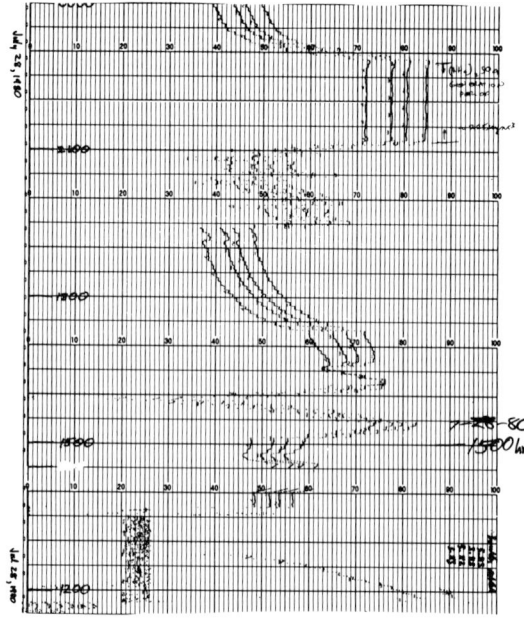

FIGURE 1. Continuous chart of aerosol scattering coefficients recorded during the workshop. The four channels represent the four wavelengbths of light given in Table 1, and Eq. (2) can be used to convert chart ordinate values to the scattering coefficients. The top edge above continues with the bottom edge on the next page, with similar continuations on subsequent pages

GMCC FOUR-WAVELENGTH NEPHELOMETER

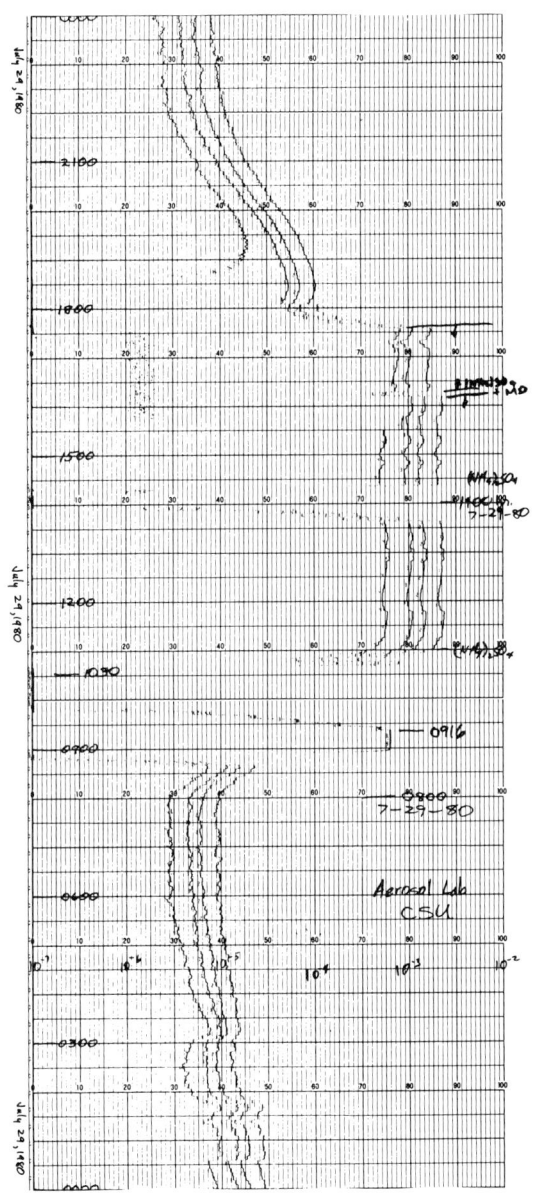

FIGURE 1. Continued

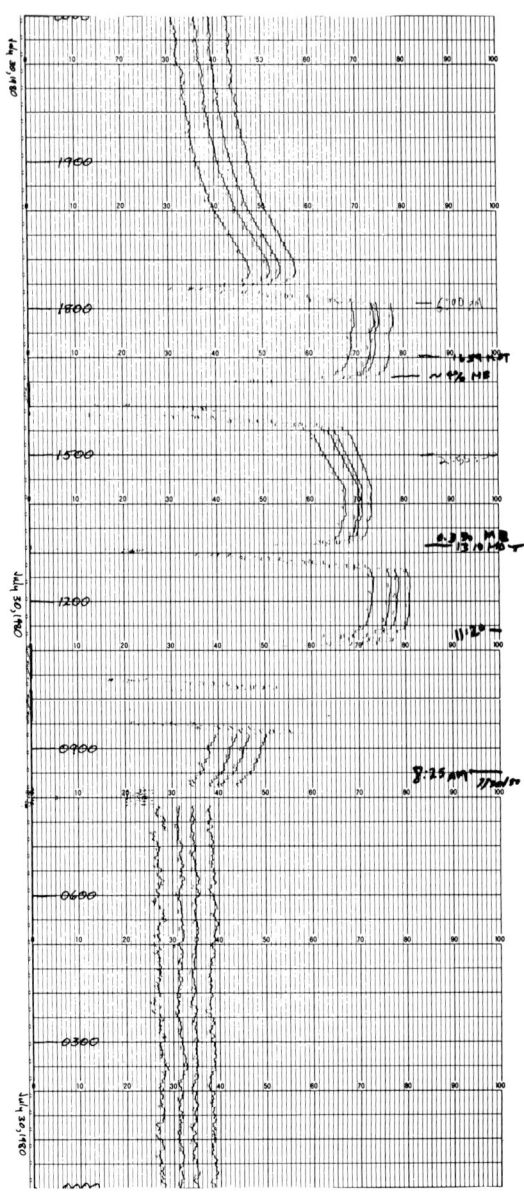

FIGURE 1. Continued

GMCC FOUR-WAVELENGTH NEPHELOMETER 155

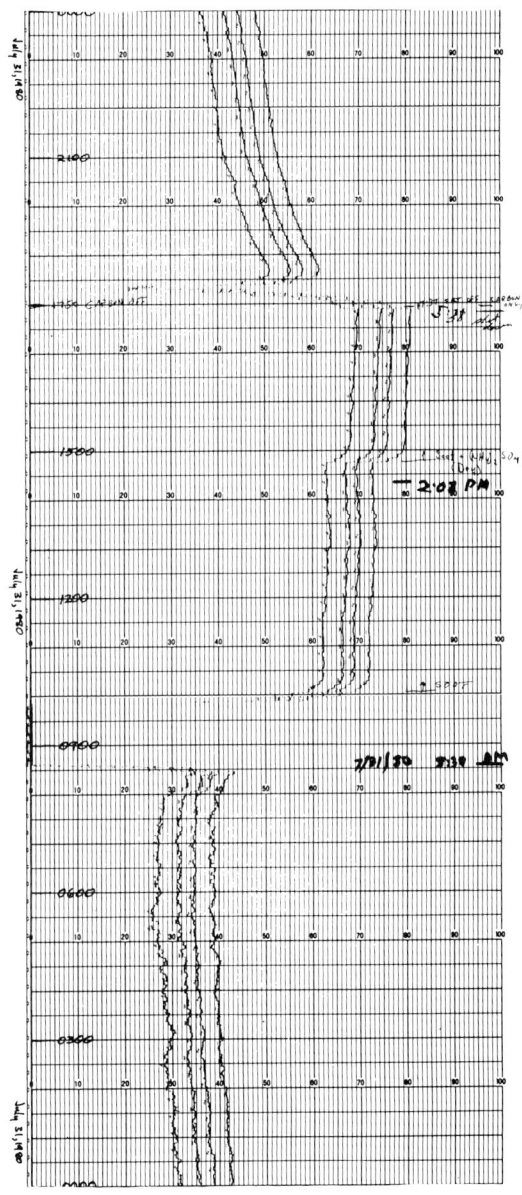

FIGURE 1. Continued

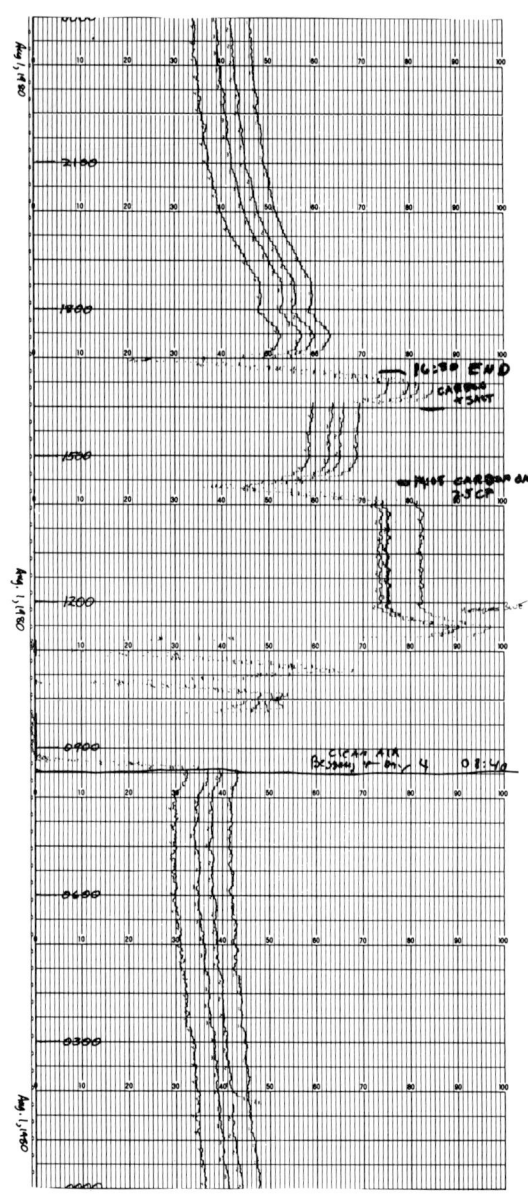

FIGURE 1. Continued

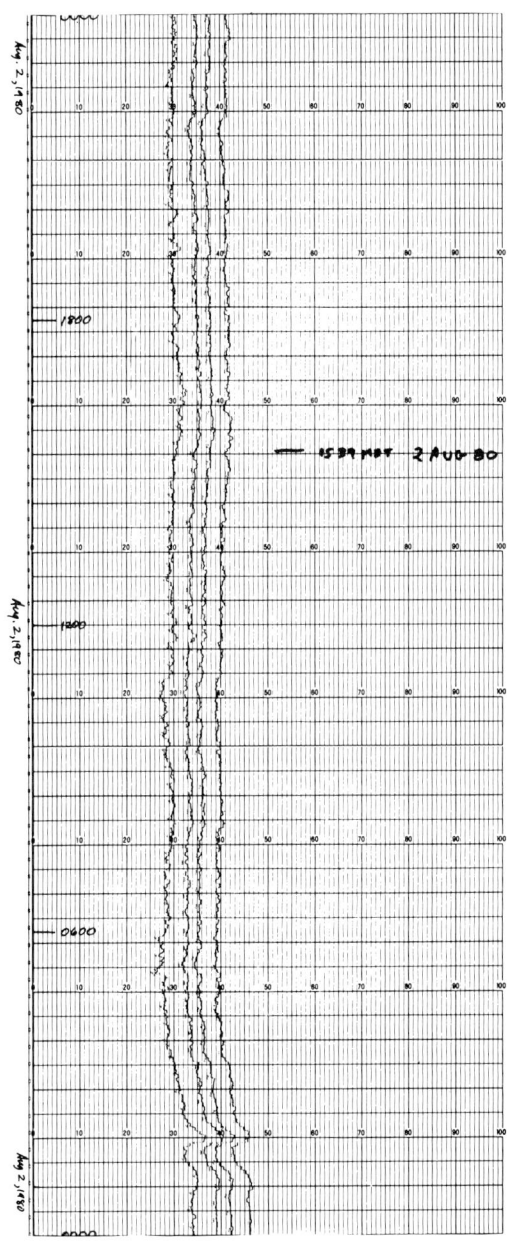

FIGURE 1. Continued

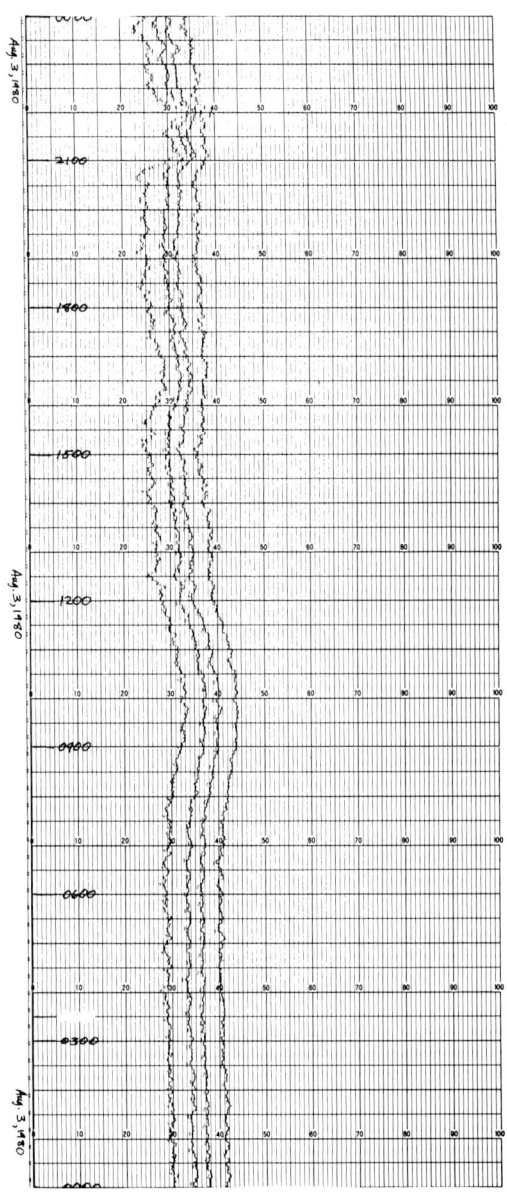

FIGURE 1. Continued

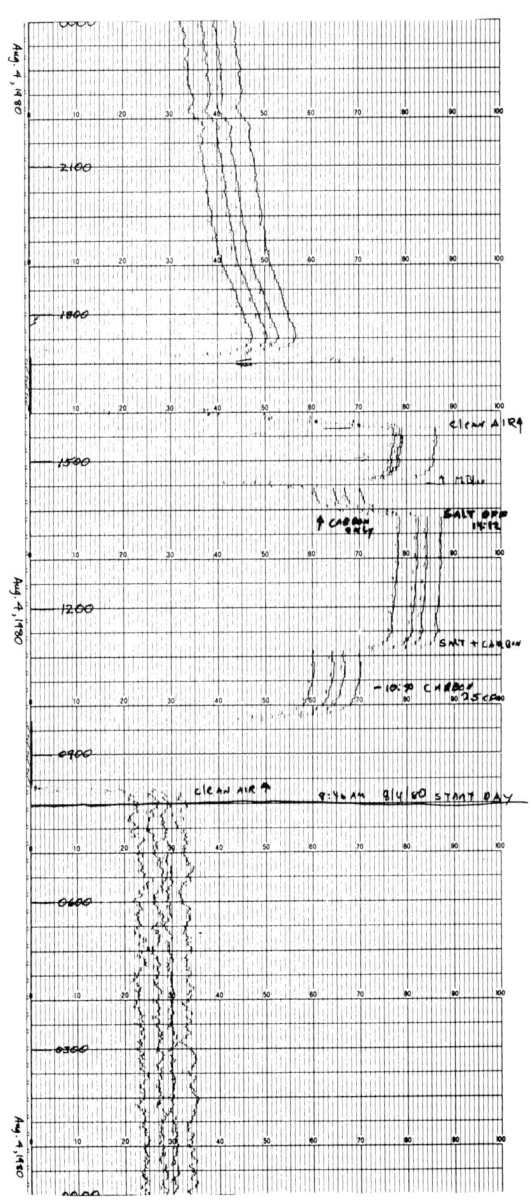

FIGURE 1. Continued

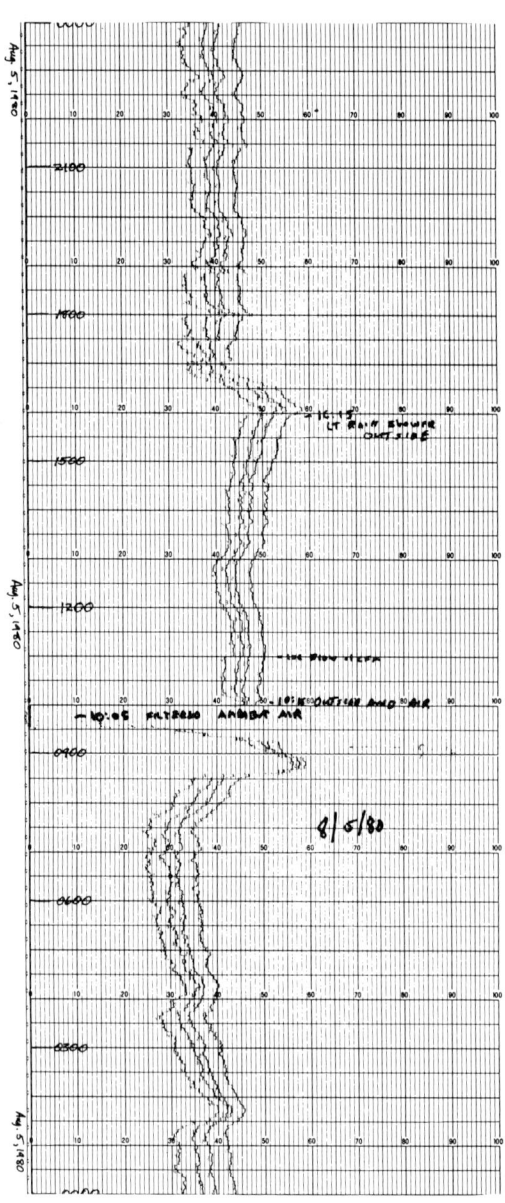

FIGURE 1. Continued

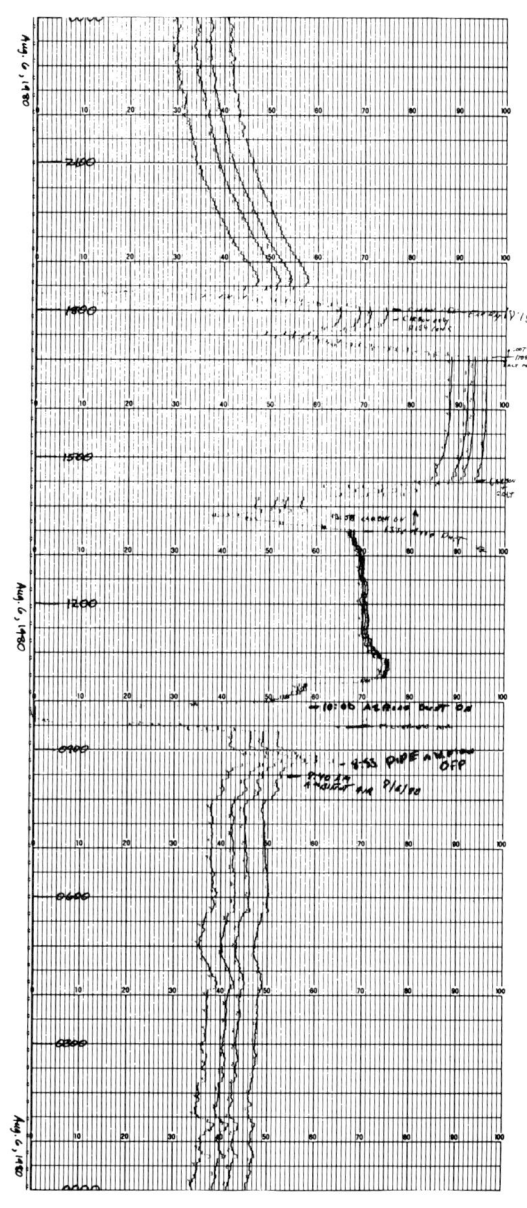

FIGURE 1. Continued

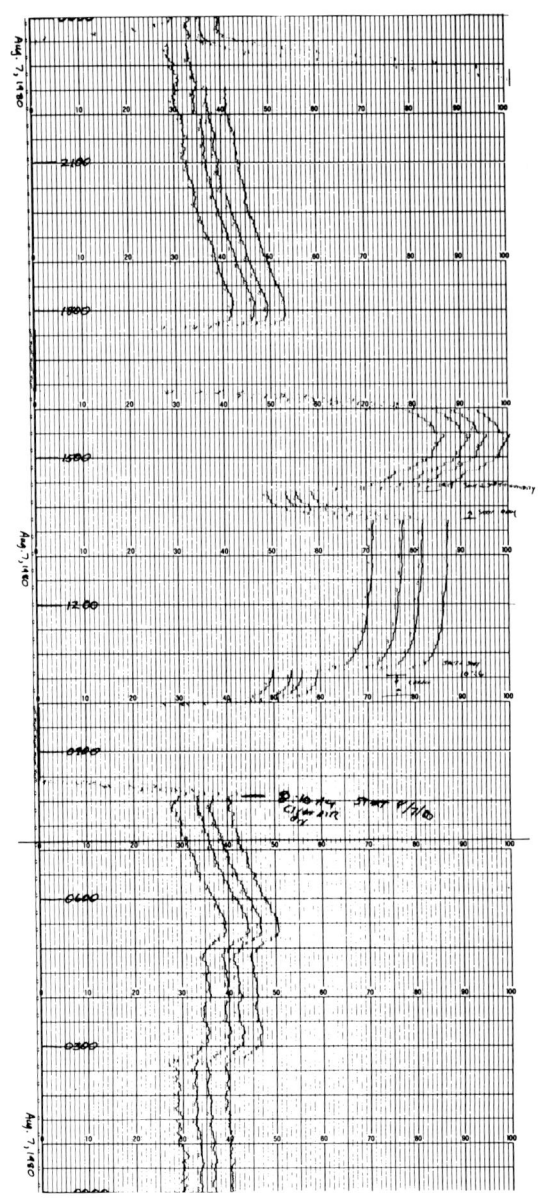

FIGURE 1. Continued

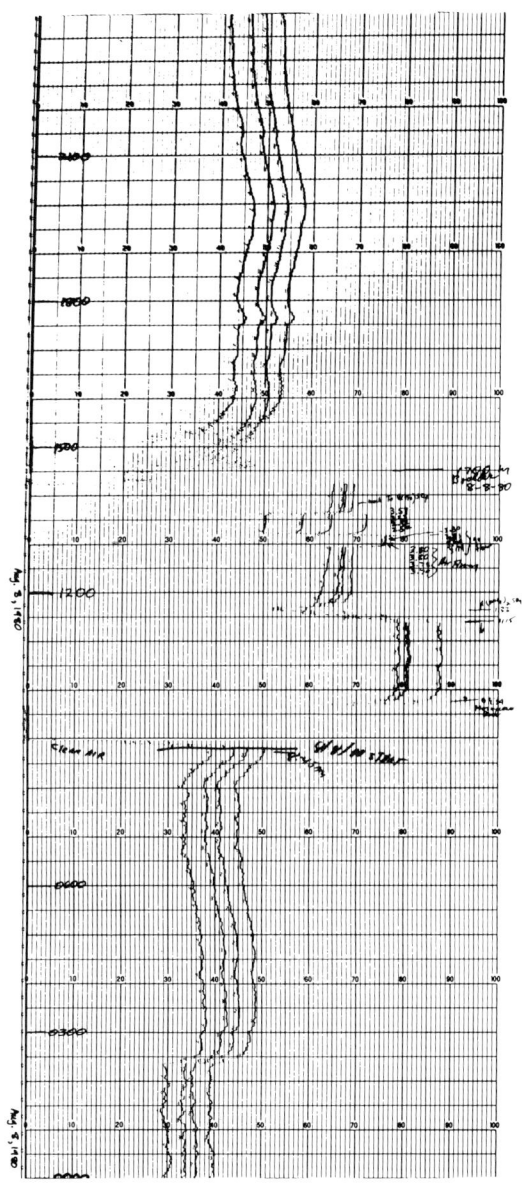

FIGURE 1. Continued

on the chart correspond to those in Table 1. The scale is logarithmic and is approximately 20 ordinate units per decade with zero equal to 10^{-7} m^{-1}. However, more accurate values may be obtained by using Eq. (2).

Light scattering and Angstrom exponent for selected time periods throughout the workshop are given in Table 2. Angstrom exponents were calculated using Eq. (3) for successive wavelength pairs of data and may be considered as a three-segment approximation of the aerosol size distribution using the relation $v = \alpha + 3$, where v is the slope of the size distribution expressed as $dN/dr = Kr^{-v}$. It should be remembered that α is a numerical approximation to the first derivative of light scatter with respect to wavelength and therefore exhibits more uncertainty than the light scatter measurements themselves.

It is interesting to consider the Angstrom exponent as size distribution information for the selected time periods given in Table 2. The size distribution of the ammonium sulfate aerosol is determined by the geometry of the bubbler and the properties of the solution (such as concentration, surface tension, etc.). It is apparent that an adjustment to the aerosol generator or the addition of methylene blue to the solution could change the aerosol size distribution. Similarly, an adjustment to the soot generator could change the size distribution of the soot aerosol, although that does not seem to be the case in the examples given in Table 2.

Pure ammonium sulfate aerosols appear to consist primarily of small particles with an α between about 1.5 and 2.5. The addition of methylene blue appears to introduce larger particles into the system and tends to reduce α_{23}. A solution of pure methylene blue (e.g., 1 August 1200 hr) produces the strange result of large negative values of α. A narrow peak in the size distribution at about 1 μm may account in part for the negative α; however, absorption as a function of wavelength may partially explain the large negative values of α_{34}.

The values of α for ambient air (5 August) are fairly typical for a rural area and may contain a contribution of larger-sized dust aerosol which could account for the small values of α_{23}. The Arizona road dust data (6 August) are typical of a large-dust aerosol at least 2 μm in diameter to give small Angstrom exponents, i.e., nearly equal scattering at all wavelengths.

The Angstrom exponent data for the pure soot aerosol are quite consistent with an average of $\alpha_{12} = 1.9$, $\alpha_{23} = 1.6$, $\alpha_{34} = 2.1$ for the soot experiments of 31 July and 1-, 4-, 6-, and 7-August. This suggests an aerosol size distribution which falls off rapidly from the small sizes to about 0.55 μm diameter, levels off slightly to about 0.70 μm, and then falls off rapidly again for the larger sizes.

In the case of the absorbing aerosols (and mixtures), however, one must await detailed size distribution and absorption coefficient information in order

TABLE 2. Light-Scattering Data for Selected Time Intervals.

	Ordinate	$\sigma_s(m^{-1})$	α
29 July 1980			
1130-1330	86.8	1.76×10^{-3}	2.01
$(NH_4)_2SO_4$	83.3	1.18×10^{-3}	1.85
	80.5	8.54×10^{-4}	2.72
	75.2	4.64×10^{-4}	
1430-1600	86.4	1.69×10^{-3}	2.18
$(NH_4)_2SO_4$	82.6	1.09×10^{-3}	1.86
	76.6	7.88×10^{-4}	2.77
	74.4	4.23×10^{-4}	
1630-1730	84.0	1.28×10^{-3}	2.35
$(NH_4)_2SO_4 + 1\%\ MB$	79.9	7.97×10^{-4}	1.78
	77.2	5.84×10^{-4}	
	M	---	
30 July 1980			
1130-1230	80.8	8.84×10^{-4}	1.49
$(NH_4)_2SO_4$	78.2	6.56×10^{-4}	1.12
	76.5	5.39×10^{-4}	2.00
	72.6	3.44×10^{-4}	
1330-1430	72.6	3.44×10^{-4}	1.09
$(NH_4)_2SO_4 + 0.3\%\ MB$	70.7	2.76×10^{-4}	0.527
	69.9	2.52×10^{-4}	1.38
	67.2	1.85×10^{-4}	
1430-1530	70.2	2.61×10^{-4}	1.49
$(NH_4)_2SO_4 + 0.3\%\ MB$	67.6	1.93×10^{-4}	0.923
	66.2	1.65×10^{-4}	1.84
	62.6	1.09×10^{-4}	
1700-1800	77.3	5.91×10^{-4}	1.72
$(NH_4)_2SO_4 + \sim 4\%\ MB$	74.3	4.18×10^{-4}	0.659
	73.3	3.73×10^{-4}	2.20
	69.0	2.27×10^{-4}	
31 July 1980			
1100-1430	72.7	3.48×10^{-4}	1.84
Soot	69.5	2.41×10^{-4}	1.58
	67.1	1.83×10^{-4}	2.20
	62.8	1.11×10^{-4}	
1500-1730	80.3	8.35×10^{-4}	2.18
Soot + $(NH_4)_2SO_4$	76.5	5.39×10^{-4}	1.78
	73.8	3.95×10^{-4}	2.51
	68.9	2.25×10^{-4}	
1 August 1980			

TABLE 2. Continued.

	Ordinate	$\sigma_s(m^{-1})$	α
1 August 1980			
1200-1330	75.3	4.69×10^{-4}	0.975
MB	73.6	3.86×10^{-4}	-0.988
	75.1	4.59×10^{-4}	-3.74
	82.4	1.06×10^{-3}	
1500-1600	68.9	2.25×10^{-4}	2.01
Soot	65.4	1.50×10^{-4}	1.52
	63.1	1.15×10^{-4}	2.15
	58.9	7.11×10^{-5}	
1615-1630	84.9	1.42×10^{-3}	1.72
Soot + $(NH_4)_2SO_4$	81.9	1.00×10^{-3}	1.45
	79.7	7.79×10^{-4}	2.25
	75.3	4.69×10^{-4}	
4 August 1980			
1000-1100	70.0	2.55×10^{-4}	1.89
S o o t	66.7	1.74×10^{-4}	1.71
	64.1	1.29×10^{-4}	2.10
	60.0	8.07×10^{-5}	
1200-1330	87.1	1.83×10^{-3}	1.72
Soot + $(NH_4)_2SO_4$	84.1	1.29×10^{-3}	1.32
	82.1	1.03×10^{-3}	2.00
	78.2	6.56×10^{-4}	
1500-1530	78.9	7.11×10^{-4}	2.75
MB	77.1	4.09×10^{-4}	-2.83
	78.4	6.71×10^{-4}	-3.95
	86.1	1.63×10^{-3}	
5 August 1980			
1000-1200	49.6	2.44×10^{-5}	1.78
Ambient air	46.5	1.70×10^{-5}	0.988
	45.0	1.43×10^{-5}	1.23
	42.6	1.09×10^{-5}	
1200-1400	49.0	2.27×10^{-5}	1.95
Ambient air	45.6	1.54×10^{-5}	1.19
	43.8	1.25×10^{-5}	1.23
	41.4	9.48×10^{-6}	
6 August 1980			
1130-1300	70.8	2.80×10^{-4}	0.287
Arizona road dust	70.3	2.64×10^{-4}	-0.329
	70.8	2.80×10^{-4}	0.512
	69.8	2.49×10^{-4}	

TABLE 2. Continued.

	Ordinate	$\sigma_s(m^{-1})$	α
1500-1700	95.8	4.97×10^{-3}	1.43
Soot + $(NH_4)_2SO_4$	93.3	3.73×10^{-3}	0.988
	91.8	3.14×10^{-3}	1.95
	88.0	2.03×10^{-3}	
1745-1800	74.5	4.28×10^{-4}	1.89
Soot	71.2	2.93×10^{-4}	1.58
	68.8	2.22×10^{-4}	2.10
	64.7	1.39×10^{-4}	
7 August 1980			
1015-1030	59.4	7.53×10^{-5}	1.89
Soot	56.1	5.15×10^{-5}	1.58
	53.7	3.91×10^{-5}	2.00
	49.8	2.49×10^{-5}	
1130-1330	86.9	1.78×10^{-3}	3.10
Soot + $(NH_4)_2SO_4$	81.5	9.59×10^{-4}	2.83
	77.2	5.84×10^{-4}	3.23
	70.9	2.83×10^{-4}	
1500-1530	99.3	7.44×10^{-3}	2.75
Soot + $(NH_4)_2SO_4$			
+ moisture	94.5	4.28×10^{-3}	2.31
	91.0	2.86×10^{-3}	2.92
	85.3	1.48×10^{-3}	
8 August 1980			
1000-1100	80.6	8.64×10^{-4}	1.09
MB	78.7	6.94×10^{-4}	-1.45
	80.9	8.95×10^{-4}	-3.48
	87.7	1.96×10^{-3}	
1200-1400	68.9	2.25×10^{-4}	0.975
$(NH_4)_2SO_4$	67.2	1.85×10^{-4}	0.461
	66.5	1.70×10^{-4}	1.43
	63.7	1.23×10^{-4}	

to give a more accurate interpretation of the nephelometer light scattering data. The data collected by GMCC at the aerosol absorption workshop are available upon request.

ACKNOWLEDGMENTS

The author wishes to thank E. Hindman for his help in setting up the equipment and A. Waggoner for his help in interpreting four-wavelength nephelometer data. Participation in the workshop was supported, in part, by NSF grant ATM 8005356.

REFERENCES

1. Mendonca, B. G., ed., *Geophysical Monitoring for Climatic Change, No. 7, Summary Report,* NOAA/ARL, US Department of Commerce (1978).
2. Ahlquist, N. C., and Charlson, R. J., Measurement of the wavelength dependence of atmospheric extinction due to scatter, *Atmos. Environ. 3,* 551-564 (1969).
3. Bodhaine, B. A., and Mendonca, B. G., Preliminary four-wavelength nephelometer measurements at Mauna Loa Observatory, *Geopys. Res. Lett. 1,* 119-122 (1974).
4. Bodhaine, B. A., Measurement of the Rayleigh scattering properties of some gases with a nephelometer, *Appl. Opt. 18,* 121-125 (1979).
5. Charlson, R. J., Ahlquist, N. C., Selvidge, H., and MacCready, P. B., Jr., Monitoring of atmospheric aerosol parameters with the integrating nephelometer, *J. Air. Poll. Contr. Assoc. 19,* 937-942 (1969).
6. Waggoner, A. P., Ahlquist, N. C., and Charlson, R. J., Recent developments in nephelometers, In *Atmospheric Aerosols: Their Optical Properties and Effects,* NASA CP-2004 (1976).

REMOTE DETERMINATION OF AEROSOL INDEX OF ABSORPTION: THE DIFFUSE-TO-DIRECT METHOD[1]

Michael A. Box
J. Douglas Copp

Institute of Atmospheric Physics
University of Arizona
Tucson, Arizona

The theory for the diffuse-to-direct method of determining the aerosol index of absorption is outlined, inherent assumptions are discussed, and the necessary atmospheric measurements are described. A lack of suitable atmospheric conditions prevented measurements during the First International Workshop on Light Absorption of Aerosol Particles (Fort Collins, Colorado, August 1980).

[1] Participation in this workshop was supported in part by National Science Foundation Grant No. ATM 8005326.

I. INTRODUCTION

Aerosols in the atmosphere scatter and absorb solar radiation. The combination of scattering plus absorption, extinction, clearly reduces the direct solar beam at the earth's surface. However, much of the light scattered by the aerosols will reappear as diffuse light. Between these two quantities, therefore, there must exist some information concerning aerosol absorption. By taking the ratio it is not necessary to know the extraterrestrial solar flux, eliminating a major source of error. Throughout this discussion all quantities will be assumed monochromatic although the wavelength dependence will not be specifically indicated.

II. THEORY

King and Herman (1) have shown that if the atmosphere is bounded below by a Lambertian surface of albedo A, the diffuse-to-direct ratio R for a solar zenith cosine μ_o may be written as

$$R(\mu_o) = \frac{a(\mu_o) + \bar{s} \mu_o A}{1 - \bar{s} A} \quad (1)$$

where a and $\bar{s}$ depend on the relevant atmospheric parameters [see King and Herman (1) for the full definition of a and $\bar{s}$].

In order to locate the relevant parameters affecting this ratio, a and $\bar{s}$ have been evaluated using single-scattering approximation. It was found that

$$a(\mu_o) = \frac{1}{2} \delta_R + \delta_P \tilde{\omega}[1 - \beta_P(\mu_o)] \quad (2)$$

where δ_R is the Rayleigh optical thickness, δ_P is the particulate optical thickness, $\tilde{\omega}$ is the particulate single scattering albedo, and $\beta_P(\mu_o)$ is the so-called backscattered fraction, which is dependent on the aerosol phase function (2), and particularly on the asymmetry factor g [see also Russell et al. (3)].

For the range of g values for typical aerosol phase functions, β_P, and hence a, can be approximated as follows

$$a(\mu_o) \cong \frac{1}{2} \delta_R + \frac{1}{2} \delta_R \tilde{\omega}(1 + \mu_o) \quad (3)$$

Similarly, for $\bar{s}$

$$\bar{s} = \delta_R + 2 \delta_P \tilde{\omega} \bar{\beta} \quad (4)$$

$$\cong \delta_R + \frac{1}{2} \delta_P \tilde{\omega} \tag{5}$$

From Eqs. (2) through (5) it is clear that R is expected to be basically linear in $\tilde{\omega}$. Correct values for β_P will also require a knowledge of the aerosol phase function, which can be obtained from the size distribution.

In this analysis two unknowns are assumed—ground albedo A and aerosol albedo $\tilde{\omega}$ (or imaginary part of the refractive index n_2). Thus, one cannot expect to extract either from a single measurement. Instead, measurements are taken throughout the day, corresponding to varying μ_0, and a best fit solution is sought (1). This approach clearly requires temporal stability, which can be partially monitored using a radiometer (which is needed to obtain δ_P, and thence the aerosol size distribution).

An alternative approach is to take measurements at a series of wavelengths during a period of a minute or so. This procedure requires that n_2 and A be essentially constant over the wavelength range involved. If the implementation of the first method during stable days suggests that this is indeed the case, then this method may be used on less stable days. Alternately, it may be assumed that ground albedo does not vary significantly over a short period, and the results from a nearby stable day can be used. This reduces the problem to only one unknown, which can be extracted from a single measurement.

III. IMPLEMENTATION

A hemispheric flux sensor has recently been constructed at the University of Arizona, and a program of measurements has commenced. In routine operation, measurements of both global and diffuse (by occulting the sun) radiation at a total of eight wavelengths between 0.4 μm and 1.0 μm are made within a 2-minute period. These measurements are made at regular intervals throughout the day. The diffuse-to-direct ratios may then be obtained.

Because the dependence of R on n_2 is nonlinear, the minimization process to obtain the best fit to a set of data cannot be performed using a standard linear least squares approach. Although nonlinear least squares techniques could be used, these are iterative and hence expensive. Instead, a library has been produced of tapes of all relevant Mie parameters for four different refractive indices (1.54, 1.54—i0.002, 1.54—i0.005, 1.54—i0.01). These may be integrated over the appropriate aerosol size distribution to provide the phase function and single scattering albedo which serve as input to the radiative transfer flux code. This code computes $a(\mu_0)$ and $\bar{s}$. These results are sufficiently linear that one may interpolate between different values of μ_0 or n_2, to within the experimental errors anticipated. Thus the problem is reduced to a linear one, which may be solved using standard techniques.

IV. ASSUMPTIONS

Throughout the analysis, a number of assumptions are made (4). Two of the most important assumptions concern the properties of the atmospheric aerosol itself. First, the aerosol size distribution is obtained by the inversion of multispectral extinction measurements taken at the same time as the flux measurements. From the size distribution, the phase function is obtained. As the effects of varying phase function are usually small, this procedure should be sufficiently accurate.

The second major assumption is that the atmospheric aerosol may be characterized by a single refractive index. This is almost certainly not the case. The absorption index obtained in this way may not be representative of any specific aerosol component present at the time and thus is clearly only an effective index. Its use is thus limited to calculations of the radiative properties of the atmospheric aerosol, which, of course, are very important climatologically. Provided these results are not used out of context, they are quite significant.

V. WORKSHOP MEASUREMENTS

The opportunity for measurements during the workshop week of 4-8 August 1980 was limited to the mornings of 4 and 6 August when cloud-free skies permitted use of the technique. Evaluation of the measurements showed that aerosol loading was very low on both days. On 4 August, in fact, the loading was consistent with zero, while on 6 August the flux ratio data showed large temporal fluctuations and were thus unsuitable for further analysis.

REFERENCES

1. King, M. D., and Herman, B. M., Determination of the ground albedo and the index of absorption of atmospheric particulates by remote sensing—Part I: Theory, *J. Atmos. Sci. 36*, 163-173 (1979).
2. Wiscombe, W. J., and Grams, G. W., The backscattered fraction in two-stream approximations, *J. Atmos. Sci. 33*, 2440-2451 (1976).
3. Russell, P. B., Livingston, J. M., and Uthe, E. E., Aerosol-induced albedo change: Measurement and modeling of an incident, *J. Atmos. Sci. 36*, 1587-1608 (1979).
4. Herman, B. M., Browning, R. S., and DeLuisi, J. J., Determination of the effective imaginary term of the complex refractive index of atmospheric dust by remote sensing: The diffuse-direct radiation method, *J. Atmos. Sci. 32*, 918-925 (1975).

RESULTS FROM UNIVERSITY OF WASHINGTON PARTICIPATION IN FIRST INTERNATIONAL WORKSHOP ON LIGHT ABSORPTION BY AEROSOL PARTICLES[1]

Antony D. Clarke
Alan P. Waggoner

Department of Civil Engineering
University of Washington
Seattle, Washington

The modified integrating plate method was used in conjunction with ancillary equipment to obtain the absorption coefficient, specific absorption, and single scattering albedo for a variety of generated aerosols. A computer driven multichannel optical particle counter also provided real time output for particle size distributions. Size segregated sampling was done for appropriate aerosols, and inferences were made on values for the complex refractive index. A new technique was also used for samples having low values of absorption, and results for these samples are included. These new results are considered the most accurate, and they take precedence over values obtained with the original method (retained for completeness and intercomparison purposes).

[1]*This work was supported in part by the Environmental Protection Agency under Grant CR807376, the National Science Foundation under Grant ATM 8005356, and the National Ocean and Atmospheric Administration under Grant NA79RAD00018.*

I. INTRODUCTION

Participation by the University of Washington Air Resources Group in the First International Workshop on Light Absorption by Aerosol Particles (Fort Collins, Colorado, August 1980) focused on the integrating plate method [so-called Charlson technique developed by Lin et al. (1) and modified by Weiss et al. (2) and others]. Modifications to this technique that have not been reported in the literature are mentioned. Additional instrumentation was also employed in order to characterize the aerosol independent of the system operated by Colorado State University (CSU).

It must be stressed that the results of the present measurements may not exemplify the typical accuracy or confidence normally expected. This is a consequence of the unavailability of a suitably accurate microbalance at the workshop. The use of a balance accurate to ± 2 μg is a critical element in the standard measurement technique. In view of this problem most filter loadings had to be obtained from a combination of CSU aerosol loadings, rotameter sample-flow values (never intended for more than a backup flow check), and load correction factors obtained from the four-wavelength-nephelometer σ_s values (3). In view of this limitation several samples taken at the workshop have not been reported because of either too short a sample time (flow uncertainty) or unstable σ_s (uncertain load).

With the exception of very short sample times a *posteriori* measurements suggest that rotameter flow uncertainty was 10%. However, filter loadings obtained from flows and mass concentrations measured by CSU for the low concentration line are suspect. The three methods employed by CSU [particle mass monitor (PMM), integrated size distribution, and filter weight] to determine mass loadings are discussed elsewhere and reveal large differences and inconsistencies for certain aerosol particles (4). With the exception of the directly measured absorption coefficient (σ_a) this uncertainty will be directly reflected in the particle-mass-dependent variables obtained for this report.

II. INSTRUMENTS

The fundamental equipment required for the measurements consists of 47-mm (0.4-μm) Nuclepore filters (preweighed and premeasured for light transmission), a microbalance (± 2 μg), an opal-glass transmission photometer ($\lambda = 550$ nm), and standard pumps, vacuum gauges, and flowmeters. For this experiment a modified nephelometer (MRI 1560 series), a modified Royco 220 optical particle counter, and an Apple II Plus computer were also utilized. The latter two instruments were coupled via hardware and software in order to provide real-time readout of particle size distributions [$dN/d(\log d)$ vs $\log d$] over the size range 0.25 μm $\leqslant$ d $\leqslant$ 10 μm. Various keyboard controls allowed choices of output displays or scaling options, and hardcopy printouts were retained for all runs for which samples were taken.

UNIVERSITY OF WASHINGTON RESULTS

In this experiment the results of the integrating plate method were obtained with a modification of the methodology reported elsewhere (1,2). It has been found that in the configuration which was previously used, reflections from the opal glass contributed to the transmitted flux and they are a strong function of the sample filter load at light loading. The instrument has been modified to accommodate a Kodak no. 96 0.8-NDF (neutral density filter) which is inserted between the Nuclepore filter and the substrate. When analyzed in the reverse configuration, with source and detector interchanged, the effect is reduced but is still significant. Not using the NDF results in a 5% to 10% overestimate of the absorption coefficient and dependent parameters for significantly absorbing species.

An integrating nephelometer was also brought to the workshop, although some damage during shipping apparently affected the calibration and caused intermittent failure, making it unreliable for the workshop. Fortunately, this nephelometer was adjacent to the four-wavelength nephelometer of Bodhaine (3) which provided the σ_s values referred to in this paper.

III. PROCEDURE

The measurements are aimed at obtaining the specific absorption coefficient $B_a(m^2/g)$ for an aerosol sample. This is generally accomplished by sampling on a 0.4-μm Nuclepore filter and measuring the change in mass and in optical transmission before and after particle collection. Details of this method are discussed elsewhere with the exception of the refinement of using a neutral density filter (5,6). The following relationships can be obtained from the Beer-Lambert law in the absence of scattering:

$$I = I_o \exp(-\sigma_a X) \tag{1}$$

where I_o is the incident light intensity, I is the transmitted light intensity, and

$$X = \frac{\text{Vol passed}}{\text{Filter area}} = \frac{V}{A}$$

$$\ln(I/I_o) = -\sigma_a \frac{V}{A} \tag{2}$$

$$M = M_v V \tag{3}$$

where M is the total particle mass and M_v is the mass concentration. Combining Eqs. (1) to (3) gives

$$B_a = \sigma_a/M_v = \frac{[\ln(I/I_o)]A/V}{M/V} \qquad (4)$$

or for light loading where

$$\ln I/I_o \cong \Delta I/I_o$$

Eq. (4) reduces to

$$B_a \cong \frac{\Delta I}{I_o} \frac{A}{M} \qquad (5)$$

To properly apply this method for strong absorbers the sample load (M) should be known to within ± 2 μg. Although this was attempted on the CSU microbalance, lack of reproducibility rendered this impossible. This is illustrated in Fig. 1 where the difference between final filter weights measured at CSU and at the University of Washington (U. of W.) are plotted as a histogram. As a consesquence of this difficulty the present results were obtained from the following analysis.

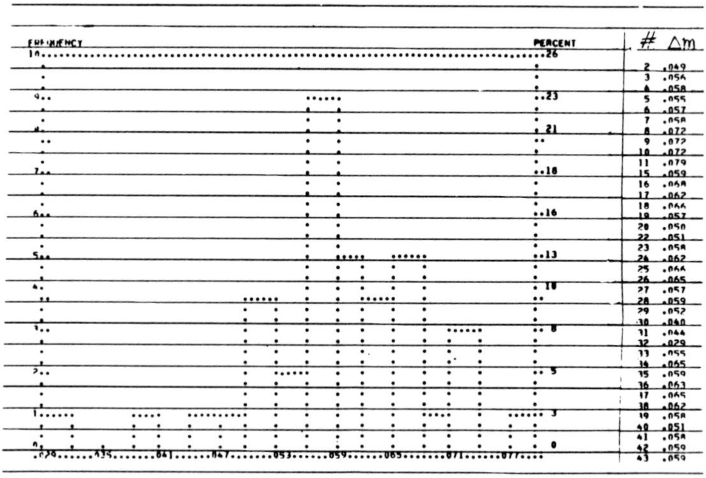

FIGURE 1. *Histogram of mass differences between filters loaded with particles as measured at CSU and at U. of W. (μg).*

Samples were collected over a known time interval and flow rates were recorded at the beginning and end of each collection. The flow meters (rotameters) were removed during the intervening time to avoid possible particle losses. Due to the rotameters operating under ambient pressure at CSU, flow values were adjusted in accordance with the well-known formula

$$F_{actual} = F_{read} \sqrt{\frac{P_{sea\ level}}{P_{CSU}}} \qquad (6)$$

Laboratory tests indicated that this correction may be a slight overestimate, but it is approximately valid for the meters used.

Next, sample load values were adjusted by using the linear relationship generally observed between σ_s and the fine particle mass loading (7); changes in the σ_s values ($\lambda = 550$ nm) of Bodhaine (3) were used to scale the CSU mass loading values for the times that the U. of W. samples were taken. This procedure proved valuable in reducing the variance in the values obtained for the optical constants of a given aerosol type.

In addition to these procedures, steps were taken to size-segregate the ambient and the Arizona road dust samples by simultaneous sample collections with and without a cyclone. The cyclone was a Bendix model no. 18 with a modified sleeve insert to provide a 50% cut at d $\cong$ 1.4 μm for a flow of 23 lpm.

Additional data of the aerosol number and volume (mass) distributions as a function of particle diameter were recorded using a modified Royco 220 optical particle counter interfaced with an Apple microcomputer (Fig. 2). This afforded a means of estimating the mass fraction in the fine particle mode and it was used to obtain mass fractions for the "cycloned" samples of filters no. 33 and no. 35. Examples of this output are provided in Fig. 3.

IV. RESULTS

The results of the analysis of particles collected on 40 filters (see Table 1 for sampling periods) are listed in Tables 2 and 3 (reference notes follow these tables and should be consulted in conjunction with their use). In view of the uncertainties in mass loadings previously described, an effort was made to arrive at the most reliable value for each filter. This was based on examining mass concentrations provided by CSU (4) and filter weighings at the workshop. Those values judged most reliable are boxed in Table 2 and are summarized in Table 3 after averaging over sample periods. The greatest uncertainties in these values, due in general to the mass measurements, should be less than 30% and hopefuly better in most cases of the selected data (Table 3).

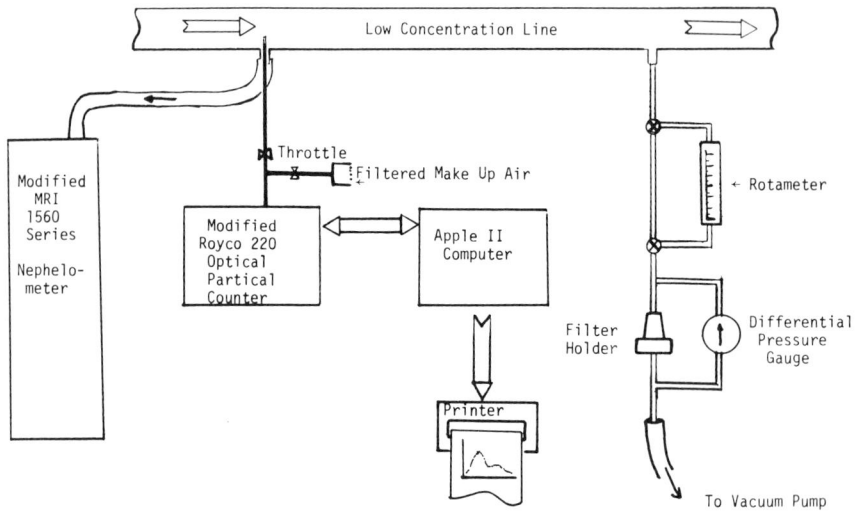

FIGURE 2. *Schematic for University of Washington sampling system.*

The complex index of refraction (n_2), provided for comparison purposes, is only a very rough estimate due in part to assumptions of a homogeneous aerosol of assumed density and real index of refraction. It is further assumed that the sample consisted of particles in the fine particle mode (approximately true for most samples) and it is determined only for one wavelength ($\sim$ 550 nm). With these assumptions it can be shown that the imaginary index of refraction can be written (6)

$$n_2 = \sigma_a(n_1^2 + 2)^2/(18 K n_1 V_T) \tag{7}$$

where

$$K = 2\pi/\lambda$$

$$V_T = \frac{\text{particle volume}}{\text{volume of air}}$$

Equation (7) was used to obtain the values of n_2 listed in this report.

Table 3 also includes σ_s values of Bodhaine (3) averaged over the U. of W. collection intervals. These values of σ_s (λ = 550 nm) and the σ_a values listed were used to obtain the single scattering albedos $\tilde{\omega}$ which are also given in Table 3.

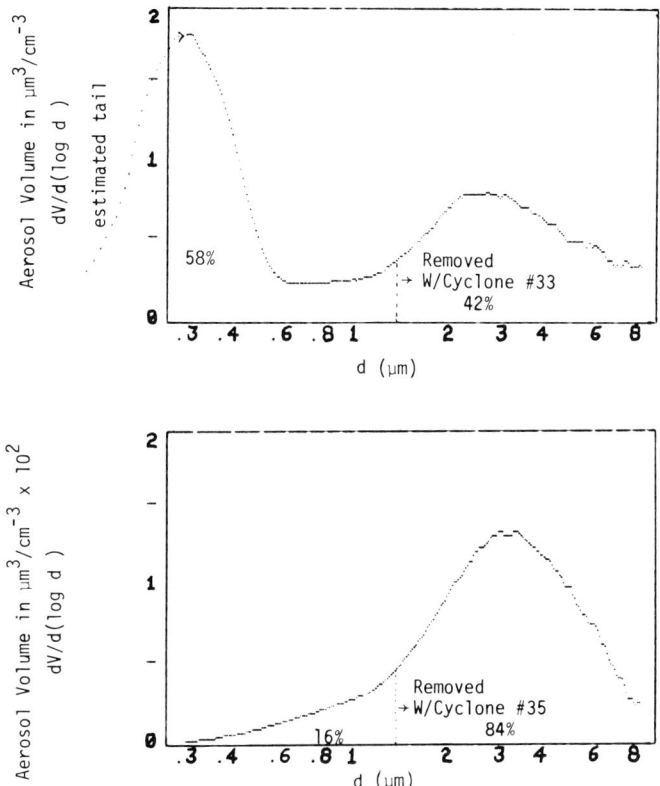

FIGURE 3. Plots of dV/d(log d) for samples 32 through 35 from Royco-Apple system. (a) Plot of dV/d(log d) for samples no. 32 and no. 33 (ambient night sample). (b) Plot of dV/d(log d) for samples no. 34 and no. 35 (Arizona road dust).

Samples no. 32 and no. 33 (ambient) were simultaneous collections as were samples no. 34 and no. 35 (Arizona road dust). The aerosol for no. 33 and no. 35 was passed through a cyclone (modified Bendix no. 18) allowing collection of particles with diameter below 1.4 μm. This collection along with the optical particle counter output (Fig. 3) allowed characterization of the optical properties for the fine particle fraction in these samples. The percentage values indicated in Fig. 3 are estimated values from the plots of the total volume (mass) above and below the 1.4-μm cut point. Agreement with measured mass fractions is excellent for no. 33 and no. 34, while weighed fine particle mass values

TABLE 1. *Sampling Periods for University of Washington Group.*

Date	Data collection period (MDT)	Sample number
7/29/80	1630-1703	2
7/30/80	1109-1139	3
7/30/80	1147-1202	4
7/30/80	1350-1426	5
7/30/80	1430-1528	6
7/30/80	1633-1648	7
7/30/80	1652-1734	8
7/31/80	1022-1034	9
7/31/80	1043-1103	10
7/31/80	1110-1115	11
7/31/80	1318-1333	13
7/31/80	1343-1347	15
7/31/80	1403-1426	16
7/31/80	1521-1541	17
7/31/80	1546-1637	18
7/31/80	1641-1651	19
7/31/80	1654-1657	20
7/31/80	1729-1735	21
8/1/80	1133-1205	22
8/1/80	1220-1328	23
8/1/80	1333-1345	24
8/1/80	1443-1520	25
8/1/80	1523-1544	26
8/4/80	1042-1114	27
8/4/80	1143-1214	28
8/4/80	1304-1400	29
8/4/80	1536-1557	30
8/5/80	1017-1146	31
8/5-6/80	1803-0831	32 & 33
8/6/80	1050-1241	34 & 35
8/6/80	1511-1623	36
8/6/80	1625-1708	37
8/6/80	1744-1804	38
8/7/80	1152-1244	39
8/7/80	1250-1302	40
8/8/80	1005-1039	41
8/8/80	1043-1104	42
8/8/80	1217-1333	43

TABLE 2. Workshop Data

Nominal mix	Number ID	$MX_v \mu g\, m^{-3}$	$MY_v \mu g\, m^{-3}$	MX/MY	$\sigma_a\, m^{-1}$	$XB_a m^2 g^{-1}$	Xn_2	$YB_a m^2 g^{-1}$	Yn_2
1% MB	1	No data							
	2 MBAS	2.93×10^2	5.42×10^2	5.42×10^{-1}	1.14×10^{-4}	3.87×10^{-1}	3.00×10^{-2}	2.10×10^{-1}	1.63×10^{-2}
	3 ASUF	1.29×10^2	1.80×10^2	7.13×10^{-1}	5.02×10^{-5}	3.90×10^{-1}	3.03×10^{-2}	2.78×10^{-1}	2.16×10^{-2}
	4 ASUF	1.86×10^2	1.43×10^2	1.30	6.42×10^{-5}	3.45×10^{-1}	2.67×10^{-2}	4.49×10^{-1}	3.48×10^{-2}
0.3% MB	5 MBAS	7.32×10^1	5.76×10^1	1.27	5.79×10^{-6}	7.91×10^{-2}	6.13×10^{-3}	1.00×10^{-1}	7.78×10^{-3}
"	6 MBAS	5.19×10^1	8.13×10^1	6.38×10^{-1}	5.14×10^{-6}	9.90×10^{-2}	7.67×10^{-3}	6.32×10^{-2}	4.90×10^{-3}
4.0% MB	7 MBAS	1.01×10^2	9.36×10^1	1.08	8.83×10^{-5}	8.70×10^{-1}	6.74×10^{-2}	9.44×10^{-1}	7.31×10^{-2}
"	8 MBAS	1.03×10^2	7.04×10^1	1.46	1.14×10^{-4}	1.11	8.57×10^{-2}	1.62	1.26×10^{-1}
	9 Soot	3.88×10^1	1.06×10^2	3.65×10^{-1}	9.79×10^{-4}	2.53×10^1	1.96	9.21	7.14×10^{-1}
	10 Soot	6.65×10^1	1.06×10^2	6.26×10^{-1}	9.90×10^{-4}	1.49×10^1	1.15	9.31	7.21×10^{-1}
	11 Soot		1.06×10^2		1.30×10^{-3}			1.23×10^1	9.50×10^{-1}
	12	No data							
	13 Soot	1.13×10^2	9.97×10^1	1.13	1.11×10^{-3}	9.81	7.60×10^{-1}	1.11×10^1	8.62×10^{-1}
	14	No data							
	15 Soot	1.14×10^2	9.49×10^1	1.20	1.22×10^{-3}	1.07×10^1	8.28×10^{-1}	1.28×10^1	9.94×10^{-1}
	16 Soot	1.12×10^2	9.97×10^1	1.12	1.06×10^{-3}	9.50	7.36×10^{-1}	1.07×10^1	8.26×10^{-1}
	17 SOAS	1.30×10^2	1.62×10^2	7.98×10^{-1}	3.64×10^{-4}	2.81	2.18×10^{-1}	2.24	1.74×10^{-1}
	18 SOAS	1.70×10^2	1.59×10^2	1.06	3.80×10^{-4}	2.24	1.74×10^{-1}	2.38	1.85×10^{-1}
	19 SOAS	2.20×10^2	1.51×10^2	1.46	3.97×10^{-4}	1.80	1.40×10^{-1}	2.63	2.04×10^{-1}
	20 SOAS		1.52×10^2		4.76×10^{-4}			3.13	2.42×10^{-1}
	21 SOAS	8.42×10^2	1.42×10^2	5.92	4.27×10^{-4}	5.07×10^{-1}	2.93×10^{-2}	3.01	2.33×10^{-1}
	22 MBLU	2.64×10^2	4.05×10^2	6.51×10^{-1}	4.68×10^{-4}	1.77	1.37×10^{-1}	1.15	8.95×10^{-2}
	23 MBLU	2.98×10^2	4.05×10^2	7.37×10^{-1}	4.96×10^{-4}	1.66	1.29×10^{-1}	1.22	9.49×10^{-2}
	24 MBLU	2.44×10^2	4.05×10^2	6.03×10^{-1}	4.05×10^{-4}	1.66	1.28×10^{-1}	1.00	7.75×10^{-2}

TABLE 2. Continued

Nominal mix	Number ID	$MX_v\mu g\ m^{-3}$	$MY_v\mu g\ m^{-3}$	MX/MY	$\sigma_a\ m^{-1}$	$XB_am^2g^{-1}$	Xn_2	$YB_am^2g^{-1}$	Yn_2
	25 Soot	7.21×10^1	6.07×10^1	1.19	7.18×10^{-4}	9.96	7.72×10^{-1}	1.18×10^1	9.17×10^{-1}
	26 Soot	9.01×10^1	6.07×10^1	1.48	7.27×10^{-4}	8.07	7.25×10^{-1}	1.20×10^1	9.28×10^{-1}
	27 Soot	1.17×10^2	6.93×10^1	1.68	8.01×10^{-4}	6.87	5.33×10^{-1}	1.16×10^1	8.96×10^{-1}
50% Soot	28 SOAS	3.72×10^2	2.42×10^2	1.54	9.30×10^{-4}	2.50	1.94×10^{-1}	3.85	2.98×10^{-1}
,,	29 SOAS	3.76×10^2	2.42×10^2	1.56	9.71×10^{-4}	2.58	2.00×10^{-1}	4.02	3.12×10^{-1}
	30 MBLU	5.04×10^2	1.04×10^3	4.87×10^{-1}	6.79×10^{-4}	1.35	1.04×10^{-1}	6.56×10^{-1}	5.08×10^{-2}
	31 AMBI	1.98×10^1	1.51×10^1	1.32	8.42×10^{-6}	4.25×10^{-1}	3.29×10^{-2}	5.59×10^{-1}	4.33×10^{-2}
	32 AMBI	6.91	6.01	1.15	2.86×10^{-6}	3.70×10^{-1}	2.87×10^{-2}	4.27×10^{-1}	3.34×10^{-2}
Cyclone	33 AMBI	3.40	3.49	9.76×10^{-1}	2.54×10^{-6}	7.45×10^{-1}	5.77×10^{-2}	7.27×10^{-1}	5.64×10^{-2}
	34 AZRD	1.15×10^2	9.93×10^2	1.16×10^{-1}	2.92×10^{-5}	2.53×10^{-1}	1.96×10^{-2}	2.94×10^{-2}	2.28×10^{-3}
Cyclone	35 AZRD	3.81×10^1	1.59×10^2	2.40×10^{-1}	1.40×10^{-5}	3.67×10^{-1}	2.85×10^{-2}	8.81×10^{-2}	6.83×10^{-3}
4.0% Soot	36 SOAS	9.82×10^2	2.08×10^3	4.71×10^{-1}	5.09×10^{-4}	5.18×10^{-1}	4.02×10^{-2}	2.44×10^{-1}	1.89×10^{-2}
,,	37 SOAS	9.24×10^2	1.90×10^3	4.87×10^{-1}	4.89×10^{-4}	5.29×10^{-1}	4.10×10^{-2}	2.58×10^{-1}	2.00×10^{-2}
	38 Soot	1.93×10^2	7.03×10^1	2.74	1.41×10^{-3}	7.31	5.66×10^{-1}	2.00×10^1	1.55
5.0% Soot	39 SOAS	6.12×10^2	1.55×10^2	3.96	4.48×10^{-4}	7.31×10^{-1}	5.67×10^{-2}	2.89	2.24×10^{-1}
,,	40 SOAS	7.12×10^2	1.46×10^2	4.87	4.73×10^{-4}	6.64×10^{-1}	5.15×10^{-2}	3.24	2.51×10^{-1}
	41 MBLU	3.61×10^2	1.35×10^3	2.68×10^{-1}	6.91×10^{-4}	1.91	1.48×10^{-1}	5.13×10^{-1}	3.98×10^{-2}
	42 MBLU	4.50×10^2	1.29×10^3	3.50×10^{-1}	7.43×10^{-4}	1.65	1.28×10^{-1}	5.78×10^{-1}	4.48×10^{-2}
	43 ASUF	5.19×10^1	2.85×10^1	1.82	7.33×10^{-6}	1.41×10^{-1}	1.09×10^{-2}	2.57×10^{-1}	1.99×10^{-2}

Reference notes for Tables 2 and 3 follow Table 3.

UNIVERSITY OF WASHINGTON RESULTS

TABLE 3. Averaged and/or Most Representative Preferred Values for Samples Selected from Previous Data Listing.

	Day 1980	Time period	Filter number	Sample	σ_s†	$\sigma_a(m^{-1})$	$B_a(m^2g^{-1})$	n_2	$\tilde{\omega}$
*	7/29	1630-1703	2	MBAS	1.09×10^{-3}	1.14×10^{-4}	0.387	0.03	0.905
*	7/30	1109-1202	3,4	ASUF	1.12×10^{-3}	5.02×10^{-5}	0.367	0.057	0.957
*	7/30	1350-1528	5,6	MBAS	2.34×10^{-4}	5.46×10^{-6}	0.089	0.0069	0.977
*	7/30	1633-1734	7,8	MBAS	4.18×10^{-4}	1.01×10^{-4}	0.99	0.76	0.805
	7/31	1022-1426	9-11,13,15,16	Soot	2.41×10^{-4}	1.11×10^{-3}	9.71	0.784	0.178
	7/31	1521-1735	17-21	SOAS	5.39×10^{-4}	4.08×10^{-4}	2.60	0.202	0.569
	8/1	1133-1345	22-24	MBLU	3.86×10^{-4}	4.56×10^{-4}	1.69	0.131	0.458
	8/1	1443-1544	25,26	Soot	1.5×10^{-4}	7.22×10^{-4}	11.9	0.922	0.172
	8/4	1042-1114	27	Soot	1.74×10^{-4}	8.01×10^{-4}	11.6	0.896	0.178
	8/4	1143-1400	28,29	SOAS 50%	1.29×10^{-3}	9.5×10^{-4}	2.54	0.197	0.575
	8/4	1536-1557	30	MBLU	4.09×10^{-4}	6.79×10^{-4}	1.35	0.104	0.376
	8/5	1017-1146	31	AMBI	1.7×10^{-5}	8.42×10^{-6}	0.559	0.043	0.668
	8/5	1803-1831	32	AMBI	1.54×10^{-5}	2.86×10^{-6}	0.427	0.033	0.843
⌐	8/5	1803-1831	33	AMBI (d≤1.4 μm)		2.54×10^{-6}	0.727	0.056	
*	8/6	1050-1241	34	AZRD	2.64×10^{-4}	2.92×10^{-5}	0.253	0.0196	0.900
⌐*	8/6	1050-1241	35	AZRD (d≤1.4 μm)		1.40×10^{-5}	0.367	0.0285	
	8/6	1511-1708	36,37	SOAS 4%	3.73×10^{-3}	4.99×10^{-4}	0.525	0.0406	0.882
	8/6	1744-1804	38	Soot	2.93×10^{-4}	1.41×10^{-3}	7.31	0.566	0.219
	8/7	1152-1302	39,40	SOAS 5%	9.59×10^{-4}	4.6×10^{-4}	0.697	0.054	0.675
	8/8	1005-1104	41,42	MBLU	6.94×10^{-4}	7.17×10^{-4}	1.79	0.138	0.491
*	8/8	1217-1333	43	ASUF	1.85×10^{-4}	7.33×10^{-6}	0.257	0.020	0.962

* See Section V for revised values; † from Bodhaine's nephelometer @ λ = 550 nm; [simultaneous collection with and without cyclone. Reference notes for Tables 2 and 3 follow.

Reference notes for Tables 2 and 3.

The appearance of X or Y in conjunction with the table headings indicate values based on direct mass weighings at CSU (X), or calculated mass loadings based on total mass values from the CSU particle sizing system (Y).

The complex refractive index as it appears in the tables represents a rough estimate only based upon numerous assumptions including the following values for particle density and real refractive index:

Aerosol	ID	Density	Real refractive index
Ammonium sulfate	ASUF	1.77	1.53
Methylene blue	MBLU	1	1.5
Soot	Soot	1.8	1.45
Ambient aerosol	AMBI	2.3	1.45
Arizona road dust	AZRD	2.3	1.45
Soot and ammonium sulfate	SOAS	1.8	1.51
Methylene blue and ammonium sulfate	MBAS	1.7	1.5

The latter two aerosol mixtures have not had their density and real refractive index adjusted for relative component masses, so that the imaginary index of such samples is to be ignored.

Boxed values (Table 2) are those considered to be the most reliable. These were chosen based on the mass measurement believed to have the least uncertainty for a given filter sample.

are higher than estimated from the plot for no. 35. As the cut on the cyclone is not sharp, the difference for no. 35 may be due to the steep increase of the distribution with diameter at d = 1.4 μm. Variation of density of the particles as a function of size may also affect this difference.

Comparisons of samples with and without the use of the cyclone for both ambient and Arizona road dust collections reveal substantially higher specific absorption for the cycloned samples. For the ambient sample, the Royco-Apple size distribution is illustrated in Fig. 3(a) and suggests that somewhat over half (58% estimate only) of the mass resides in the cycloned (< 1.4 μm) fraction. It is of interest to note that this percentage of the mass multiplied by the specific absorption for the fine particle fraction yields the absorption of the total. This suggests that virtually all of the absorption is due to the fine particle mode for this ambient sample. Implications are that the fine and coarse modes differ in both nature and origin.

Similar observations can be made for the Arizona road dust sample by referring to Fig. 3(b). Once again higher specific absorption is found for the fine mode. However, in this instance the fine mode (estimated 16% by volume) accounts for approximately 26% of the absorption of the total. Given the uncertainties associated with mass measurements, density, cyclone size cut, etc., it is believed that in contrast to the ambient sample, Arizona road dust is largely homogeneous in composition regardless of particle size.

It should be noted that the volume (mass) distribution in Fig. 3(b) for Arizona road dust does not agree with the analysis provided at the workshop, in that the large particle fraction in Fig. 3(b) appears to be decreasing rapidly for d > 3 μm. This is not surprising as the fall speed for a 6-μm-diameter particle is on the order of 2 mm/sec. The U. of W. sampling post was located at approximately 25 m from the generator and about 15 sec away in time. Consequently, most particles in this size range would have had an opportunity to settle out. Particles much larger than this would not be observed at all [as seen in Fig. 3(b)].

V. CONCLUDING COMMENTS

In addition to the reflections by the opal glass, which are reduced with the neutral density filter, other optical effects are present in Nuclepore membrane filters which require additional corrections. These are most significant for very low light absorption values (e.g., with ammonium sulfate) which result in < 5% light attenuation for the filter sample. These effects have been the subject of extensive analysis in the laboratory, and an alternate more sensitive technique has been developed specifically for this situation following the workshop. As this new technique was not available for the CSU workshop, the results described in this report are based primarily on values obtained using the opal-glass technique with the 0.8 ND filter.

It is believed that the values for the absorption coefficient and the specific absorption coefficient reported in Table 3 for absorbing particles may be systematically high by about 5% to 15%. This is a consequence of absorbing particles perturbing the optical properties of the Nuclepore filter. For nonabsorbing particles (e.g., ammonium sulfate) the reported values are excessively high and are considered to be in error. However, these values have been retained in this report for the purpose of comparison with values obtained by others at the workshop using the integrating plate method.

Using the integrating plate method for particles with low light absorption tends to significantly overestimate the absorption coefficient due to both backscatter and particle/filter surface effects. Following the CSU workshop, a new technique, the "integrating sandwich method," was developed. This technique effectively eliminates those problems by inserting the particles between two diffusively highly reflecting surfaces, and measuring the light transmission through the particles before and after insertion between the surfaces. This technique shares features in common with the integrating sphere such that the results are independent of forward and backward scattering effects, and absorption is predictably enhanced allowing calculation of the absorption coefficient. A paper on the new technique, materials, calibration, and field tests is currently in preparation.

The integrating sandwich was used to reanalyze filter samples no. 3, no. 4, and no. 43 (ammonium sulfate), nos. 5-8 (methylene blue and ammonium sulfate), and no. 34 and no. 35 (Arizona road dust). These samples showed little light attenuation using the integrating plate method and thus were in the region of spurious effects mentioned in the which is indicated in Table 3. The reanalyzed values are listed in Table 4.

Pure ammonium sulfate (with the exception of no. 43 which was contaminated) showed no absorption upon remeasurement with the improved technique, although calibration uncertainty for these cases only guarantee that B_a lies below 0.004 m^2 gm^{-1}.

Intercomparisons of relative absorption for ammonium sulfate containing various amounts of soot or methylene blue showed mixed results, although trends were consistent. Specific absorption B_a for MBAS (4%) was 16 times that of MBAS (0.3%). The MBAS (1%) value was low by comparison, but mass uncertainty for that case was much larger. In all cases these values exceeded those expected compared to the pure methylene blue value (B_a) by approximately a factor of three. The reason for this is uncertain, but a contributing factor may be the different morphology (SEM photos) for the pure and mixed aerosol. Samples of MBAS (4%) also had a pinkish hue of unknown origin.

Similar comparisons for soot and ammonium sulfate (SOAS) were closer to expected values. The ratio of B_a values for SOAS (4%) to soot was 4.4% and

TABLE 4. Remeasured Values for Low Absorbers Using The Integrating Sandwich Method.

Filter no.	Sample	$\sigma_a(m^{-1})$	$B_a(m^2gm^{-1})$	$\sigma_s(m^{-1})$	$\tilde{\omega}$
2	MBAS (1.0%)[c]	1.7×10^{-5}	0.058	1.09×10^{-3}	0.984[a]
3	ASUF	$<3 \times 10^{-7}$	<0.004	1.12×10^{-3}	0.9999
4	ASUF	$<3 \times 10^{-7}$	<0.004	1.12×10^{-3}	0.9999
43	ASUF	$\sim 4.5 \times 10^{-6}$	~ 0.16	1.8×10^{-4}	0.975[b]
5	MBAS (0.3%)	2×10^{-6}	0.027	2.5×10^{-4}	0.992
6	MBAS (0.3%)	1.3×10^{-6}	0.025	1.65×10^{-4}	0.992
7	MBAS (~4%)	4.2×10^{-5}	0.415	4.18×10^{-4}	0.909
8	MBAS (~4%)	5.3×10^{-5}	0.514	4.18×10^{-4}	0.887
34	AZRD (no cyclone)	5.4×10^{-6}	0.047	2.64×10^{-4}	0.980
35	AZRD (cyclone)	2.6×10^{-6}	0.068	---	---

[a]*Large mass uncertainty.*

[b]*The relatively large absorption present for filter no. 43 was found to be due to contamination by large particles containing methylene blue, the origin of which is uncertain.*

[c]*Numbers in parentheses refer to fraction of methylene blue present by mass relative to ammonium sulfate.*

for SOAS (5%) to soot was 5.8%. For SOAS (50%) the value for B_a was only half of that expected; however, mass determinations were again in conflict for this sample.

For all samples the optical measurements are believed to have an accuracy better than ± 20%. However, reliance on approximate flow values and uncertain mass determinations has increased this in some extreme cases to as much as ± 50% for derived quantities. The remeasured values taken with the alternative technique have an accuracy of better than ± 30%, provided that the values are for the low absorbers.

REFERENCES

1. Lin, C. I., Baker, M., B., and Charlson, R. J., Absorption coefficient of atmospheric aerosols: A method for measurement, *Appl. Opt. 12(6)*, 1356-1363 (1973).
2. Weiss, R. E., Waggoner, A. P., Thorsell, D. L., Hall, J. S., Riley, L. A., and Charlson, R. J., On the nature of light absorbing aerosol, *Proc. Conf. on Carbonaceous Particles in the Atmosphere*, Report LDL-9037, Lawrence Berkeley Laboratory, Berkeley, California (1978).

3. Bodhaine, B. A., The GMCC four-wavelength nephelometer, *Light Absorption by Aerosol Particles* (H. E. Gerber and E. E. Hindman, eds.), Spectrum Press, Hampton, Virginia (1982).
4. Hindman, E. E., II, Horn, R. D., and Finnegan, W. G., Generation and characterization of aerosol particles, *Light Absorption by Aerosol Particles* (H. E. Gerber and E. E. Hindman, eds.), Spectrum Press, Hampton, Virginia (1982).
5. Lin. C. I., "Light Absorption by Atmospheric Aerosol and Its Global Climatic Implications," Ph.D. Thesis, University of Washington, Seattle (1973).
6. Weiss, R. E., "The Optical Absorption Properties of Suspended Particles in the Lower Troposphere at Visible Wavelengths," Ph.D. Thesis, University of Washington, Seattle (1980).
7. Waggoner, A. P., and Weiss, R. E., Comparison of fine particle mass concentration and light scattering in ambient aerosol, *Atmos. Environ.* *14,* 623-626 (1980).

LIGHT ABSORPTION MEASUREMENTS USING INTEGRATING PLATE METHOD[1]

Stan Cowen

Meteorology Research, Inc.
Altadena, California

David S. Ensor

Research Triangle Institute
Research Triangle Park, North Carolina

Leslie E. Sparks

US Environmental Protection Agency
Industrial Environmental Research Laboratory
Research Triangle Park, North Carolina

A double-beam spectrophotometer is used with the Integrating Plate Method (1) to measure the absorption coefficient of aerosol particles generated at the First International Workshop on Light Absorption by Aerosol Particles (Fort Collins, Colorado, August 1980). The imaginary refractive index is estimated for mixtures of ammonium sulfate and methylene blue.

[1]This work was supported by US Environmental Protection Agency Cooperative Agreement R806718010.

I. INTRODUCTION AND SCOPE

This report is a brief summary of light absorption measurements of aerosols generated at the First International Workshop on Light Absorption by Aerosol Particles held at Colorado State University, Fort Collins, Colorado. The experimental technique which was used is called the Integrating Plate Method (IPM) which compares light absorption through a clean Nuclepore filter to absorption through one with a single layer of aerosol. An opal glass diffuses the light so only absorption is measured. The aerosols sampled at the workshop include ammonium sulfate (AS), AS plus 0.3% methylene blue (MB), AS plus 1% MB, AS plus 4% MB, soot, and soot mixed with AS.

The ability to accurately measure absorption or the imaginary refractive index using the IPM technique depends on careful measurement of the aerosol size distribution. If the sample contains only submicron aerosol, then the absorption can be easily obtained and is proportional to the aerosol volume. The use of Mie theory to reduce the data assumes that a filter coated with aerosol can represent the optical effects of a suspension of aerosol.

II. EXPERIMENTAL PROCEDURE

The experimental technique is an application of the Integrating Plate Method (IPM) as developed by Lin et al. (1). Weiss et al. (2) further tested and calibrated this technique with monodisperse aerosol of methylene blue. Aerosol samples are collected on a Nuclepore filter with pore size equal to 0.2 μm.

IPM is a simple direct measurement of aerosol light absorption based on comparing the transparency of a clean Nuclepore filter with one containing a thin layer of particles. The particles are on the side of the filter facing the incident light. A Hitachi model 60 double-beam spectrophotometer with a specially designed filter holder was used for the comparison. Opal glass behind each filter integrates the transmitted light as a function of angle so that only absorption is recorded. Back scattering probably can be ignored when the real refractive indices of particles and filters are similar. The double-beam automatically gives the ratio of the aerosol absorption to the blank filter. The optical arrangement for the beam measuring the sample is shown in Fig. 1.

The real part of the index of refraction is measured with oil-immersion techniques. The sample is spread over a microscope slide so that individual particles can be identified. The particle is then immersed in an oil of known refractive index. The Becke line test indicates whether the particle index of refraction is lower or higher than that of the immersion oil.

The quantity of material collected on each filter ranged for 10 μg to 700 μg as measured by a Cahn 25 balance. A dry gas meter downstream of the filter measured gas volume through the filter. The calculation of the absorption

INTEGRATING PLATE METHOD

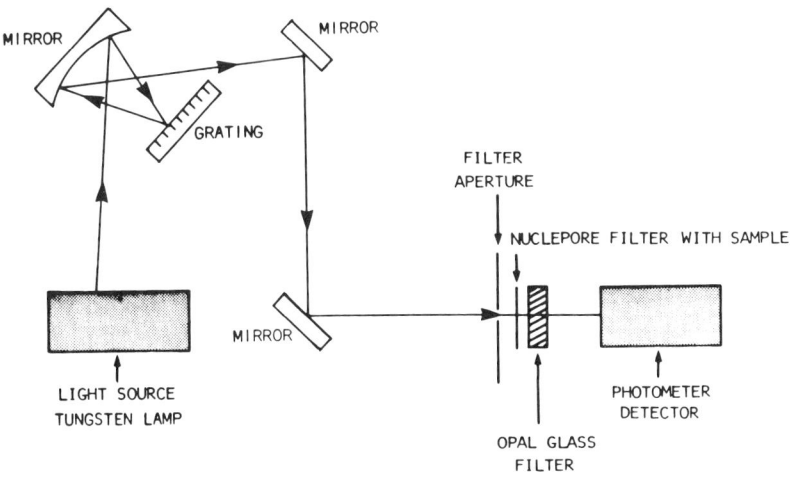

FIGURE 1. Schematic diagram of the optical system incorporating the integrating plate method.

coefficient σ_a required measurement of an effective pathlength which was determined from the gas volume divided by the area of the filter.

III. DATA ANALYSIS

Data reduction for the Integrating Plate Method uses Mie theory to determine absorption effects. It is assumed that the filter does not interfere with the light absorption of the particles. Nuclepore membrane filters were used because the particles are on the surface of the filter and not embedded as they would be in a fiberglass filter.

The transmission of light through a volume containing an aerosol is described by the Beer-Lambert law

$$\text{Light transmittance} = I/I_o = \exp(-\sigma_e L) \tag{1}$$

where I is the transmitted light, I_o is the incident light, σ_e is the light extinction coefficient, and L is the illumination path length. The light extinction coefficient is the sum of the scattering coefficient (σ_s) and the absorption coefficient (σ_a).

The ratio of intensity (I') of light transmitted through the opal glass and a filter coated with aerosol to the intensity (I_o') transmitted through an opal glass and blank filter can be used to calculate the absorption coefficient

$$I'/I_o' = \exp(-\sigma_a L) \tag{2}$$

The specific absorption coefficient B_a is calculated by determining the ratio of the absorption coefficient (m^{-1}) to the mass concentration of the aerosol ($\mu g/m^3$).

The imaginary refractive index is not a direct measurement obtained from the IPM technique; it must be inferred based on Mie theory. The imaginary refractive index is obtained from calibration curves using methylene blue aerosol. Figure 2 shows the calibration curves for methylene blue in ammonium sulfate aerosol versus a volumetric specific absorption coefficient (S_{va}). Before discussing the use of this calibration curve, this absorption coefficient should be defined.

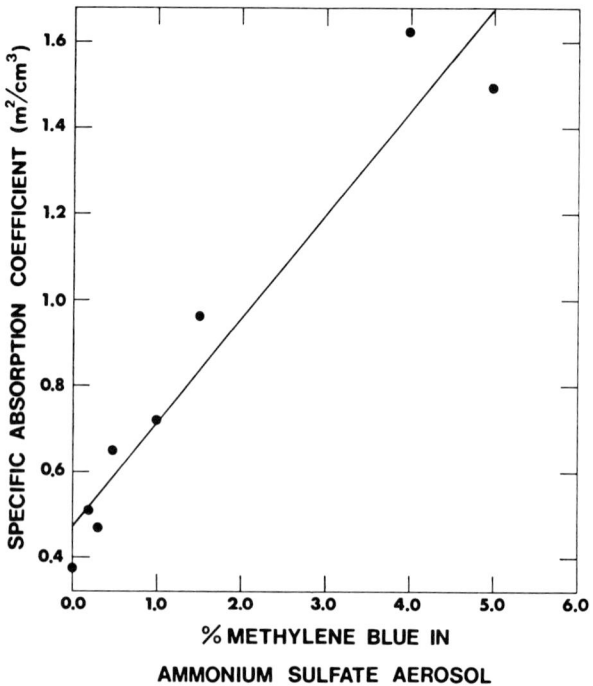

FIGURE 2. *The absorption per unit volume (S_{va}) as a function of methylene blue content.*

Ensor and Pilat (3) have shown that the aerosol mass concentration is related to optical transmittance through a modified form of the Beer-Lambert law

$$\frac{I}{I_o} = \exp - \left(\frac{M_v L S_{ve}}{\varrho} \right) \qquad (3)$$

where M_v is the particle mass concentration (g/m^3), S_{ve} (m^2/cm^3) is the ratio of the light extinction coefficient (m^{-1}) to the specific particulate volume (ratio of particle volume to aerosol volume, cm^3/m^3; reciprocal of "K" in the original paper), ϱ is the average particle specific gravity (g/m^3), and L is the optical pathlength (m). Optical absorption can similarly be formulated by

$$\frac{I'}{I_o'} = \exp - \left(\frac{M_v L S_{va}}{\varrho} \right) \qquad (4)$$

where S_{va} (m^2/cm^3) is the ratio of the light absorption coefficient to the specific particulate volume. The optical pathlength is determined by the length of the volume of sampled air of cross-sectional area equal to the filter.

The specific absorption coefficient (S_{va}) is converted to an imaginary refractive index (n_2) by calibrating the instrument with a submicron aerosol. This was accomplished with an aerosol of spherical ammonium sulfate particles containing various amounts of methylene blue dye to obtain various degrees of light absorption. The S_{va} is then proportional to the product of imaginary refractive index (n_2) of methylene blue and its weight percent in the ammonium sulfate aerosol, a nonabsorbing material. The imaginary refractive index of the samples are calculated in the following manner: The specific absorption coefficient (S_{va}) of the sample is plotted on the methylene-blue ammonium-sulfate calibration curve (Fig. 2). The fraction of methylene blue corresponding to this plotted point multipled by the imaginary refractive index of methylene blue (0.1 in this case) equals the imaginary refractive index of the sample. There is very little dependence on the real part of the refractive index, because the Integrating Plate Method diminishes scattering effects by comparing a blank filter to one with a layer of aerosol where the real refractive indices of aerosol and filter are similar.

IV. RESULTS

Absorption measurements obtained at the workshop from 100 μg to 500 μg samples of methylene-blue/ammonium-sulfate aerosol particles are shown in Fig. 2. The linear relationship found in Fig. 2 was also obtained by using Mie theory with particle sizes measured by scanning electron microscopy and an Electrical Aerosol Analyzer. However, in additional measurements on similar aerosols at Meteorological Research, Inc., the linear relationship was found to apply only to aerosols with a low methylene-blue content. As shown in Fig. 3 an asymptotic relationship is found between the specific absorption coefficient of the aerosol and the increasing methylene-blue content of the ammonium-sulfate particles.

The summary of absorption measurements made at the workshop is presented in Table 1. Where more than one sample was taken, the standard deviation is computed. These samples were analyzed at the wavelength equal to 0.56 μm.

The measurement of finite absorption for ammonium sulfate, a nonabsorbing species, indicates some interference effects due to back scattering or multiple scattering. These effects put the lower bound on the sensitivity of the technique at 10^{-4} m^{-1} for the absorption coefficient.

The mass specific absorption coefficient for "graphitic" carbon ranged from 9.1 to 12.8. These values, somewhat higher then expected, probably are a result of scattering "interference" effects.

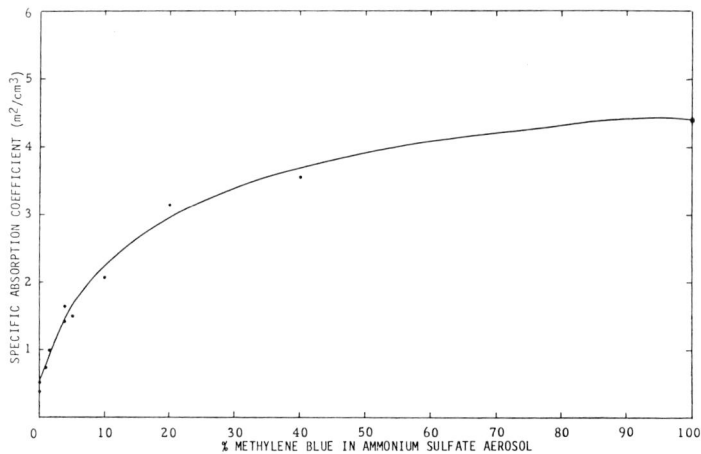

FIGURE 3. Methylene-blue/ammonium-sulfate calibration of the integrating plate method.

TABLE 1. Absorption Measurement Summary (High Aerosol Concentration Sampling Line).

Aerosol	Date/time	Absorption coefficient[a] σ_a (m^{-1})	Mass specific absorption coefficient, B_a ($m^2 \, g^{-1}$)	Single scattering albedo[b] $\tilde{\omega}$	Volume-specific absorption coefficient S_{va} (m^2/cm^3)	Imaginary refractive index[c] n_2
$(NH_4)_2SO_4$	7/29/80 1100-1600	$5.88 \pm 0.57 \times 10^{-4}$	0.111 ± 0.009	0.021	0.190 ± 0.025	--
$NH_4)_2SO_4$	7/30/80 1130-1210	$5.5 \pm 0.44 \times 10^{-4}$	0.114 ± 0.004	0.024	0.205 ± 0.007	--
$(NH_4)_2SO_4$ + 1% MB	7/29/80 1625-1715	$1.605 \pm 0.03 \times 10^{-3}$	0.406 ± 0.016	0.104	0.720 ± 0.028	0.001
$(NH_4)_2SO_4$ + 0.3% MB	7/30/80 1330-1530	$5.3 \pm 0.6 \times 10^{-4}$	0.27 ± 0.020	0.075	0.470 ± 0.037	0.0003
$(NH_4)_2SO_4$ + 4% MB	7/30/80 1630-1730	$2.69 \pm 0.15 \times 10^{-3}$	0.92 ± 0.045	0.135	1.620 ± 0.08	0.004
Carbon (soot)	7/31/80 1024-1448	$2.46 \pm 9.36 \times 10^{-2}$	9.64 ± 1.86	0.764	9.700 ± 1.9	--
Carbon (soot) $(NH_4)_2SO_4$	7/31/80 1502-1713	$1.08 \pm 0.04 \times 10^{-2}$	2.15 ± 0.29	0.365	4.100 ± 0.52	--
MB	8/1/80 1140-1337	$1.38 \pm 0.083 \times 10^{-2}$	1.81 ± 0.107	0.444	2.660 ± 0.16	0.12
MB	8/4/80 1519-1538	$2.16 \pm 0.047 \times 10^{-2}$	1.37 ± 0.11	0.370	2.030 ± 0.19	0.10
Carbon (soot)	8/1/80 1414-1536	$1.76 \pm 0.19 \times 10^{-2}$	9.1 ± 1.9	0.756	15.000 ± 1.6	--
Carbon (soot)	8/4/80 1032-1110	$2.01 \pm 0.17 \times 10^{-2}$	12.8 ± 1.9	0.835	17.100 ± 1.4	--
$NH_4)_2SO_4$ + soot	8/4/80 1214-1355	$2.52 \pm 0.16 \times 10^{-2}$	3.21 ± 0.13	0.379	6.000 ± 0.2	--

[a] Wavelength of incident light = 0.56 μm.
[b] σ_s measurements from Bodhaine nephelometer using a correction factor equal to the ratio of the high concentration to low concentration.
[c] Mie theory calculation gives n_2 of methylene blue (MB) at 0.56 μm equal to approximately 0.1.

REFERENCES

1. Lin, C., Baker, M., and Charlson, R. J., Absorption coefficient of atmospheric aerosol: A method for measurement, *Appl. Opt. 12,* 1356-1363 (1973).
2. Weiss, R. E., Waggoner, A. P., Charlson, R. J., Thorsell, D. L., Hall, J. S., and Riley, L. A., Studies of the optical, physical, and chemical properties of light absorbing aerosols, *Proceedings Conference Carbonaceous Particles in the Atmosphere, Report LDL-9037,* Lawrence Berkeley Laboratory, Berkeley, California (1979).
3. Ensor, D. S., and Pilat, M. J., Calculation of smoke plume opacity from particulate air pollutant properties, *J. Air Poll. Contr. Assn. 21,* 496 (1971).

OPTICAL PROPERTIES OF STANDARD AEROSOLS: A REPORT OF MEASUREMENTS FOR THE FIRST INTERNATIONAL WORKSHOP ON LIGHT ABSORPTION BY AEROSOL PARTICLES[1]

W. G. Egan

Research Department
Grumman Aerospace Corporation
Bethpage, New York

The absorption portions of the optical complex index of refraction of various standard aerosols were determined, following collection, by the application of the Kubelka-Munk radiative transfer theory to total diffuse transmission and reflection measurements on potassium bromide pellets containing the samples. The Brewster angle technique was used on bulk samples to determine the refractive portion. Ellipsometry was also used as a check on the absorption of bulk samples. The experimental investigation included propane soot, $(NH_4)_2SO_4$, and mixtures of these with methylene blue, graphite, Arizona dust, and ambient air. The results are presented tabularly and graphically, and may be used in Mie theory calculations to model visibility for remote sensing in the atmosphere.

[1]This research was supported in part by National Science Foundation Grant ATM 8005356.

I. INTRODUCTION

The earth's atmosphere is known to contain numerous micrometer-sized aerosol particles that are composed of silicate dust, ammonium sulfate [$(NH_4)_2SO_4$], amorphous carbon, sodium chloride, water, and solutions of nitric, sulfuric, and hydrochloric acids. In order to characterize the optical behavior of these particles so that their effect on the atmospheric optical transmission, scattering, and absorption may be determined, the fundamental optical properties (i.e., the complex index of refraction) must be determined.

There have been many techniques described in the literature for the determination of the complex index of refraction (1). The refractive (real) and absorptive (imaginary) portions are usually measured separately as a function of wavelength. The refractive portion of a bulk or compressed powder sample of an inhomogeneous material can be determined from the Brewster angle, and the absorption portion from the application of the basic or modified Kubelka-Munk theories (1). The Kubelka-Munk approach which will be discussed requires that a finely dispersed powder of the material under investigation be embedded in a potassium bromide (KBr) pellet.

The majority of the absorption measurement techniques determine the absorption portion alone, although the Kramers-Kronig technique produces both the refractive portion and the absorption portion. However, this technique is not valid when scattering is present as in the ultraviolet, visual, and near-infrared portions of the spectrum (1). It is desirable and necessary to compare the results of the determinations of the complex index of refraction by various techniques in order to assess their validity.

II. EXPERIMENTAL APPROACH

Aerosol samples were collected during the period 4 August to 7 August 1980 from the high-density pipe-line source furnished by the Cloud Simulation and Aerosol Laboratory of the Department of Atmospheric Science, Colorado State University, Fort Collins, Colorado. The collections were made using Millipore type PVC5 filters supported by AP10 MF support pads, in the 37-mm, three-section, in-line monitor. A Millipore vacuum pressure pump was used at 10 liter/minute flow during the collection runs. The samples acquired and the collection durations are listed in Table 1. The samples furnished were soot (amorphous carbon), $(NH_4)_2SO_4$, methylene blue, Arizona dust, ambient air, and mixtures of soot and $(NH_4)_2SO_4$. The Millipore filters were weighed before and after the collections to determine the quantity of aerosol collected during each run; the filters were reinserted into the filter housing for storage after the final weighing, with a small amount of silica gel dessicant. The dessicant was separated from the sample by a partition made from the blue filter pad spacers, and the holder openings were sealed with plastic plugs; in this way

TABLE 1. *List of Sample Collections Made from the High-Concentration Line Located in the Downstairs Laboratory (except as noted) Using Vacuum Pump, and Millipore PVC5 Filters on Pad in Holders.*

Date	Aerosol material	Time on hours	Time off hours	Sample holder no.
8/4/80	Propane soot	1030	1045	3
8/4/80	Propane soot	1045	1125	2
8/4/80	Propane soot + $(NH_4)_2SO_4$	1152	1212	4
8/4/80	Propane soot + $(NH_4)_2SO_4$	1212	1400	5
8/4/80	Methylene blue	1500	1545	6
8/4/80	Methylene blue	1525	1527	7
8/5-6/80	Ambient air	1035	0830	8
8/6/80	Arizona road dust	1040	1300	1
8/6/80	Propane soot + $(NH_4)_2SO_4$	1445	1700	9
8/7/80	Propane soot	1000	1040	10
8/7/80	Propane soot + $(NH_4)_2SO_4$	1045	1340	11
8/7/80	Humidified soot + $(NH_4)_2SO_4$ (upstairs)	1420	1550	12
8/8/80	$(NH_4)_2SO_4$	1130	1200	13
8/8/80	$(NH_4)_2SO_4$	1200	1535	14

the samples could be carried back to the Grumman Optics Laboratory for sample preparation for total diffuse reflectance and transmission measurements with a $BaSO_4$-coated integrating sphere (1). Also, samples of bulk $(NH_4)_2SO_4$, methylene blue, and Arizona dust (which were used to generate the aerosols) were brought back for determinations of the refractive portion of the index of refraction using the Brewster angle technique on compressed pellets of these materials. Soot was obtained (in the form of lampblack) from Fischer (stock no. C-198). The soot was amorphous, as verified by an X-ray diffraction measurement. However, the soot particles were coated with a layer of sulfur; the sulfur presumably condensed on the particles during the burning of high sulfur oil under the conditions of oxygen-deficient combustion necessary to produce soot. The soot was also used in a Brewster angle determination of the refractive portion of the index of refraction. As a point of comparison, spectroscopically pure graphite ("graphite 1") was also investigated using ellipsometry to determine the refractive and absorptive portions; also a powder of the graphite was prepared in a KBr pellet to determine the absorptive portion using the Kubelka-Munk theory for comparison.

The sample collection filters were removed from their holder upon arrival at the Grumman Optics Laboratory, and the material collected on the filter surface was mixed with a small measured quantity of KBr and dispersed using an agate mortar and pestle. This sample was then inserted into a Barnes 13-mm KBr die and another thicker layer of KBr powder placed above it. Compression of the composite at a pressure of 3.3×10^7 dyne/mm^2 resulted in a thin layer of powder sample in the KBr pellet. It was found that the $(NH_4)_2SO_4$ aerosol samples could not be prepared because the sample imbedded itself into the filter, because the filter pore sizes were too large. Future collections of $(NH_4)_2SO_4$ aerosol would require the use of glass fiber collection filters such as Millipore type AP40.

Optical measurements of total diffuse transmission and reflection were made using a modified Perkin Elmer model 13U spectrograph with a BaSO$_4$-coated integrating sphere. Initial measurements (reported here) were made at selected wavelengths between 0.185 μm and 1.105 μm. From the measurements, the Kubelka-Munk (K-M) or modified Kubelka-Munk (MKM) theories were used to calculate the absorption coefficient of the samples independent of the scattering (1). For the application of the MKM theory, the real portion of the refractive index of the sample is necessary. This index was determined from Brewster angle measurements on compressed pellets composed completely of powders of $(NH_4)_2SO_4$, methylene blue, lampblack, and Arizona dust; a sample block of spectroscopically pure graphite was also used for measurements of the real index. For the highly opaque samples (lampblack, graphite, and methylene blue), ellipsometric measurements were made to determine the absorption portion of the index of refraction.

Calculations of the absorption portion of the complex index of refraction were made on an IBM 370/168 computer, and the refractive portions of the index were made using a least squares curve fit program on an HP-9810-A computer plotter. Comparisons of the present set of measurements to previously published data on $(NH_4)_2SO_4$, graphite, and atmospheric aerosols are made.

III. RESULTS AND DISCUSSION

The results of the measurements of total diffuse reflectance and transmission (betweeen wavelengths of 0.185 μm and 1.105 μm) on the samples, as well as the calculated imaginary index (n_2) are presented in Figs. 1 through 11 and Tables 2 through 12; also included is the Brewster angle determination of the refractive portion of the index (n_1). For the high opacity samples (graphite 1, methylene blue, and lampblack), the results of the ellipsometric measurements are presented separately in Figs. 12 through 14 and Tables 13 through 15. A similar set of results of measurements on pure $(NH_4)_2SO_4$ in a KBr pellet is shown in Fig. 15 and Table 16. In the figures, the logarithm (base 10) of the total diffuse transmission (T), the total diffuse reflection (R), and the refractive and absorption portions of the refractive index (n_1 and n_2, respectively) are plotted as a function of the wavelength in micrometers. The effective thickness specified is the equivalent thickness of a layer of the aerosol material that is suspended in the KBr pellet. The tables give numerical values (estimated error $<\pm$ 10%) for the data presented in the figures. It is to be noted that in the notation of Egan and Hilgeman (1) $n = n_1$ and $K_O = n_2$.

The optical properties n_1 and n_2 are the significant results, whereas the diffuse transmission and reflection are presented mainly as a point of information; in the event that better theories for obtaining the optical complex index of refraction become available, they may be applied to these raw data.

A comparison of the optical properties n_1 and n_2 of Arizona dust aerosol (Fig. 1 and Table 2) with samples prepared from the bulk sample (Figs. 2 and 3; Tables 3 and 4) indicate that the aerosol sample has about one order of magnitude higher absorption than the bulk samples. The aerosol sample would generally consist of smaller particles with a lower real refractive index, which is inferred to be the result of quantum size effects, QSE [see for instance Egan and Hilgeman (2)]. This lower effective index would increase scatter in the KBr matrix and increase absorption as concentration is lowered, as was found.

Propane soot (Figs. 4 and 5; Tables 6 and 7) shows a general agreement in magnitude of n_2 between the two samples, with the lower concentration sample (Fig. 4) showing a slightly higher absorption; but both absorptions are lower than that for bulk soot (Fig. 14 and Table 15), which is also attributed to QSE and possibly to the effect of the sulfur contaminant.

The mixture of propane soot and $(NH_4)_2SO_4$ (Figs. 6 and 7; Tables 7 and 8) reveals an absorption n_2 dominated by soot, the higher absorber, as expected, with the lower concentration sample showing a slightly higher absorption.

Methylene blue (Figs. 8 and 9; Tables 9 and 10) shows an anomalous refractive effect on n_1 in the blue, with a slight corresponding increase in n_2. The sample prepared from the bulk (Fig. 9) shows an order of magnitude larger value for n_2 than the finer aerosol sample (Fig. 8). The bulk sample analyzed with the ellipsometric technique (Fig. 13; Table 14) has even a much higher absorption; this too is inferred to be the result of QSE.

The ambient aerosol collected over a 22-hour period had the optical absorption properties shown in Fig. 10 and Table 11. The absorption is comparable to a "dirty" silicate.

A comparison of absorption n_2 for particulate graphite (Fig. 11, Table 12) with absorption for the same material in bulk form (Fig. 12, Table 13) shows a two order of magnitude decrease in the absorption portion, which is also presumed to be caused by QSE. The characteristics of the bulk graphite are

TABLE 2. Arizona Dust (Fig. 1).

(μm)	$Log_{10}T$	R	$Log_{10}\, n_2$	n_1
0.185	−0.783	0.125	-0.213×10^1	1.390
0.190	−0.706	0.134	-0.219×10^1	1.387
0.200	−0.710	0.129	-0.215×10^1	1.383
0.210	−0.695	0.126	-0.212×10^1	1.380
0.215	−0.678	0.131	-0.214×10^1	1.377
0.220	−0.668	0.136	-0.215×10^1	1.374
0.225	−0.664	0.134	-0.213×10^1	1.370
0.233	−0.650	0.137	-0.213×10^1	1.367
0.240	−0.652	0.136	-0.212×10^1	1.364
0.260	−0.545	0.152	-0.220×10^1	1.361
0.280	−0.523	0.153	-0.219×10^1	1.357
0.300	−0.463	0.159	-0.222×10^1	1.354
0.325	−0.446	0.153	-0.219×10^1	1.351
0.360	−0.387	0.149	-0.220×10^1	1.347
0.370	−0.378	0.151	-0.220×10^1	1.344
0.400	−0.364	0.152	-0.219×10^1	1.341
0.433	−0.338	0.145	-0.217×10^1	1.340
0.466	−0.327	0.144	-0.215×10^1	1.339
0.500	−0.310	0.139	-0.213×10^1	1.338
0.533	−0.298	0.137	-0.211×10^1	1.337
0.566	−0.293	0.137	-0.209×10^1	1.336
0.600	−0.280	0.133	-0.208×10^1	1.335
0.633	−0.277	0.133	-0.206×10^1	1.334
0.666	−0.255	0.123	-0.206×10^1	1.333
0.700	−0.276	0.124	-0.199×10^1	1.333
0.817	−0.241	0.111	-0.195×10^1	1.332
0.907	−0.223	0.102	-0.191×10^1	1.331
1.000	−0.215	0.106	-0.190×10^1	1.330
1.105	−0.196	0.112	-0.193×10^1	1.329

OPTICAL PROPERTIES OF STANDARD AEROSOLS

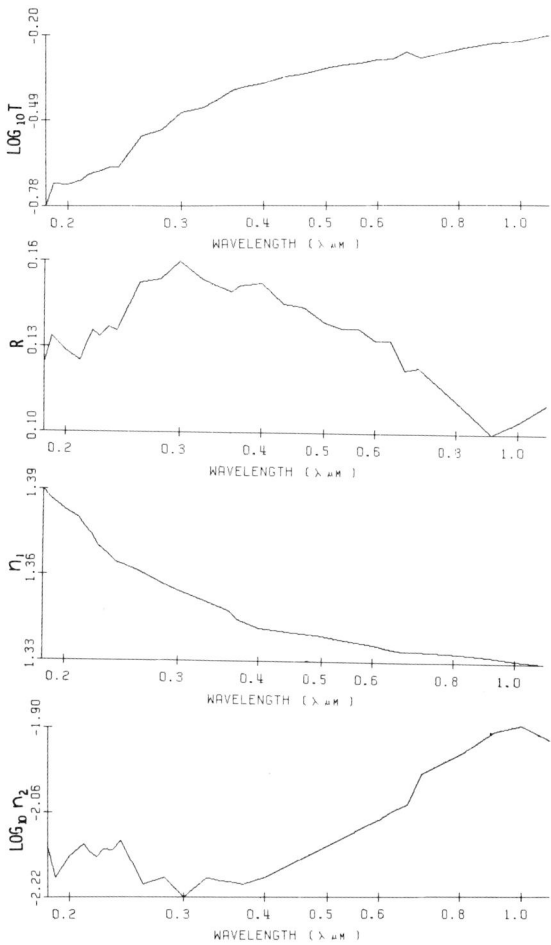

FIGURE 1. Arizona dust, CSU aerosol sample no. 1, compressed KBr pellet (effective thickness 0.000149 cm); n_1 from Brewster angle measurement.

TABLE 3. Arizona Dust (Fig. 2).

(μm)	$Log_{10}T$	R	$Log_{10} n_2$	n_1
0.185	-0.201×10^1	0.073	-0.281×10^1	1.390
0.190	-0.198×10^1	0.069	-0.278×10^1	1.387
0.200	-0.208×10^1	0.062	-0.271×10^1	1.383
0.210	-0.207×10^1	0.060	-0.269×10^1	1.380
0.215	-0.200×10^1	0.062	-0.270×10^1	1.377
0.220	-0.195×10^1	0.067	-0.272×10^1	1.374
0.225	-0.195×10^1	0.064	-0.270×10^1	1.370
0.233	-0.192×10^1	0.066	-0.270×10^1	1.367
0.240	-0.187×10^1	0.066	-0.270×10^1	1.364
0.260	-0.159×10^1	0.075	-0.277×10^1	1.361
0.280	-0.147×10^1	0.075	-0.277×10^1	1.357
0.300	-0.140×10^1	0.080	-0.277×10^1	1.354
0.325	-0.132×10^1	0.084	-0.278×10^1	1.351
0.360	-0.115×10^1	0.087	-0.280×10^1	1.347
0.370	-0.110×10^1	0.092	-0.283×10^1	1.344
0.400	-0.102×10^1	0.094	-0.283×10^1	1.341
0.433	-0.917	0.101	-0.286×10^1	1.340
0.466	-0.866	0.104	-0.286×10^1	1.339
0.500	-0.807	0.113	-0.289×10^1	1.338
0.533	-0.752	0.122	-0.291×10^1	1.337
0.566	-0.684	0.141	-0.298×10^1	1.336
0.600	-0.650	0.149	-0.299×10^1	1.335
0.633	-0.618	0.152	-0.300×10^1	1.334
0.666	-0.573	0.150	-0.300×10^1	1.333
0.700	-0.569	0.156	-0.300×10^1	1.333
0.817	-0.520	0.151	-0.296×10^1	1.332
0.907	-0.499	0.155	-0.294×10^1	1.331
1.000	-0.469	0.143	-0.289×10^1	1.330
1.105	-0.440	0.148	-0.289×10^1	1.329

OPTICAL PROPERTIES OF STANDARD AEROSOLS

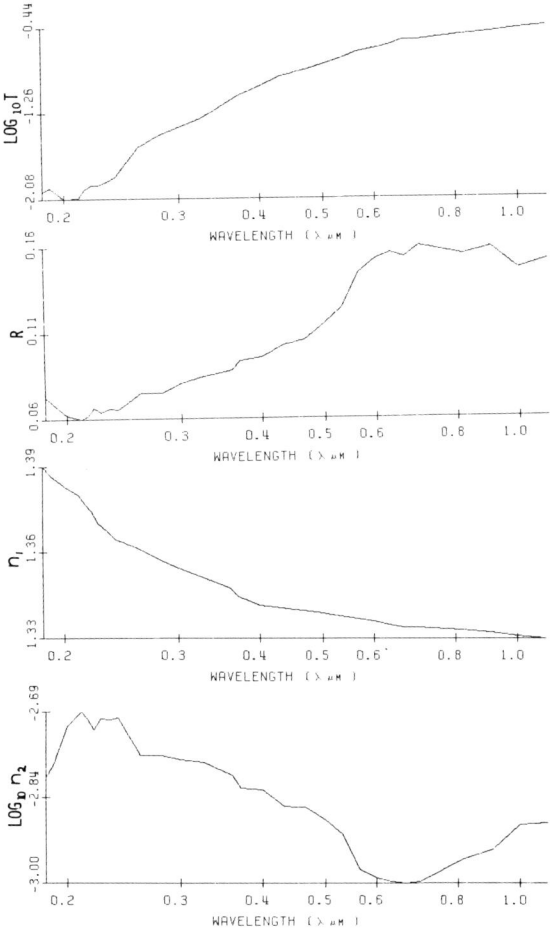

FIGURE 2. Arizona dust, prepared from raw dust, CSU sample no. 1A, compressed KBr pellet (effective thickness 0.00263 cm); n_1 from Brewster angle measurement.

TABLE 4. Arizona Dust (Fig. 3).

(µm)	$Log_{10}T$	R	$Log_{10}\ n_2$	n_1
0.185	-0.262×10^1	0.061	-0.287×10^1	1.390
0.190	-0.306×10^1	0.069	-0.283×10^1	1.387
0.200	-0.294×10^1	0.068	-0.282×10^1	1.383
0.210	-0.293×10^1	0.059	-0.276×10^1	1.380
0.215	-0.280×10^1	0.067	-0.281×10^1	1.377
0.220	-0.279×10^1	0.071	-0.282×10^1	1.374
0.225	-0.277×10^1	0.070	-0.280×10^1	1.370
0.233	-0.273×10^1	0.071	-0.280×10^1	1.367
0.240	-0.267×10^1	0.070	-0.280×10^1	1.364
0.260	-0.229×10^1	0.081	-0.287×10^1	1.361
0.280	-0.219×10^1	0.081	-0.285×10^1	1.357
0.300	-0.201×10^1	0.087	-0.288×10^1	1.354
0.325	-0.190×10^1	0.091	-0.288×10^1	1.351
0.360	-0.172×10^1	0.095	-0.289×10^1	1.347
0.370	-0.166×10^1	0.099	-0.291×10^1	1.344
0.400	-0.158×10^1	0.104	-0.291×10^1	1.341
0.433	-0.143×10^1	0.115	-0.295×10^1	1.340
0.466	-0.135×10^1	0.123	-0.297×10^1	1.339
0.500	-0.127×10^1	0.119	-0.295×10^1	1.338
0.533	-0.117×10^1	0.143	-0.302×10^1	1.337
0.566	-0.107×10^1	0.169	-0.310×10^1	1.336
0.600	-0.102×10^1	0.182	-0.312×10^1	1.335
0.633	-0.967	0.185	-0.313×10^1	1.334
0.666	-0.914	0.185	$-0.31\ 3 \times 10^1$	1.333
0.700	-0.936	0.192	-0.311×10^1	1.333
0.817	-0.810	0.198	-0.312×10^1	1.332
0.907	-0.783	0.195	-0.308×10^1	1.331
1.000	-0.728	0.192	-0.306×10^1	1.330
1.105	-0.690	0.191	-0.303×10^1	1.329

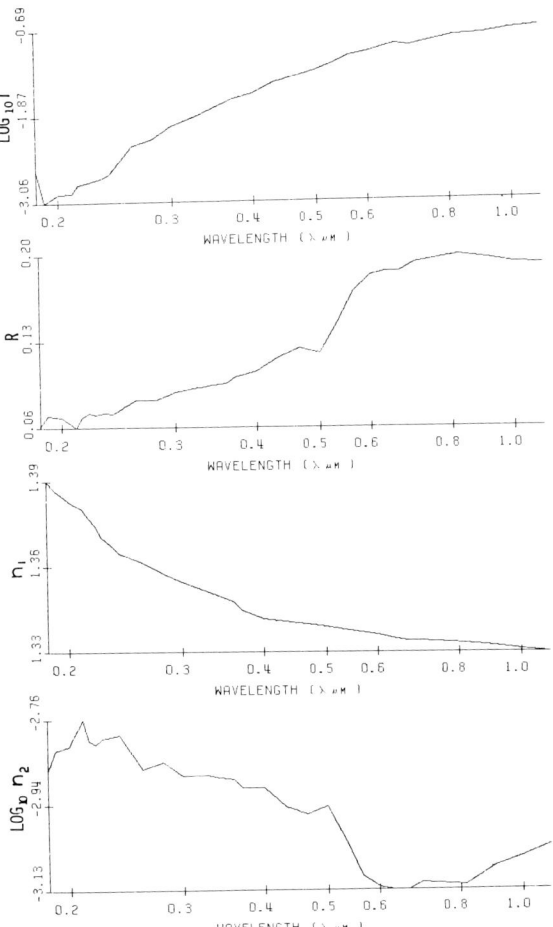

FIGURE 3. Arizona dust, prepared from raw dust, CSU sample no. 1B, compressed KBr pellet (effective thickness 0.00448 cm); n_1 from Brewster angle measurement.

TABLE 5. Propane Soot (Fig. 4).

(μm)	$Log_{10}T$	R	$Log_{10}\, n_2$	n_1
0.185	-0.207×10^1	0.083	-0.108×10^1	0.472
0.190	-0.176×10^1	0.080	-0.113×10^1	1.331
0.200	-0.180×10^1	0.081	-0.110×10^1	1.409
0.210	-0.179×10^1	0.078	-0.108×10^1	1.367
0.215	-0.176×10^1	0.080	-0.108×10^1	1.508
0.220	-0.175×10^1	0.084	-0.107×10^1	1.491
0.225	-0.175×10^1	0.081	-0.106×10^1	1.633
0.233	-0.174×10^1	0.081	-0.105×10^1	1.659
0.240	-0.173×10^1	0.079	-0.104×10^1	1.684
0.260	-0.159×10^1	0.080	-0.104×10^1	1.678
0.280	-0.157×10^1	0.078	-0.102×10^1	1.684
0.300	-0.152×10^1	0.078	-0.100×10^1	1.737
0.325	-0.148×10^1	0.074	-0.976	1.697
0.360	-0.139×10^1	0.072	-0.959	1.674
0.370	-0.137×10^1	0.073	-0.951	1.665
0.400	-0.133×10^1	0.075	-0.933	1.668
0.433	-0.128×10^1	0.075	-0.915	1.675
0.466	-0.124×10^1	0.075	-0.894	1.687
0.500	-0.119×10^1	0.072	-0.882	1.701
0.533	-0.116×10^1	0.073	-0.866	1.689
0.566	-0.113×10^1	0.074	-0.851	1.731
0.600	-0.109×10^1	0.076	-0.843	1.716
0.633	-0.106×10^1	0.076	-0.829	1.729
0.666	-0.102×10^1	0.069	-0.827	1.775
0.700	-0.102×10^1	0.071	-0.805	1.753
0.817	-0.932	0.069	-0.776	1.762
0.907	-0.886	0.078	-0.753	1.750
1.000	-0.857	0.076	-0.725	1.753
1.105	-0.790	0.079	-0.717	1.757

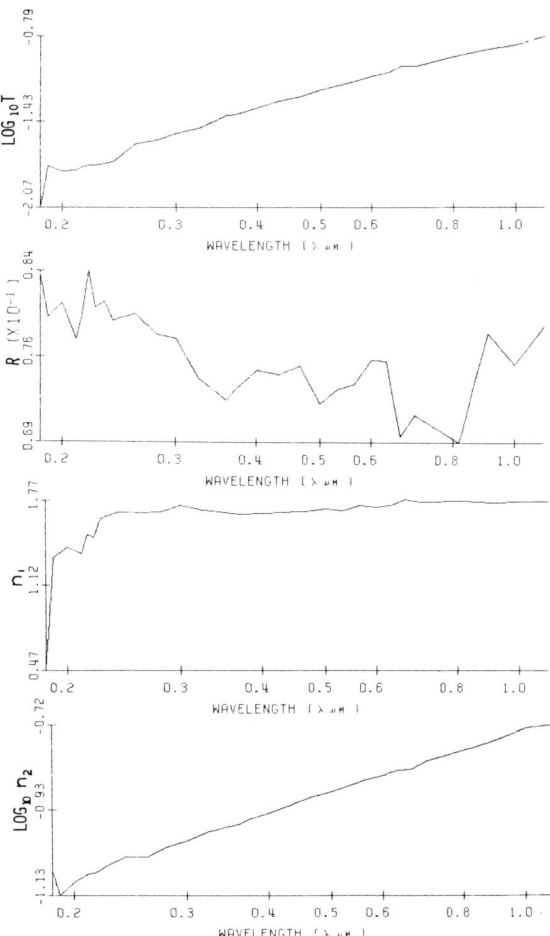

FIGURE 4. Propane soot, CSU aerosol sample no. 2, compressed KBr pellet (effective thickness 0.0000831 cm); n_1 from Fig. 14.

TABLE 6. Propane Soot (Fig. 5).

(μm)	$Log_{10}T$	R	$Log_{10}\ n_2$	n_1
0.185	-0.221×10^1	0.074	-0.149×10^1	0.472
0.190	-0.267×10^1	0.076	-0.139×10^1	1.331
0.200	-0.252×10^1	0.076	-0.139×10^1	1.409
0.210	-0.245×10^1	0.072	-0.139×10^1	1.367
0.215	-0.238×10^1	0.076	-0.139×10^1	1.508
0.220	-0.238×10^1	0.077	-0.138×10^1	1.491
0.225	-0.249×10^1	0.075	-0.135×10^1	1.633
0.233	-0.246×10^1	0.076	-0.134×10^1	1.659
0.240	-0.249×10^1	0.075	-0.132×10^1	1.684
0.260	-0.229×10^1	0.076	-0.132×10^1	1.678
0.280	-0.225×10^1	0.074	-0.130×10^1	1.684
0.300	-0.222×10^1	0.075	-0.127×10^1	1.737
0.325	-0.216×10^1	0.072	-0.125×10^1	1.697
0.360	-0.201×10^1	0.070	-0.124×10^1	1.674
0.370	-0.201×10^1	0.071	-0.123×10^1	1.665
0.400	-0.188×10^1	0.073	-0.122×10^1	1.668
0.433	-0.180×10^1	0.073	-0.120×10^1	1.675
0.466	-0.172×10^1	0.073	-0.119×10^1	1.687
0.500	-0.170×10^1	0.070	-0.117×10^1	1.701
0.533	-0.166×10^1	0.071	-0.115×10^1	1.689
0.566	-0.161×10^1	0.074	-0.114×10^1	1.731
0.600	-0.159×10^1	0.073	-0.112×10^1	1.716
0.633	-0.154×10^1	0.074	-0.111×10^1	1.729
0.666	-0.153×10^1	0.069	-0.109×10^1	1.775
0.700	-0.150×10^1	0.071	-0.108×10^1	1.753
0.817	-0.142×10^1	0.069	-0.103×10^1	1.762
0.907	-0.134×10^1	0.075	-0.101×10^1	1.750
1.000	-0.132×10^1	0.073	-0.977	1.753
1.105	-0.130×10^1	0.078	-0.941	1.757

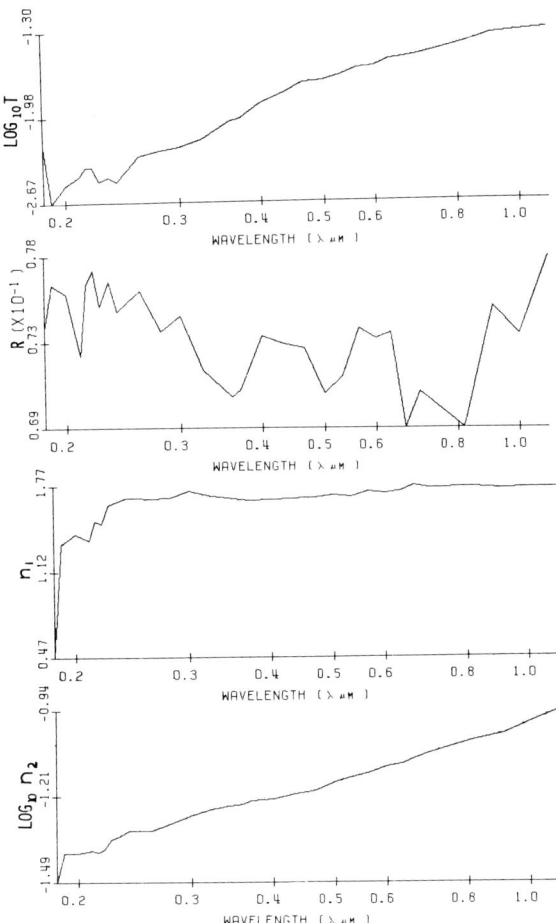

FIGURE 5. Propane soot, CSU aerosol sample no. 2A, prepared from CSU sample no. 2, compressed KBr pellet (effective thickness 0.000229 cm); n_1 from Fig. 12.

TABLE 7. Propane Soot + $(NH_4)_2SO_4$ (Fig. 6).

(μm)	$Log_{10}T$	R	$Log_{10} n_2$	n_1
0.185	-0.141×10^1	0.093	-0.137×10^1	1.005
0.190	-0.135×10^1	0.072	-0.138×10^1	1.432
0.200	-0.135×10^1	0.074	-0.136×10^1	1.468
0.210	-0.136×10^1	0.071	-0.133×10^1	1.444
0.215	-0.134×10^1	0.073	-0.133×10^1	1.511
0.220	-0.133×10^1	0.075	-0.132×10^1	1.500
0.225	-0.129×10^1	0.072	-0.132×10^1	1.568
0.233	-0.129×10^1	0.072	-0.131×10^1	1.578
0.240	-0.126×10^1	0.070	-0.131×10^1	1.588
0.260	-0.119×10^1	0.071	-0.130×10^1	1.582
0.280	-0.116×10^1	0.069	-0.128×10^1	1.582
0.300	-0.107×10^1	0.070	-0.128×10^1	1.606
0.325	-0.106×10^1	0.066	-0.125×10^1	1.583
0.360	-0.979	0.064	-0.124×10^1	1.569
0.370	-0.955	0.064	-0.124×10^1	1.561
0.400	-0.967	0.067	-0.120×10^1	1.560
0.433	-0.943	0.065	-0.118×10^1	1.563
0.466	-0.910	0.066	-0.116×10^1	1.568
0.500	-0.879	0.056	-0.114×10^1	1.575
0.533	-0.764	0.071	-0.118×10^1	1.569
0.566	-0.839	0.066	-0.111×10^1	1.590
0.600	-0.812	0.066	-0.110×10^1	1.582
0.633	-0.790	0.067	-0.109×10^1	1.588
0.666	-0.770	0.060	-0.108×10^1	1.610
0.700	-0.762	0.063	-0.106×10^1	1.599
0.817	-0.699	0.056	-0.103×10^1	1.602
0.907	-0.654	0.065	-0.101×10^1	1.596
1.000	-0.728	0.057	-0.925	1.596
1.105	-0.606	0.063	-0.962	1.598

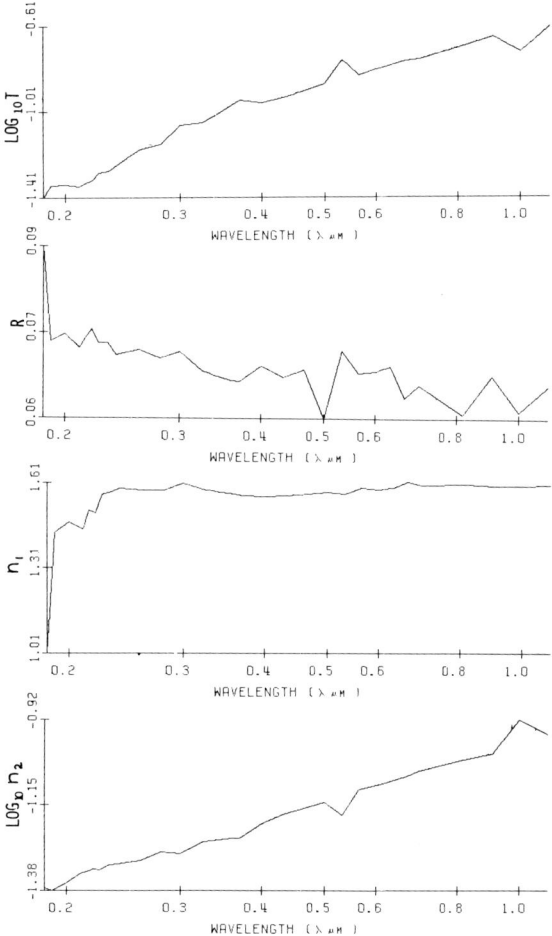

FIGURE 6. Propane soot + $(NH_4)_2SO_4$, CSU aerosol sample no. 5, compressed KBr pellet (effective thickness 0.000112 cm); n_1 is average from Figs. 12 and 15.

TABLE 8. Propane Soot + $(NH_4)_2SO_4$ (Fig. 7).

(μm)	$Log_{10}T$	R	$Log_{10}\ n_2$	n_1
0.185	-0.222×10^1	0.072	-0.105×10^1	1.005
0.190	-0.229×10^1	0.065	-0.103×10^1	1.432
0.200	-0.234×10^1	0.069	-0.997	1.468
0.210	-0.239×10^1	0.063	-0.967	1.444
0.215	-0.235×10^1	0.067	-0.964	1.511
0.220	-0.235×10^1	0.069	-0.954	1.500
0.225	-0.235×10^1	0.067	-0.945	1.568
0.233	-0.233×10^1	0.067	-0.931	1.578
0.240	-0.230×10^1	0.066	-0.925	1.588
0.260	-0.220×10^1	0.066	-0.910	1.582
0.280	-0.215×10^1	0.064	-0.886	1.582
0.300	-0.205×10^1	0.065	-0.878	1.606
0.325	-0.197×10^1	0.063	-0.859	1.583
0.360	-0.185×10^1	0.060	-0.842	1.569
0.370	-0.183×10^1	0.060	-0.837	1.561
0.400	-0.170×10^1	0.057	-0.835	1.560
0.433	-0.163×10^1	0.056	-0.818	1.563
0.466	-0.156×10^1	0.060	-0.805	1.568
0.500	-0.151×10^1	0.057	-0.790	1.575
0.533	-0.147×10^1	0.059	-0.771	1.569
0.566	-0.143×10^1	0.058	-0.760	1.590
0.600	-0.139×10^1	0.057	-0.744	1.582
0.633	-0.136×10^1	0.058	-0.732	1.588
0.666	-0.132×10^1	0.053	-0.723	1.610
0.700	-0.131×10^1	0.051	-0.703	1.599
0.817	-0.120×10^1	0.052	-0.677	1.602
0.907	-0.113×10^1	0.058	-0.656	1.596
1.000	-0.108×10^1	0.053	-0.632	1.596
1.105	-0.104×10^1	0.050	-0.606	1.598

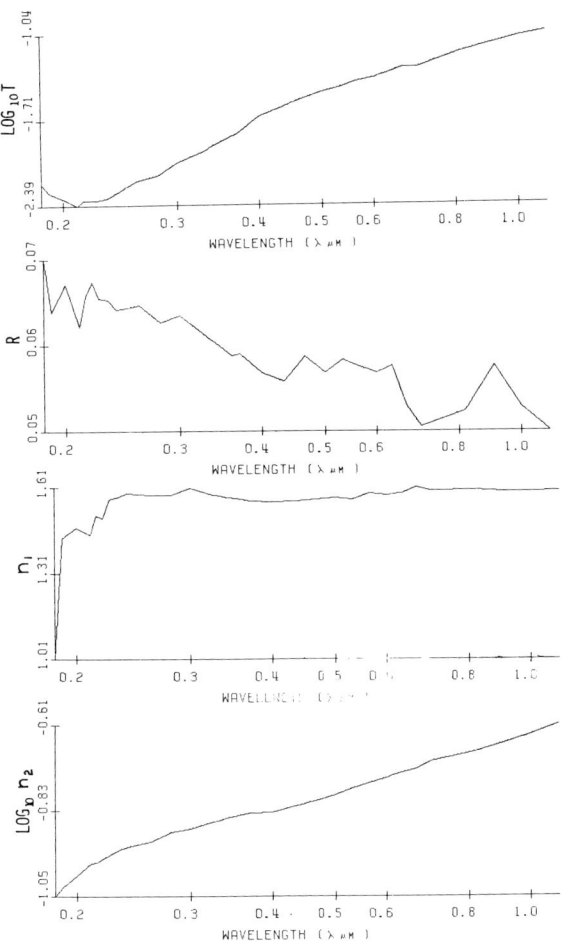

FIGURE 7. Propane soot + $(NH_4)_2SO_4$, CSU aerosol sample no. 5A, prepared from CSU sample no. 5, compressed KBr pellet (effective thickness 0.0000850 cm); n_1 is average from Figs. 12 and 15.

TABLE 9. Methylene Blue (Fig. 8).

(µm)	$Log_{10}T$	R	$Log_{10}\ n_2$	n_1
0.185	-0.224×10^1	0.080	-0.120×10^1	1.535
0.190	-0.200×10^1	0.072	-0.124×10^1	1.536
0.200	-0.209×10^1	0.067	-0.119×10^1	1.537
0.210	-0.211×10^1	0.064	-0.117×10^1	1.539
0.215	-0.213×10^1	0.067	-0.116×10^1	1.540
0.220	-0.216×10^1	0.073	-0.114×10^1	1.541
0.225	-0.217×10^1	0.074	-0.113×10^1	1.543
0.233	-0.217×10^1	0.075	-0.111×10^1	1.544
0.240	-0.216×10^1	0.074	-0.110×10^1	1.545
0.260	-0.191×10^1	0.076	-0.112×10^1	1.547
0.280	-0.182×10^1	0.072	-0.111×10^1	1.548
0.300	-0.171×10^1	0.072	-0.111×10^1	1.549
0.325	-0.146×10^1	0.078	-0.114×10^1	1.551
0.360	-0.117×10^1	0.094	-0.119×10^1	1.552
0.370	-0.116×10^1	0.092	-0.119×10^1	1.510
0.400	-0.105×10^1	0.110	-0.120×10^1	1.467
0.433	-0.114×10^1	0.085	-0.113×10^1	1.356
0.466	-0.138×10^1	0.057	-0.101×10^1	1.243
0.500	-0.170×10^1	0.061	-0.887	1.190
0.533	-0.180×10^1	0.077	-0.835	1.451
0.566	-0.177×10^1	0.085	-0.815	1.750
0.600	-0.172×10^1	0.085	-0.802	1.854
0.633	-0.150×10^1	0.082	-0.840	1.829
0.666	-0.132×10^1	0.084	-0.873	1.803
0.700	-0.117×10^1	0.092	-0.903	1.778
0.817	-0.712	0.232	-0.107×10^1	1.752
0.907	-0.602	0.302	-0.111×10^1	1.727
1.000	-0.577	0.300	-0.109×10^1	1.701
1.105	-0.548	0.286	-0.107×10^1	1.676

OPTICAL PROPERTIES OF STANDARD AEROSOLS

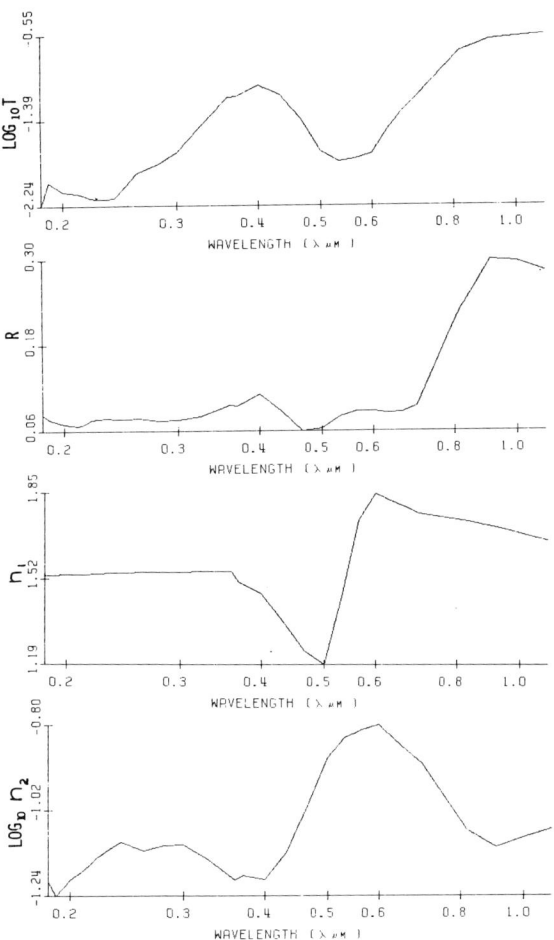

FIGURE 8. Methylene blue, CSU aerosol sample no. 6, compressed KBr pellet (effective thickness 0.00012 cm); n_1 from Fig. 13.

TABLE 10. Methylene Blue (Fig. 9).

(μm)	$Log_{10}T$	R	$Log_{10}\ n_2$	n_1
0.185	-0.135×10^1	0.070	-0.207×10^1	1.535
0.190	-0.130×10^1	0.076	-0.207×10^1	1.536
0.200	-0.131×10^1	0.070	-0.205×10^1	1.537
0.210	-0.130×10^1	0.068	-0.203×10^1	1.539
0.215	-0.128×10^1	0.071	-0.203×10^1	1.540
0.220	-0.128×10^1	0.075	-0.202×10^1	1.541
0.225	-0.130×10^1	0.075	-0.200×10^1	1.543
0.233	-0.144×10^1	0.076	-0.194×10^1	1.544
0.240	-0.130×10^1	0.075	-0.197×10^1	1.545
0.260	-0.116×10^1	0.075	-0.199×10^1	1.547
0.280	-0.112×10^1	0.071	-0.197×10^1	1.548
0.300	-0.107×10^1	0.071	-0.196×10^1	1.549
0.325	-0.983	0.0672	-0.197×10^1	1.551
0.360	-0.845	0.065	-0.198×10^1	1.552
0.370	-0.824	0.064	-0.198×10^1	1.510
0.400	-0.770	0.069	-0.198×10^1	1.467
0.433	-0.815	0.061	-0.192×10^1	1.356
0.466	-0.879	0.057	-0.185×10^1	1.243
0.500	-0.939	0.061	-0.180×10^1	1.190
0.533	-0.971	0.075	-0.175×10^1	1.451
0.566	-0.983	0.079	-0.172×10^1	1.750
0.600	-0.951	0.080	-0.171×10^1	1.854
0.633	-0.900	0.076	-0.171×10^1	1.829
0.666	-0.830	0.068	-0.172×10^1	1.803
0.700	-0.812	0.070	-0.171×10^1	1.778
0.817	-0.572	0.097	-0.180×10^1	1.752
0.907	-0.471	0.158	-0.185×10^1	1.727
1.000	-0.438	0.163	-0.184×10^1	1.701
1.105	-0.400	0.176	-0.184×10^1	1.676

OPTICAL PROPERTIES OF STANDARD AEROSOLS 219

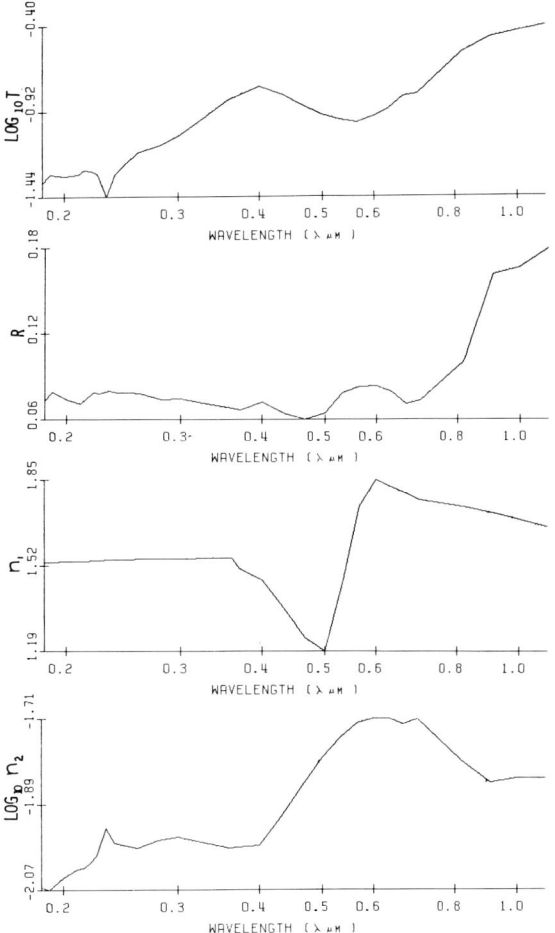

FIGURE 9. Methylene blue, prepared from bulk methylene blue, CSU sample no. 6A. compressed KBr pellet (effective thickness 0.000536 cm); n_1 from Fig. 13.

TABLE 11. Ambient Air, Fort Collins, Colorado (Fig. 10).

(μm)	$Log_{10}T$	R	$Log_{10}\ n_2$	n_1
0.185	−0.807	0.131	−0.829	1.390
0.190	−0.750	0.141	−0.875	1.387
0.200	−0.767	0.134	−0.822	1.383
0.210	−0.788	0.122	−0.755	1.380
0.215	−0.767	0.127	−0.770	1.377
0.220	−0.759	0.133	−0.780	1.374
0.225	−0.664	0.133	−0.825	1.370
0.233	−0.638	0.135	−0.831	1.367
0.240	−0.693	0.130	−0.769	1.364
0.260	−0.623	0.133	−0.786	1.361
0.280	−0.604	0.132	−0.763	1.357
0.300	−0.585	0.148	−0.789	1.354
0.325	−0.548	0.152	−0.791	1.351
0.360	−0.491	0.140	−0.760	1.347
0.370	−0.469	0.136	−0.757	1.344
0.400	−0.421	0.132	−0.757	1.341
0.433	−0.386	0.130	−0.756	1.340
0.466	−0.359	0.123	−0.737	1.339
0.500	−0.354	0.137	−0.753	1.338
0.533	−0.334	0.118	−0.695	1.337
0.566	−0.318	0.113	−0.677	1.336
0.600	−0.310	0.113	−0.664	1.335
0.633	−0.288	0.106	−0.652	1.334
0.666	−0.279	0.102	−0.632	1.333
0.700	−0.276	0.106	−0.629	1.333
0.817	−0.241	0.087	−0.563	1.332
0.907	−0.223	0.087	−0.557	1.331
1.000	−0.211	0.094	−0.564	1.330
1.105	−0.196	0.085	−0.525	1.329

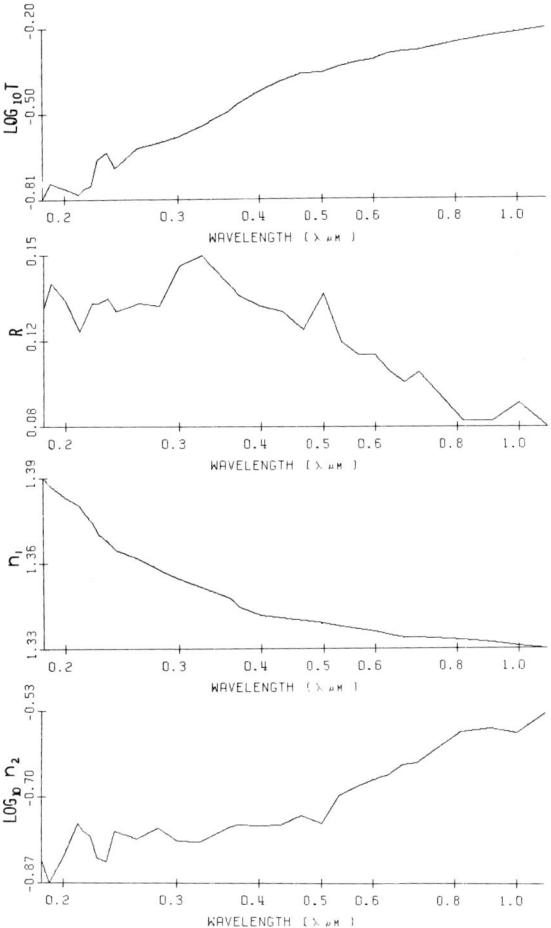

FIGURE 10. Ambient air, Fort Collins, Colorado, CSU aerosol sample no. 8, compressed KBr pellet (effective thickness 0.00000737 cm); n_1 from Table 2.

TABLE 12. Spectroscopically Pure Graphite (Fig. 11).

(μm)	$Log_{10}T$	R	$Log_{10}\, n_2$	n_1
0.185	-0.229×10^1	0.181	-0.243×10^1	0.879
0.190	-0.274×10^1	0.159	-0.234×10^1	1.797
0.200	-0.278×10^1	0.163	-0.231×10^1	1.767
0.210	-0.262×10^1	0.155	-0.232×10^1	1.856
0.215	-0.270×10^1	0.161	-0.229×10^1	1.867
0.220	-0.263×10^1	0.156	-0.230×10^1	1.947
0.225	-0.243×10^1	0.150	-0.232×10^1	1.939
0.233	-0.240×10^1	0.149	-0.231×10^1	1.951
0.240	-0.241×10^1	0.138	-0.229×10^1	1.975
0.260	-0.240×10^1	0.131	-0.226×10^1	2.031
0.280	-0.236×10^1	0.124	-0.224×10^1	2.059
0.300	-0.240×10^1	0.094	-0.220×10^1	2.062
0.325	-0.242×10^1	0.113	-0.216×10^1	2.065
0.360	-0.223×10^1	0.109	-0.215×10^1	2.035
0.370	-0.222×10^1	0.102	-0.214×10^1	2.026
0.400	-0.225×10^1	0.104	-0.210×10^1	2.000
0.433	-0.218×10^1	0.104	-0.208×10^1	2.010
0.466	-0.219×10^1	0.077	-0.205×10^1	2.020
0.500	-0.218×10^1	0.099	-0.202×10^1	2.030
0.533	-0.214×10^1	0.101	-0.200×10^1	2.038
0.566	-0.212×10^1	0.101	-0.198×10^1	2.045
0.600	-0.210×10^1	0.106	-0.196×10^1	2.053
0.633	-0.208×10^1	0.101	-0.194×10^1	2.054
0.666	-0.204×10^1	0.096	-0.192×10^1	2.055
0.700	-0.203×10^1	0.095	-0.190×10^1	2.057
0.817	-0.207×10^1	0.108	-0.183×10^1	2.060
0.907	-0.207×10^1	0.114	-0.178×10^1	2.064
1.000	-0.205×10^1	0.117	-0.174×10^1	2.067
1.105	-0.203×10^1	0.111	-0.171×10^1	2.071

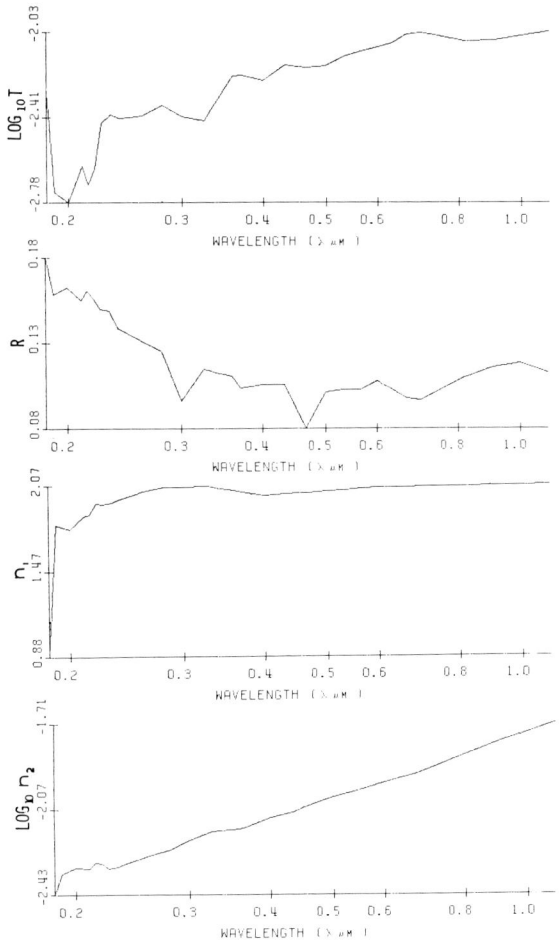

FIGURE 11. Spectroscopically pure graphite, CSU sample "graphite," compressed KBr pellet (effective thickness 0.00208 cm).

TABLE 13. Spectroscopically Pure Graphite (Fig. 12).

(μm)	n_1	n_2
0.185	0.472	0.338
0.190	0.133×10^1	0.834
0.200	0.141×10^1	0.616
0.210	0.137×10^1	0.897
0.215	0.151×10^1	0.659
0.220	0.149×10^1	0.881
0.225	0.163×10^1	0.550
0.233	0.166×10^1	0.511
0.240	0.168×10^1	0.519
0.260	0.168×10^1	0.711
0.280	0.168×10^1	0.771
0.300	0.174×10^1	0.653
0.325	0.170×10^1	0.758
0.360	0.167×10^1	0.731
0.370	0.166×10^1	0.728
0.400	0.167×10^1	0.649
0.433	0.168×10^1	0.658
0.466	0.169×10^1	0.656
0.500	0.170×10^1	0.652
0.533	0.169×10^1	0.703
0.566	0.173×10^1	0.616
0.600	0.172×10^1	0.680
0.633	0.173×10^1	0.650
0.666	0.177×10^1	0.608
0.700	0.175×10^1	0.591
0.817	0.176×10^1	0.576
0.907	0.175×10^1	0.623
1.000	0.175×10^1	0.624
1.105	0.176×10^1	0.626

OPTICAL PROPERTIES OF STANDARD AEROSOLS

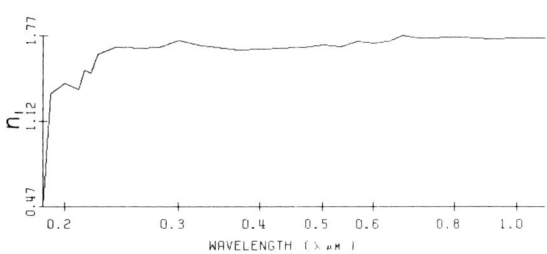

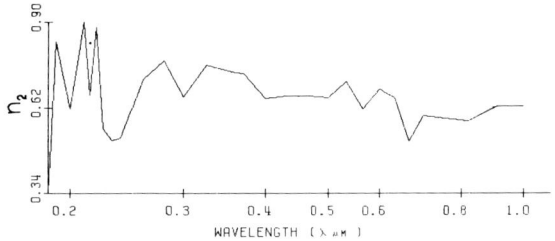

FIGURE 12. Spectroscopically pure graphite, bulk sample, CSU sample "graphite I."

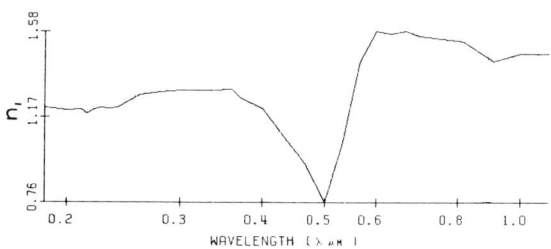

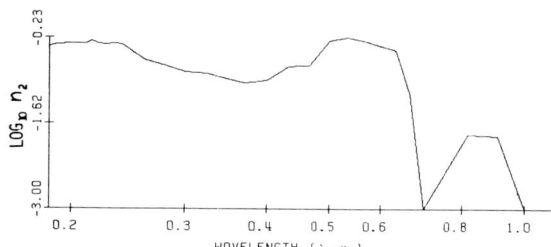

FIGURE 13. Bulk methylene blue.

TABLE 14. Bulk Methylene Blue (Fig. 13).

(μm)	n_1	n_2
0.185	0.121×10^1	0.431
0.190	0.120×10^1	0.453
0.200	0.120×10^1	0.474
0.210	0.120×10^1	0.468
0.215	0.118×10^1	0.517
0.220	0.120×10^1	0.473
0.225	0.121×10^1	0.459
0.233	0.121×10^1	0.467
0.240	0.121×10^1	0.456
0.260	0.127×10^1	0.254
0.280	0.128×10^1	0.205
0.300	0.129×10^1	0.163
0.325	0.130×10^1	0.154
0.360	0.130×10^1	0.116
0.370	0.126×10^1	0.108
0.400	0.121×10^1	0.118
0.433	0.107×10^1	0.201
0.466	0.946	0.204
0.500	0.756	0.512
0.533	0.105×10^1	0.586
0.566	0.143×10^1	0.522
0.600	0.158×10^1	0.426
0.633	0.156×10^1	0.364
0.666	0.158×10^1	0.716×10^{-1}
0.700	0.155×10^1	0.100×10^{-2}
0.817	0.152×10^1	0.160×10^{-1}
0.907	0.143×10^1	0.150×10^{-1}
1.000	0.147×10^1	0.100×10^{-2}
1.105	0.147×10^1	0.100×10^{-2}

TABLE 15. Bulk Lampblack (Fig. 14).

(μm)	n_1	n_2
0.185	0.711	0.404
0.190	0.723	0.383
0.200	0.718	0.392
0.210	0.725	0.379
0.215	0.728	0.373
0.220	0.732	0.367
0.225	0.750	0.328
0.233	0.792	0.206
0.240	0.791	0.211
0.260	0.792	0.206
0.280	0.784	0.234
0.300	0.780	0.247
0.325	0.780	0.247
0.360	0.761	0.301
0.370	0.725	0.379
0.400	0.753	0.319
0.433	0.763	0.333
0.466	0.773	0.344
0.500	0.763	0.399
0.533	0.786	0.394
0.566	0.812	0.383
0.600	0.857	0.314
0.633	0.876	0.320
0.666	0.907	0.288
0.700	0.926	0.294
0.817	0.935	0.333
0.907	0.952	0.321
1.000	0.974	0.319
1.105	0.994	0.315

different from that of amorphous carbon (Fig. 14, Table 15) which has a lower real index (the result of QSE and the sulfur coating) and an absorption decrease in the 0.26-μm region in the ultraviolet (UV). Bulk graphite has an absorption increase toward the UV (Fig. 12, Table 13) as corroborated by Wickramasinghe (3).

Also presented are the properties n_1 and n_2 for $(NH_4)_2SO_4$ (Fig. 15, Table 16), assuming the bulk sample to be representative of the aerosol, and neglecting QSE. A comparison with the results of Toon et al. (4) shows the same value for the refractive portion in the range considered, but the comparison shows quite a bit less variation in the absorption portion; these measurements show a slight decrease in n_2 with wavelength, while the observations of Toon et al. (4) indicate nearly a three order of magnitude decrease, which is suggestive of an error due to failure to eliminate the effect of scattering in the observed thin crystals.

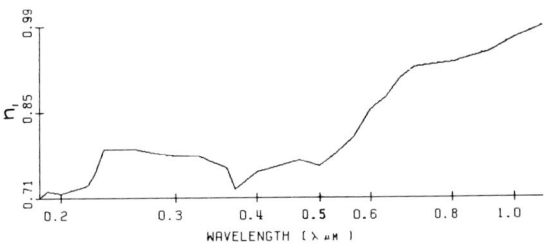

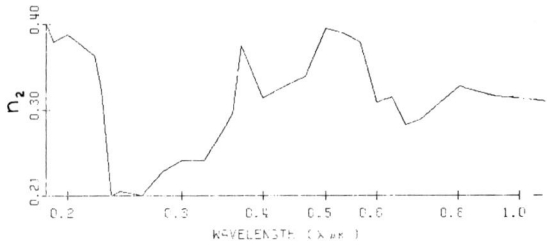

FIGURE 14. Bulk lampblack.

OPTICAL PROPERTIES OF STANDARD AEROSOLS

TABLE 16. Pure $(NH_4)_2SO_4$ (Fig. 15).

(μm)	$Log_{10}T$	R	$Log_{10} n_2$	n_1
0.185	−0.818	0.302	−0.510 x 10^1	1.537
0.190	−0.804	0.321	−0.513 x 10^1	1.532
0.200	−0.775	0.327	−0.514 x 10^1	1.526
0.210	−0.824	0.318	−0.507 x 10^1	1.520
0.215	−0.740	0.332	−0.513 x 10^1	1.514
0.220	−0.733	0.337	−0.513 x 10^1	1.509
0.225	−0.793	0.326	−0.506 x 10^1	1.503
0.233	−0.812	0.335	−0.505 x 10^1	1.497
0.240	−0.815	0.315	−0.499 x 10^1	1.491
0.260	−0.726	0.348	−0.508 x 10^1	1.486
0.280	−0.688	0.344	−0.506 x 10^1	1.480
0.300	−0.544	0.399	−0.528 x 10^1	1.474
0.325	−0.609	0.399	−0.517 x 10^1	1.468
0.360	−0.532	0.399	−0.520 x 10^1	1.463
0.370	−0.644	0.390	−0.506 x 10^1	1.457
0.400	−0.474	0.352	−0.511 x 10^1	1.451
0.433	−0.460	0.398	−0.521 x 10^1	1.450
0.466	−0.472	0.391	−0.514 x 10^1	1.449
0.500	−0.462	0.385	−0.511 x 10^1	1.448
0.533	−0.452	0.390	−0.511 x 10^1	1.448
0.566	−0.447	0.389	−0.509 x 10^1	1.448
0.600	−0.450	0.393	−0.507 x 10^1	1.448
0.633	−0.444	0.395	−0.506 x 10^1	1.447
0.666	−0.447	0.344	−0.490 x 10^1	1.445
0.700	−0.458	0.370	−0.493 x 10^1	1.444
0.817	−0.462	0.363	−0.484 x 10^1	1.442
0.907	−0.444	0.356	−0.480 x 10^1	1.441
1.000	−0.439	0.360	−0.478 x 10^1	1.439
1.105	−0.435	0.326	−0.465 x 10^1	1.438

The optical complex indices of refraction which have been presented here can be used in a Mie scattering program, employing an appropriate particle size distribution, to determine the volume scattering coefficient as well as the scattering phase function. Ultimately, an atmospheric radiative transfer calculation may be made to determine remote sensing target contrast.

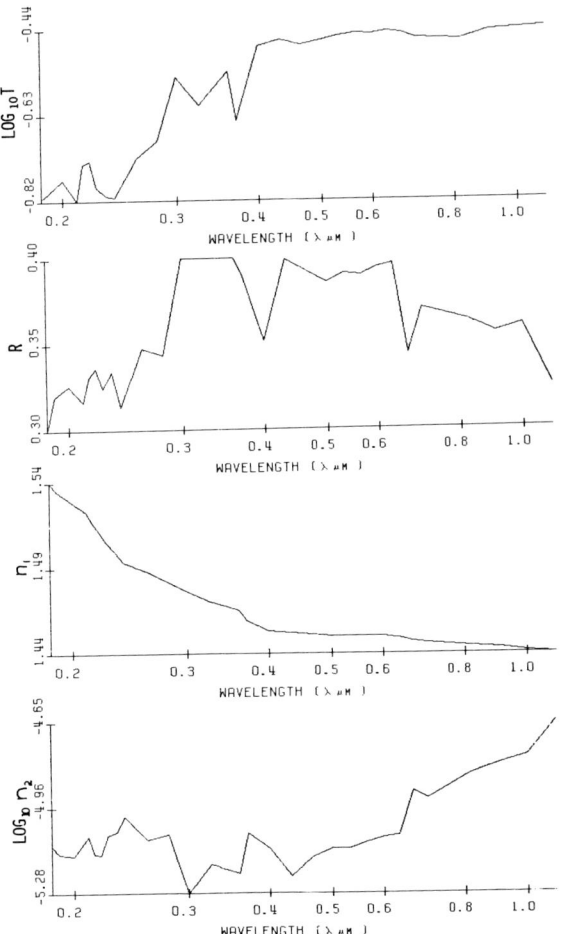

FIGURE 15. Pure $(NH_4)_2SO_4$, compressed pellet (effective thickness 0.0475 cm).

REFERENCES

1. Egan, W. G., and Hilgeman, T. W., *Optical Properties of Inhomogeneous Materials,* Academic Press, New York (1979).
2. Egan, W. G., and Hilgeman, T., Anomalous refractive index of submicron-sized particles, *Appl. Opt.* 19(22), 3724-3727 (1980).
3. Wickramasinghe, N. C., *Intersteller Grains,* Chapman and Hall, Ltd., London (1967).
4. Toon, O. B., Pollack, J. B., and Khare, B. N., Optical constants of several atmospheric aerosol species: Ammonium sulfate, aluminum oxide, and sodium chloride, *J. Geophys. Res.* 81(33), 5733-5748 (1976).

SIMULTANEOUS MEASUREMENTS OF AEROSOL SCATTERING AND EXTINCTION COEFFICIENTS IN A MULTI-PASS CELL[1]

H. E. Gerber

Naval Research Laboratory
Washington, D.C.

A multi-pass optical cell which combines a transmissiometer and a reciprocal type of integrating nephelometer was used to simultaneously measure the scattering and extinction coefficients of aerosols generated at the First International Workshop on Light Absorption by Aerosol Particles (Fort Collins, Colorado, August 1980). The instrumentation and the results of the measurements are described. A limited amount of data consisting of the aerosol absorption coefficient was obtained. Comparison of the nephelometer with another of different design present at the workshop showed good agreement.

---

[1]*This work was supported by Naval Ocean Systems Center, Project No. ZF59551002.*

I. INTRODUCTION

The multi-pass optical cell brought to the workshop for intercomparison with other techniques for measuring light absorption by aerosols was described previously (1). Transmittance measurements over the 20-m folded light path in the cell yields the aerosol extinction coefficient σ_e, and a nephelometer located in one wall of the cell gives the aerosol scattering coefficient σ_s. The difference between these coefficients is the desired quantity, the aerosol absorption coefficient σ_a.

The cell was used on a naval cruise to determine σ_a and the single-scattering albedo $\tilde{\omega}$ at a wavelength of 632.8 nm for maritime aerosols (2). At that time an aerosol concentrator which enhances the particle loading of the ambient aerosol by a factor of about 20 was used in conjunction with the cell to increase the effective length of the light path in the cell to 400 m. This extra length was necessary to achieve reasonable measurement accuracies. Subsequent improvements made to the optics of the cell have eliminated the need for the concentrator and permitted accurate transmittance measurements for moderate atmospheric turbidities.

The following report includes descriptions of the cell and of the improvements made to it. The nephelometer is also described; it is one version of the several discussed originally by Beutell and Brewer (3) and it is optically the reciprocal of the currently popular nephelometer (4). Results of workshop measurements of σ_a and $\tilde{\omega}$ with the cell are presented, and a comparison is made between the nephelometer in the cell and the four-wavelength nephelometer brought to the workshop by Bodhaine (5).

II. INSTRUMENTATION

1. Cell

The optical cell (Fig. 1) consists of a slab-like chamber 200 cm by 20 cm by 3 cm. This configuration minimizes the chamber volume (12.0 liters) and the time required to fill the clean cell with aerosol. The inside of the chamber is coated with optical-black paint, and baffles along the inner walls help collect stray light. The chamber is aluminum and weighs $\approx$ 50 kg.

During operation the long dimension of the chamber is kept vertical, so that aerosol particles introduced at the top are not lost due to sedimentation. The lower mirror in the cell is continuously flushed with clean air to prevent particle deposition. To keep the aerosol at the ambient temperature and thus help preserve the optical properties, ambient air can be flushed through the saddle chambers which flank the inner aerosol chamber.

A MULTI-PASS CELL METHOD

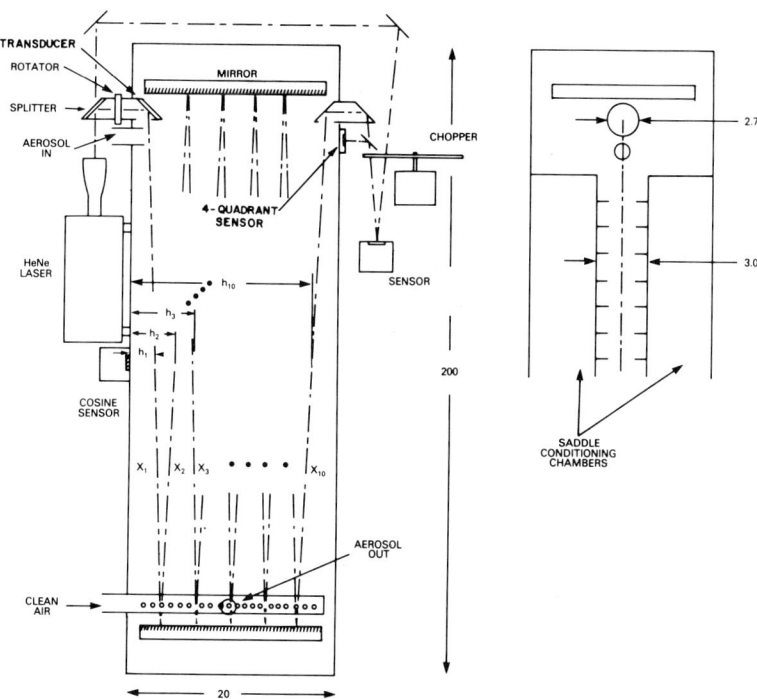

FIGURE 1. Schematic of the transmission and scattering cell showing laser beam paths, location of optics, aerosol and clean air ports, and dimensions (cm).

2. Transmissiometer

The transmissiometer consists essentially of a linearly polarized 4-mW HeNe laser, a 30-m-long collimated beam stepped through the aerosol chamber between flat mirrors located at each end of the cell, and a silicon photovoltaic sensor located near the chopper in Fig. 1. The sensor gives the transmittance

$$\tau = \frac{I}{I_o} = \exp[-(\sigma_s + \sigma_a)x] \qquad (1)$$

of the beam through the cell, where I is the light intensity for the aerosol-filled cell, I_o is the intensity for the cell filled with filtered air, and x = 20 m. The purpose of splitting the laser beam near the laser and of passing a fraction of the beam directly to the chopper and the sensor is to reference the transmittance measurement to the laser intensity. This is accomplished by permitting both the transmitted and reference portions of the beam to fall on the chopper and on the same sensor. Holes on the chopper are so arranged to cause the sensor to be illuminated alternately by each beam. A phase-locked analog demultiplexer separates the train of pulses on the sensor, and separate analog outputs of I and I_0 are obtained. This method permits precise measurements of I/I_0, since neither the drift in the laser nor the sensor comes into play. The previous work showed that I/I_0 could be resolved by $\approx \pm 0.2\%$ for the 20-m path in the cell. The major cause of this error was the drift of the laser beam on the sensor due to thermal and mechanical flexing of the cell, and turbulence-induced scintillation of the beam.

The stability of the beam was greatly improved by adding an optical servo loop to the cell which aims the transmitted laser beam to precisely the same spot on the sensor. The servo loop consists of a piezoelectric transducer (PZ-80 Burleigh Instruments, Inc.) placed near the laser, and a four-quandrant detector (PIN-SC 25, United Detector Technology, Inc.) located near the sensor. The transducer is driven by three independent high-voltage op-amps. Figure 2 shows the consequence of this addition. With the servo loop in operation the

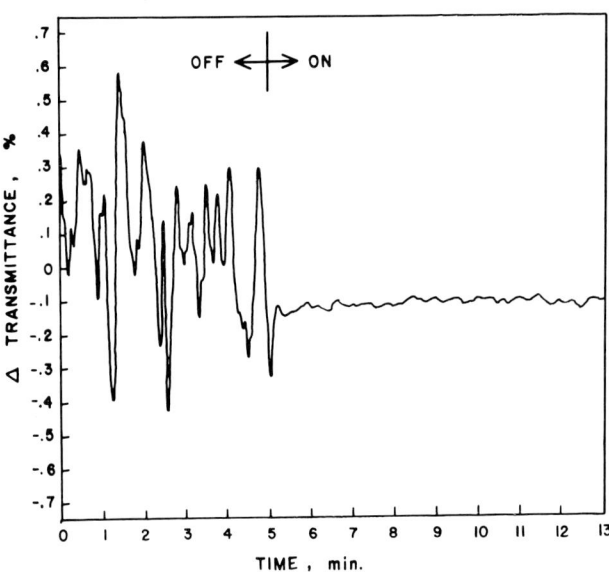

FIGURE 2. *Relative change in light transmittance over the 20-m path in the cell with the optical servo loop which steers the light beam to the sensor in the off and on condition. The beam is within the sensitive area of the sensor in both cases. The cell is filled with filtered air.*

transmittance can be measured within an error of ± 0.02%, with a drift of 0.1% per hour. This improved accuracy permits measurements of the transmittance with an accuracy of ± 10% for ambient air with a turbidity which corresponds to a visual range of about 40 km. Smaller errors result for larger turbidities.

3. Nephelometer

The integrating nephelometer consists of a radiometer with a precise cosine response (detector 700-8B and S-20 photomultiplier detector assembly 2020-10 from Gamma Scientific, Inc.) placed in the wall of the cell so that it is parallel to the laser beam in the cell. The sensor detects the light scattered by the aerosol particles and by the air molecules in the beam, and it gives an output proportional to the sum of σ_s and the molecular scattering coefficient of air σ_m.

The derivation relating cosine-sensor behavior and the scattering coefficients has not been published previously for this reciprocal type of integrating nephelometer; therefore, it will be given here: If an aerosol is irradiated by a narrow collimated beam of light of intensity i_0 as shown in Fig. 3, the irradiance of the cosine sensor by the aerosol is given by

$$dE = \int_{2\pi} B(\phi) \cos \theta \, d\omega \tag{2}$$

where

$$B(\phi) = \frac{i_0 \beta(\phi) dv}{dA'} \tag{3}$$

$$dA' = dA \cos \theta \tag{4}$$

$$dv = dA \, \Delta y \tag{5}$$

$$d\omega = \frac{d\theta \, \Delta x}{y \sec \theta} \tag{6}$$

$B(\phi)$ is the radiance of the hatched volume, $\beta(\phi)$ is the Mie angular scattering function, dA' is the area of the hatched volume dv projected in the direction of the sensor, and ω is the solid angle with its vertex at the sensor.

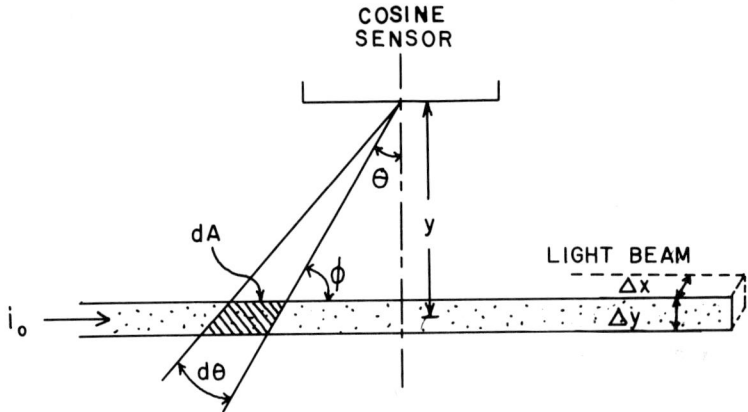

FIGURE 3. Schematic of the reciprocal type of integrating nephelometer.

Combining Eqs. (2) through (6) and changing from θ to ϕ gives

$$dE = \frac{F_o}{y} \int_0^\pi \beta(\phi) \sin \phi \, d\phi \tag{7}$$

where $F_o = i_o \, \Delta x \, \Delta y$ is the radiant flux of the beam.
By definition (Ref. 6)

$$\sigma_s + \sigma_m = 2\pi \int_0^\pi \beta(\phi) \sin \phi \, d\phi \tag{8}$$

so that Eqs. (7) and (8) give

$$dE = \frac{F_o(\sigma_s + \sigma_m)}{2\pi y} \tag{9}$$

Multiplying both sides of Eq. (9) by the area S of the sensor gives the radiant flux F received by the sensor for the case of a narrow collimated beam:

$$F = \frac{F_o S(\sigma_s + \sigma_m)}{2\pi y} \tag{10}$$

The integrating nephelometer is calibrated in the usual manner by filling the cell with Freon-12 which has a known molecular scattering coefficient σ_{sc}. The calibration flux if F_c, so that Eq. (10) becomes

$$F_c = \frac{F_o S \sigma_{sc}}{2\pi y} \tag{11}$$

and combining Eqs. (10) and (11) and solving for σ_s gives

$$\sigma_s = \sigma_{sc} \frac{F}{F_c} - \sigma_m \tag{12}$$

where $\sigma_{sc} = 9.345 \times 10^{-5}$ m^{-1} and $\sigma_m = 6.230 \times 10^{-6}$ m^{-1} (pressure is 840 mb, temperature is 20°C) are values for a wavelength of 632.8 nm which were extrapolated from values given at other wavelengths by Bodhaine (7).

Strictly, the application of Eq. (10) to the present instrument requires summing up the contributions from all 10 segments of the beam in the cell, and the complete expression for F shows a dependence on I/I_o and on the reflectivity of the mirrors in the cell (1). However, for the case where $I/I_o \geq 0.9$, Eq. (12) gives acceptable results.

Combining Eqs. (1) and (12) gives the desired expression

$$\sigma_a = \sigma_m - \frac{\sigma_{sc} F}{F_c} - \frac{1}{x} \ln \frac{I}{I_o} \tag{13}$$

III. WORKSHOP RESULTS

Simultaneous measurements of transmittance and aerosol scattering in the cell for the workshop aerosols gave the values of σ_a and $\tilde{\omega}$ shown in Table 1. The values are also shown in Fig. 4 where the error due to random noise is included with each point. Comparison of Table 1 with the workshop schedule given in Hindman et al. (9) shows that measurements were only made for a fraction of the experiments. This was due to the malfunction of the optical servo loop in the transmissiometer part of the cell. Unfortunately, this important feature could not be repaired during the workshop. The large error bars on σ_a in Fig. 4 reflect this state of affairs. The most reliable data points are for soot (measurements 4, 5, and 7) and methylene blue (measurement 11). Generally, $\tilde{\omega}$ was somewhat too large for the salt and salt mixtures; this may have been due to excessive values of I/I_o caused by scattered light from the aerosols falling onto the transmitted-light sensor. The Arizona road dust value of $\tilde{\omega}$ is much smaller than expected; here the value of σ_s may be in error due to a large nephelometer truncation error caused by the large particles in the dust.

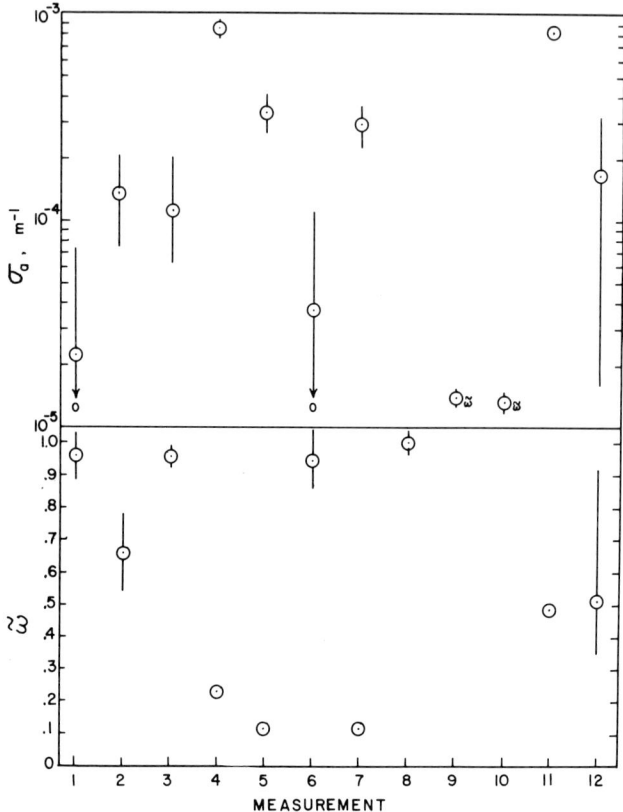

FIGURE 4. Values of the aerosol absorption coefficient σ_a and single scattering albedo $\tilde{\omega}$ measured during the workshop. Key to the measurements is given in Table 1. Line segments centered about each point give ± one sample standard deviation of the random sampling error. Systematic instrumentation errors are not included.

The integrating nephelometer in the cell wall operated satisfactorily during the workshop. This provided the opportunity to compare for the first time the two types of integrating nephelometers described originally by Beutell and Brewer (3). The two types of nephelometers differ in that the roles of the sensor and the light source are interchanged. The reciprocal type of nephelometer in the cell wall uses a cosine sensor and a collimated light beam, while the other type, which was brought to the workshop by Bodhaine (5) and which is similar to the one described by Charlson et al. (4), uses a cosine light source and a collimated field of view for the sensor. Bodhaine (5) gives values of σ_s for a wavelength λ of light at 550 nm and 655 nm; his value for $\lambda = 632.8$ nm was

A MULTI-PASS CELL METHOD

TABLE 1. Workshop Measurements of σ_a and $\tilde{\omega}$

Measurement	Date	Time	Aerosol[a]	$\sigma_a(m^{-1})$	$\tilde{\omega}$	n_2[b]
1	7/29/80	1620-1700	Salt + MB	2.20×10^{-5}	0.960	
2	8/6/80	1150	ARD	1.38×10^{-4}	0.658	
3	8/6/80	1430-1707	Salt + soot	1.14×10^{-4}	0.959	
4	8/6/80	1740-1807	Soot	8.64×10^{-4}	0.229	0.664
5	8/7/80	1025-1035	Soot	3.36×10^{-4}	0.112	0.753
6	8/7/80	1106-1330	Salt + soot	3.70×10^{-5}	0.944	
7	8/7/80	1352-1417	Soot	2.91×10^{-4}	0.113	0.653
8	8/7/80	1423-1558	Salt + soot (RH ≈ 50%)	≈ 0	1.020	
9	8/7/80	1510	Salt + soot (RH ≈ 100%)	≈ 0	1.150	
10	8/7/80	1550	Salt + soot (RH ≈ 95%)	≈ 0	1.130	
11	8/8/80	0937-1116	MB	8.34×10^{-4}	0.482	0.030
12	8/8/80	1133-1206	Salt	1.68×10^{-4}	0.513	

[a] MB = methylene blue; ARD = Arizona road dust; salt = $(NH_4)_2SO_4$.
[b] n_2 was determined from a formula given by Hänel (8).

found by linear extrapolation. Twenty-two hours of comparison between the nephelometers for various workshop aerosols are shown in Fig. 5. Good agreement was found. The nephelometer in the cell gave σ_s values which were larger by an average of 1.9% than σ_s given by Bodhaine's nephelometer, and the correlation coefficient between the two sets of measurements was 0.9992. The right-most data point in Fig. 5 was not included in those accuracies, since it corresponds to the experiment during which the aerosol was humidified in the aerosol manifold. For this one case, the larger σ_s measured by the cell nephelometer agrees with the observation that relative humidity increased with distance along the manifold thus causing changes in the aerosol size distribution, and the cell was last on line. No truncation error corrections were made for either nephelometer (see Ref. 10). The good agreement in Fig. 5 is partially due to the fact that both nephelometrs have about the same truncation error.

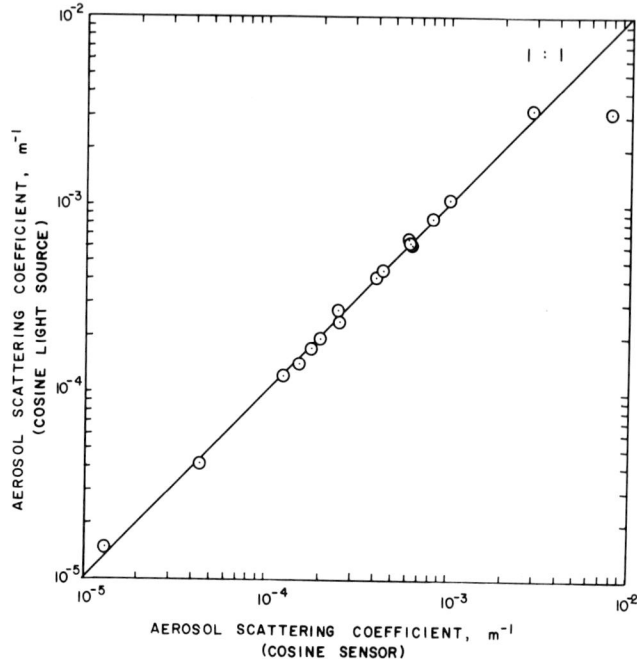

FIGURE 5. *Comparison between the integrating nephelometer (with cosine light source) brought to the workshop by Bodhaine (5) and the integrating nephelometer (with cosine sensor) located in the multi-pass optical cell. The nephelometers were exposed to the same workshop aerosols, and the comparison was done for a wavelength of light of 632.8 nm. The poor match of the right-most data point is discussed in the text.*

ACKNOWLEDGMENTS

Robert E. Daniels is thanked for his assistance during the workshop. Appreciation is expressed to the staff of the Cloud Simulation and Aerosol Laboratory for their exceptional support.

REFERENCES

1. Gerber, H. E., Portable cell for simultaneously measuring the coefficients of light scattering and extinction for ambient aerosols, *Appl. Opt. 18,* 1009-1014 (1979).
2. Gerber, H. E., Absorption of 632.8 nm radiation by maritime aerosols near Europe, *J. Atmos. Sci. 36,* 2502-2512 (1979).
3. Beutell, R. G., and Brewer, A. W., Instruments for the measurement of the visual range. *J. Sci. Instru. 26,* 357-359 (1949).
4. Charlson, R. J., Horvath, H., and Pueschel, R. F., The direct measurement of atmospheric light scattering coefficient for studies of visibility and pollution, *Atmos. Environ. 1,* 469-478 (1967).
5. Bodhaine, B. A., The GMCC four-wavelength nephelometer, *Light Absorption by Aerosol Particles* (H. E. Gerber and E. E. Hindman, eds.), Spectrum Press, Hampton, Virginia (1982).
6. Middleton, W. E. K., *Vision Through the Atmosphere*, University of Toronto Press, Toronto (1952).
7. Bodhaine, B. A., Measurement of the Rayleigh scattering properties of some gases with a nephelometer, *Appl. Opt. 18,* 121-125 (1979).
8. Hänel, G., The properties of atmospheric aerosol particles as functions of the relative humidity at thermodynamic equilibrium with the surrounding moist air, *Advances in Geophys. 19,* 73-188 (1976).
9. Hindman, E. E., Horn, R. D., and Finnegan, W. G., Generations and characterization of aerosol particles, *Light Absorption by Aerosol Particles* (H. E. Gerber and E. E. Hindman, eds.), Spectrum Press, Hampton, Virginia (1982).
10. Rabinoff, R. A., and Herman, B. M., Effect of the aerosol size distribution on the accuracy of the integrating nephelometer, *J. Appl. Meteor. 12,* 184-188 (1973).

IMAGINARY REFRACTIVE INDEX MEASUREMENTS FOR ARIZONA ROAD DUST AND METHYLENE BLUE

James B. Gillespie

Atmospheric Sciences Laboratory
White Sands Missile Range, New Mexico

Measurements of the imaginary part of the complex refractive index of Arizona road dust and methylene blue are reported for the 0.3-µm to 1.7-µm spectral region. The measurements were performed using a Cary Model 14 spectrophotometer with a custom 25-cm-diameter integrating sphere. The Kubelka-Munk diffuse reflectance theory was used to determine the absorption coefficients from which the imaginary indices were determined. The Arizona road dust had an imaginary index on the order of 10^{-4}, while the methylene blue had a value on the order of 0.5 at 0.65 µm, its peak in the visible.

I. INTRODUCTION

At the First International Workshop on Light Absorption by Aerosol Particles (Fort Collins, Colorado, August 1980), materials with well-known optical properties were generated into aerosols so that simultaneous measurements of optical properties could be performed and the results compared. Arizona road dust and methylene blue were two of the materials used in these tests. Determinations of the imaginary part of the refractive index of each of these materials are reported for the 0.3- to 1.7-μm spectral region using Kubelka-Munk diffuse reflectance methods. These methods are described by Lindberg and Snider (1) and Lindberg and Laude (2), and will therefore be described only briefly here.

II. THEORY

The complex refractive index is defined as

$$n = n_1 - in_2 \tag{1}$$

where n_1 is the real (refractive) part and is the ratio of the speed of light in vacuum to the speed of light in the material, and n_2 is the imaginary (absorptive) part. For a bulk material in which there is no scattering, the intensity I at a point is related to the initial intensity I_o by the Bouguer-Lambert law

$$I = I_o e^{-kx} \tag{2}$$

where k is the bulk absorption coefficient with units of reciprocal length and x is the pathlength in the medium. From electromagnetic theory the imaginary refractive index is related to k by

$$n_2 = \frac{k\lambda}{4\pi} \tag{3}$$

where λ is the wavelength.

The complex refractive index can be quite readily measured for liquids and for homogeneous solids which can be polished to optical flatness or which can be cut into very thin transparent slices. Specular reflectance measurements or direct transmission measurements can then be performed, and either Fresnel's equations or a Kramers-Kroenig dispersion analysis can be applied to obtain both n_1 and n_2.

Determination of the complex refractive index of particulate materials is not quite so straightforward. Attempts have been made to apply specular methods to particulates; however, except under certain conditions (see Ref. 3) these attempts must fail, since scattering and absorption are not separated out and information carried in the phase is lost. To overcome these limitations, usually a radiative transfer model is employed. One of the more successful radiative transfer theories has been the Kubelka-Munk theory of diffuse reflectance. This theory is a two-flux approximation to the radiative transfer equation. The Kubelka-Munk theory is widely used because of its relative ease of application and the general acceptance of its results. Many different equations have been derived in this theory. They relate the two basic unknown parameters K_m (the Kubelka-Munk absorption coefficient) and S (the Kubelka-Munk scattering coefficient) to measurable quantities such as R_∞ (the reflectance of a semi-infinite thick layer), R_w (the reflectance of a layer on a white background), R_B (the reflectance of a layer on a black background), T (the diffuse transmittance), etc. (See Ref. 4 for further information.) Usually two measurements can be made so that K_m and S can then be uniquely determined. The S coefficient has no direct physical interpretation; however, the K_m coefficient has been empirically shown to be directly proportional and equivalent within about 50% to the Bouguer-Lambert bulk absorption coefficient. Therefore, if Eq. (3) is judicially applied, it can be used to determine the imaginary refractive index.

For weakly absorbing powdered layers, Lindberg and Snider (1) have derived a system of Kubelka-Munk equations which allows determination of the absorption coefficient of a thin powdered sample which has been deposited onto a membrane filter. Four spectrophotometric reflectivities are required: (a) the diffuse reflectance of the clean filter on a white background, (b) the diffuse reflectance of the clean filter on a black background, (c) the diffuse reflectance of the sample on the filter on a white background, and (d) the diffuse reflectance of the sample on the filter on a black background. The thickness of the layer must also be measured.

For strongly absorbing powdered samples, Lindberg and Laude (2) have applied the Kubelka-Munk equation

$$\frac{k}{S} = \frac{(1 - R_\infty)^2}{2R_\infty} \tag{4}$$

for a mixture of a strongly absorbing sample with a nonabsorbing material (such as barium sulfate). The resulting equation is

$$k = CS^* \left[\frac{(1 - R_\infty^*)^2}{2R_\infty^*} - \frac{(1 - R_\infty)^2}{2R_\infty} \right] \tag{5}$$

where C is a concentration term, R_∞ is the absolute reflectance of the diluent, S^* is the scattering coefficient of the diluent, and R_∞^* is the absolute reflectance of the mixture. The assumption for this equation is that the percentage of sample is very much less than that of the nonabsorbing diluent and therefore the sample does all the absorbing and the diluent does all the scattering.

The Kubelka-Munk theory makes two assumptions in its derivation: (a) that the scattering is isotropic in the sample layer and (b) that the layer is diffusely illuminated. That the requirements of the first assumption are met can be readily verified by examining the sample layer in a light-scatter photometer. The second assumption is overcome by performing the diffuse reflectance measurements in an integrating sphere. An integrating sphere has several unique properties which make it ideal for use in diffuse reflectance measurements, one of which is the principle of reversibility which, combined with taking ratio measurements, approximately satisfies the requirements of the second assumption.

III. PROCEDURE

Arizona road dust and methylene blue (MB) have very different optical properties in the visible/near-infrared. Arizona road dust is a very weakly absorbing material while methylene blue is a very strongly absorbing substance. The method of Lindberg and Snider (1) was therefore applied to the Arizona road dust while the method of Lindberg and Laude (2) was applied to the methylene blue.

The Arizona road dust was suspended as an aerosol and collected as a layer onto a 47-mm-diameter 10-μm-pore-size Millipore Teflon filter using a conventional filter holder assembly and vacuum pump. The Millipore (catalog No. LCWP 04700) Teflon filter was selected because of its diffuse reflectance characteristics. Several layer samples were collected. The sample weights were on the order of 10 mg to 30 mg and the layer thicknesses were from about 0.015 cm to about 0.03 cm. The diffuse reflectivities of these filters on both black and white backgrounds were measured beforehand. After the dust samples were collected, the corresponding measurements were made for each layer on the filter sample.

Since the methylene blue was to be analyzed by the dilution method, a pure sample of several milligrams was required. Therefore, it was only necessary to obtain a bulk sample from the container and return it to the laboratory for analysis. The dilution method required only one diffuse reflectance measurement, R_∞ in Eq. (5), and the relative concentration value, C in Eq. (5). The value of C requires measurement of the specific gravity of the sample, which was found to be 2.0. A few milligrams of methylene blue were mixed with about 200 mg of Eastman White Reflectance Standard (a very highly refined $BaSO_4$). This mixture was ground in an agate vial for about 5 minutes, after which a small sample was mixed with about 11 g or $BaSO_4$ and shaken for 15 minutes to obtain a uniform mixture. The 11 g of mixture were then pressed into a 1-cm-deep stainless-steel sample dish for reflectivity measurements. Ten dilutions were made and measured. The mass concentration of methylene blue varied from about 1 part MB to 5000 parts $BaSO_4$ to about 1 part MB to 30,000 parts $BaSO_4$. The results were consistent, although differences of $\pm 10\%$ in n_2 occurred. Dilutions which produced R_∞^* values of about 0.90 were selected to avoid errors due to uncertainties in the Kubelka-Munk optical parameters described by Gillespie (5). Patterson (6) has also published similar $BaSO_4$ data, but his data are only for the 0.3- to 0.7μm spectral region. Lindberg and Gillespie (7) have pointed out that if reflectivities are about 0.90 then any errors due to the selection of K_m and S tend to offset each other and cancel out.

All of the reflectivity measurements described were performed with a Cary model 14 spectrophotometer equipped with a custom-made 25-cm integrating sphere. This was interfaced to a Varian model 620/L computer to form an SS-100 data acquisition system which was used to record and reduce the spectrophotometric data.

IV. RESULTS

The results for Arizona road dust are shown in Figs. 1 to 3. Figure 1 shows the reflectivities of the layer of dust deposited on the filter and measured on a black background R_B and on a white background R_W; note the strong absorption as the ultraviolet is approached. Figure 2 shows the spectral dependence

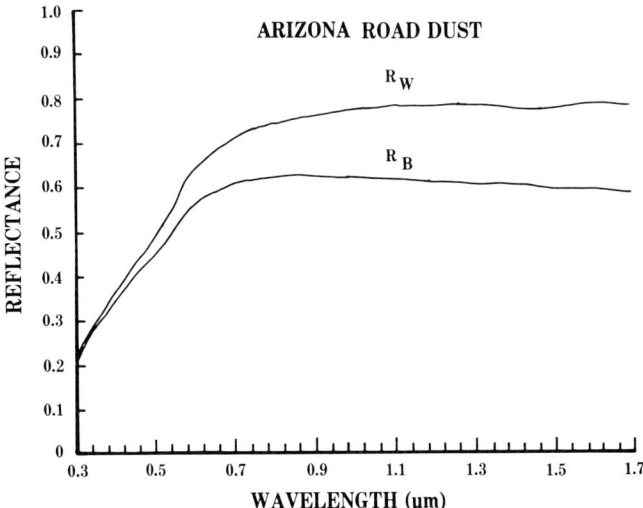

FIGURE 1. Reflectivity versus wavelength for a 0.0139-cm-thick layer of Arizona road dust collected on a Teflon filter. R_B is the reflectance on a black background and R_W is the reflectance on a white background.

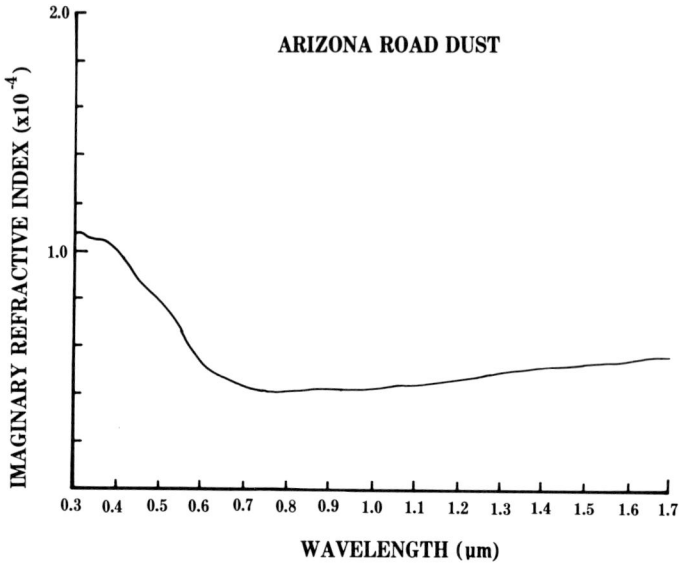

FIGURE 2. *The imaginary refractive index of Arizona road dust in the 0.3- to 1.7-μm spectral region.*

of the imaginary index n_2 in the 0.3- to 1.7-μm region. The imaginary index is on the order of 1×10^{-4} in this region. Figure 3 shows the corresponding absorption coefficient $k(\text{cm}^{-1})$. Eight different layers were measured and the results were consistent to within ± 20%. Errors in reflectance can cause 10% to 15% error with this method, and the thickness measurement of the layer was measured easily with a micrometer microscope to within a 5% error. The values of the Arizona road dust are about what one would expect for that type of material. The absorption in such a sample is probably due to the impurities, such as iron oxide, in the minerals.

The results for methylene blue are shown in Fig. 4, which gives the imaginary refractive index n_2. The curve peaks at about 0.65 μm and gives a value of about 0.6, a value close to that of carbon in the visible.

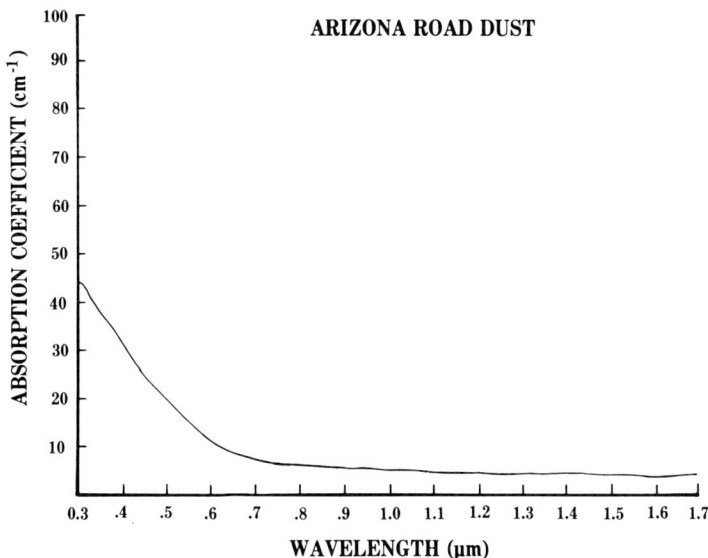

FIGURE 3. The absorption coefficient (cm^{-1}) of Arizona road dust for the 0.3- to 1.7-μm spectral region.

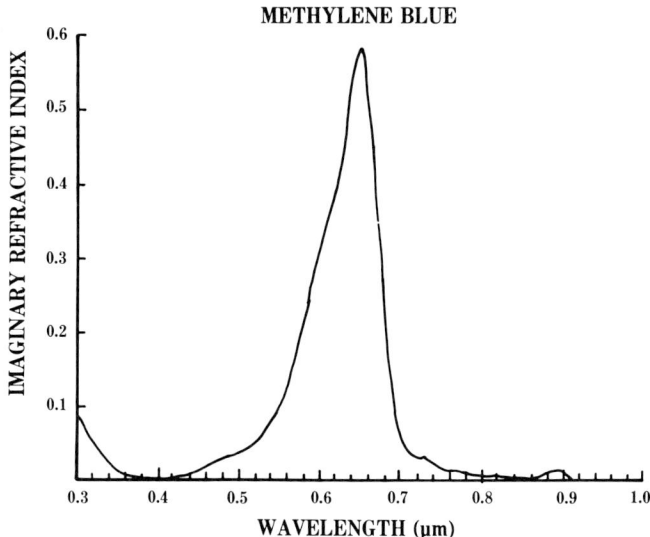

FIGURE 4. The imaginary refractive index for methylene blue for the 0.3-to 1.7-μm spectral region.

REFERENCES

1. Lindberg, J. D., and Snider, D. G., Diffuse reflectance spectra of several clay minerals, *Appl Opt.* 12, 573-578 (1973).
2. Lindberg, J. D., and Laude, L. S., A measurement of the absorption coefficient of atmospheric dust, *Appl. Opt.* 13, 1923-1927 (1974).
3. Jennings, S. G., Pinnick, R. P., and Gillespie, J. B., Relation between absorption coefficient and imaginary index of atmospheric aerosol constituents, *Appl. Opt.* 18, 1368-1371 (1979).
4. Wendlandt, W. W., and Hecht, H. G., *Reflectance Spectroscopy,* John Wiley & Sons, New York (1966)
5. Gillespie, J. B., Lindberg, J. D., and Laude, L. S., Kubelka-Munk optical coefficients for a barium sulfate white reflectance standard, *Appl. Opt.* 14, 807-809 (1975).
6. Patterson, E. M., Shelden, C. E., and Stockton, B. H., Kubelka-Munk optical properties of a barium sulfate white reflectance standard, *Appl. Opt.* 16, 729 (1977).
7. Lindberg, J. D., and Gillespie, J. B., Relationship between particle size and imaginary refractive index in atmospheric dust, *Appl. Opt.* 10, 2627-2628 (1977).

ANALYSIS OF POLAR NEPHELOMETER DATA OBTAINED AT THE FIRST INTERNATIONAL WORKSHOP ON LIGHT ABSORPTION BY AEROSOL PARTICLES[1]

Gerald W. Grams

School of Geophysical Sciences
Georgia Institute of Technology
Atlanta, Georgia

Alessandro Coletti

Instituto di Fisica dell'Atmosfera-C.N.R.
Largo Lo Sturzo 31, 00100 Roma, Italy

A polar nephelometer was exposed to the various aerosols generated at the First International Workshop on Light Absorption by Aerosol Particles (Fort Collins, Colorado, August 1980). Data consisting of the asymmetry parameters and the scattering phase functions are presented, and systematic differences for the various aerosols are noted.

---

[1]*Participation in this workshop was supported in part by National Science Foundation Grant ATM 8005356.*

I. INTRODUCTION

Measurements obtained with a laser polar nephelometer at the First International Workshop on Light Absorption by Aerosol Particles have been analyzed in terms of an aerosol volume scattering cross section σ_i and two asymmetry factors g_1 and g_2 for scattering in planes that are perpendicular and parallel, respectively, to the electric vector of an incident polarized light beam. If θ is the scattering angle and i is an index designating each of the two orthogonal scattering planes, the angular variation of intensity of the light scattered out of the beam by aerosol particles and molecules illuminated by the beam is specified by the so-called scattering phase function $P_i(\theta)$ as defined by Deirmendjian (1). The asymmetry factor is a dimensionless parameter whose value is equal to the average over a unit sphere of $\cos \theta$, weighted by the scattering phase function; it is, thereby, calculated that

$$g_i = \int_0^{180°} P_i(\theta) \cos \theta \sin \theta \, d\theta \bigg/ \int_0^{180°} P_i(\theta) \sin \theta \, d\theta \tag{1}$$

The optical parameters σ_i, g_1, and g_2 will depend on the concentration, size, shape, and refractive index (composition) of the particles. Of these three parameters, the scattering cross section σ_i is most sensitive to changes in aerosol concentration whereas the asymmetry parameters g_1 and g_2 will depend mostly on the particle size, shape, and composition. It is believed that data on temporal changes in these three parameters provide valuable information on the aerosol particles under study, especially if they were to be calculated in real time by the microprocessor that controls the polar nephelometer.

The phase function measurements carried out during the workshop were also analyzed in terms of a least-squares best fit to the so-called Henyey-Greenstein phase function (2). Asymmetry factors obtained in this way match the values calculated directly from the experimental data, suggesting the utility of using the Henyey-Greenstein formula to parameterize the optical properties of the aerosol.

II. DATA ANALYSIS PROCEDURES

The polar nephelometer (3) measures the differential scattering cross section of the aerosol particles at the helium-neon laser wavelength ($\lambda = 633$ nm) as a function of angle in 5° steps from 10° to 170°. For each scattering plane with index i, an angular function is calculated

$$S_i(n,r,\theta) = \int_0^\infty \pi r^2 Q_s(n,r) p_i(n,r,\theta) N(r) dr = \sigma_s P_i(\theta) \tag{2}$$

where n is the complex index of refraction of the particles, as defined by van de Hulst (4), r is the particle radius, Q_{sca} is the particle scattering efficiency (the scattering cross section per particle/πr^2), p_1 and p_2 are the scattering phase functions of each particle in each of the two different polarization planes, P_1 and P_2 are the particle scattering phase functions averaged over all particle sizes, N(r) is the aerosol particle size distribution, and σ_s is the volume-scattering cross-section of the aerosol particles. As a first approximation, the following parameters can be computed from the raw nephelometer signals $s_1(\theta)$ and $s_2(\theta)$ for each of the two scattering planes

$$S_i = \sum_{k=1}^{N} s_i(\theta_k)\sin(\theta_k) \qquad (3)$$

$$g_i = \sum_{k=1}^{N} s_i(\theta_k)\sin(\theta_k)\cos(\theta_k)/S_i \qquad (4)$$

The asymmetry factor g for unpolarized light can then be evaluated as

$$g = \frac{S_1 g_1 + S_2 g_2}{S_1 + S_2} \qquad (5)$$

An alternative way to compute the same quantity was also attempted by applying a nonlinear least-squares curve-fitting procedure to determine the value of g as used in the Henyey-Greenstein phase function

$$P_{HG}(\theta) = (1 - g^2)/(1 + g^2 - 2g\cos\theta)^{3/2} \qquad (6)$$

This expression was introduced many years ago in the astrophysics literature to describe observed intensity patterns for diffuse interstellar radiation (2). Hansen (5) has shown that $P_{HG}(\theta)$ can be used to replace more realistic Mie scattering phase functions in multiple-scattering calculations with no more than a few percent error in computed fluxes.

In the past, simultaneous data on scattering phase functions and particle size distributions were analyzed to determine the best-fit optical parameters such as the imaginary component n_2 of the complex refractive index for the Mie scattering theory [e.g., Grams et al. (3)]. However, the nephelometer was modified for observations of stratospheric aerosol particles on a high-altitude jet research aircraft in Poker Flat, Alaska, during July 1979 (6) so that the nephelometer operated under the control of a microprocessor data-recording system. The automated nephelometer operation substantially increased the amount of data available to process. Therefore, the Henyey-Greenstein function was used to carry out preliminary studies of the Alaska data rather than using previous analysis procedures which had been based on Mie scattering theory (and which consumed significantly more computer time). It was found, for the conditions of the Alaskan experiment, that excellent agreement was obtained using $P_{HG}(\theta)$, and the results of comparisons with the Henyey-Greenstein function are thereby used to specify realistic scattering phase functions for applications such as aerosol-climate models. An example of the results of these stratospheric observations is shown in Fig. 1

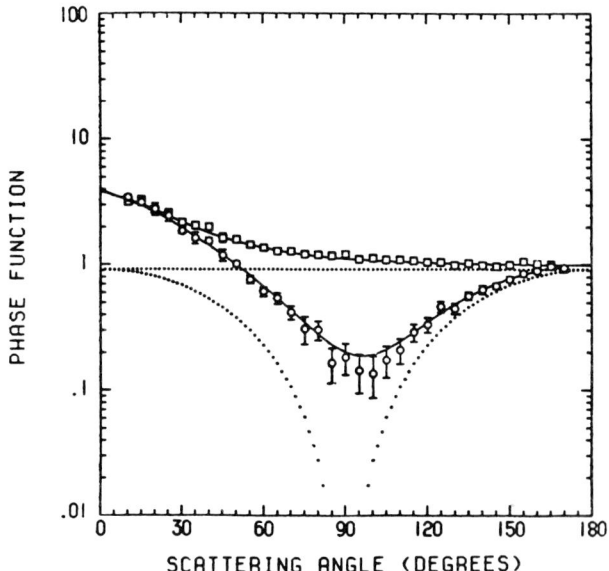

FIGURE 1. Example of polar nephelometer observations for the stratospheric aerosol layer over Poker Flat, Alaska. The data were obtained on 17 July 1979. The squares are data points for scattering in the plane perpendicular to the electric vector of incident polarized light; the circles are data points for the parallel plane. Dotted lines are relative contributions to the phase functions due to molecular scattering; solid lines represent the phase function for both aerosol and molecular scattering using the Henyey-Greenstein phase function to specify the aerosol contributions. Best-fit Henyey-Greenstein parameters are an asymmetry parameter g = 0.52 and ratio of aerosol to molecular scattering coefficient f = 0.65.

ANALYSIS OF POLAR NEPHELOMETER DATA 255

III. RESULTS

In this section a number of figures are presented showing experimental observations (with error bars) of the scattering phase functions along with the two curves for each polarization plane that represent the best-fit Henyey-Greenstein phase function for scattering by aerosol particles and air molecules. Figure 2 shows one of the five calibration curves obtained during the

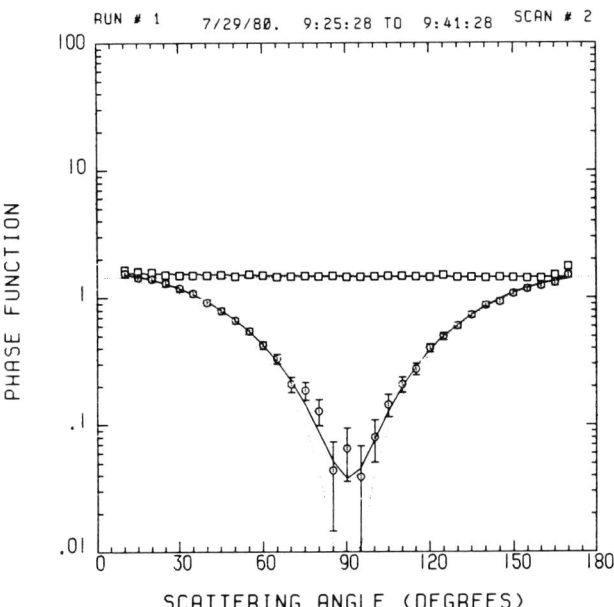

FIGURE 2. Example of phase functions observed while calibrating the polar nephelometer with filtered air during the workshop. The data were obtained on 29 July 1980. The squares are data points for scattering in the plane perpendicular to the electrical vector of incident polarized light; the circles are data points for the parallel plane. Dotted lines are relative contributions to the phase functions due to molecular scattering; solid lines represent the phase function for both aerosol and molecular scattering using the Henyey-Greenstein phase function to specify the aerosol contributions. Best-fit Henyey-Greenstein parameters gave asymmetry parameter $g = 0.34$ and a ratio of aerosol to molecular scattering coefficient $f = 0.06$. The value $f = 0.06$ indicates that the "filtered" air still has enough aerosol particles to contribute another 6% to the aerosol-free molecular scattering coefficient; the small value of $g = 0.34$ indicates the particles that are present are quite small. (Normal background aerosol particles are characterized by values of $g = 0.5$ to 0.7, whereas air molecules would have $g = 0$.)

measurements. These curves were used to calibrate the instrument through a series of observations of molecular scattering using filtered laboratory air on the first day of the workshop.

The present analysis is limited to aerosol particles generated at the workshop from 29 July to 5 August. The particle compositions included ammonium sulfate, methylene blue (MB), and soot particles, as well as mixtures of ammonium sulfate and soot and ammonium sulfate and MB (in two different concentrations, 0.3% MB and 1% MB).

1. *Ammonium Sulfate*

Figure 3 shows a typical result of the ammonium sulfate measurements (18 nephelometer scans on 29 July and five scans on 30 July). Figure 4 shows parameters S_1, S_2, g_1, and g_2 as a function of time during the instrument scan sequence on 29 July. While volume scattering parameters S_1 and S_2 (S_i values calculated separately for each polarization plane) changed from one scan to

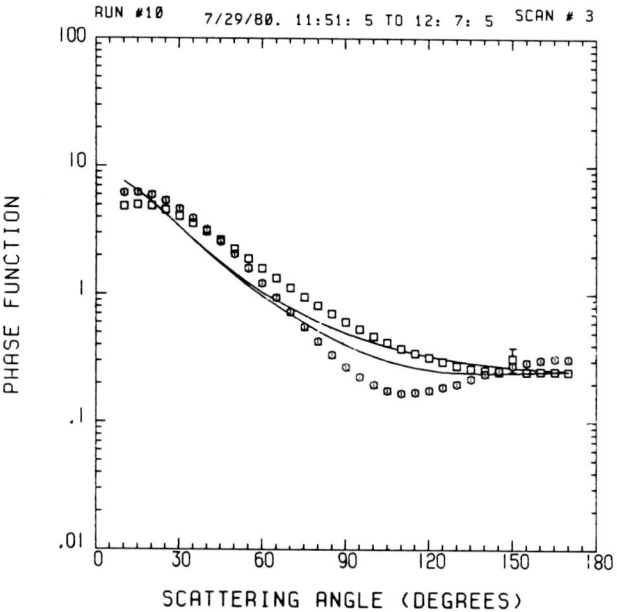

FIGURE 3. *Representative phase functions P_1 (circles) and P_2 (squares), measured for ammonium sulfate aerosol particles. As in Figs. 1 and 2, the continuous lines represent the sum of the best-fit Henyey-Greenstein phase functions for scattering by aerosol particles and the Rayleigh scattering phase functions for air molecules.*

ANALYSIS OF POLAR NEPHELOMETER DATA

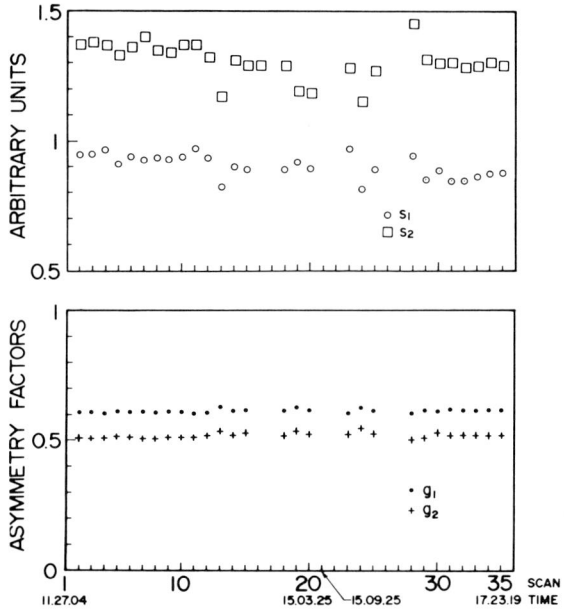

FIGURE 4. Time sequence of total scattering cross sections and asymmetry factors measured during 29 July 1980 by the polar nephelometer. (a) Values of S_1 and S_2 (each proportional to the scattering cross section for each polarization plane). (b) Asymmetry factors g_1 and g_2.

another, suggesting statistical fluctuations in particle concentration, the asymmetry parameters g_1 and g_2 appeared to be very stable. Typical values of g_1 and g_2 and the ratio f of aerosol to molecular scattering coefficient during these experiments are reported in Table 1.

2. *Methylene Blue (MB)*

The phase function data shown in Fig. 5 are representative of five scans obtained during 1 August and three scans during 4 August. The slope in the forward scattering region is steeper than it was in the previous case and the data for each of the two polarization planes are in close agreement. Both of these features are qualitatively consistent with the values calculated for g_1, g_2, and g in Table 1.

TABLE 1. *Typical Values of Aerosol to Molecular Scattering Coefficient*[a]

	g_1	g_2	g	Δg	f	Day 1980
$(NH_4)_2SO_4$	0.61	0.53	0.55	±0.01	190.00	29 July
$(NH_4)_2SO_4$	0.66	0.60	0.63	±0.007	120.00	30 July
$(NH_4)_2SO_4$ +						
0.3% MB	0.68	0.64	0.66	±0.01	50.00	30 July
$(NH_4)_2SO_4$ +						
4% MB	0.66	0.61	0.63	±0.005	80.00	30 July
Soot	0.62	0.42	0.48	±0.01	38.00	31 July
	0.61	0.41	0.48	±0.01	30.00	4 August
MB	0.74	0.70	0.72	±0.005	40.00	1 August
	0.72	0.68	0.70	±0.007	60.00	4 August
$(NH_4)_2SO_4$ +						
soot	0.63	0.54	0.58	±0.015	200.00	4 August
Ambient air	0.62	0.47	0.53	±0.03	0.55	5 August

[a] *Representative values of asymmetry parameters g_1 and g_2 for polarized light along with the asymmetry parameter g for unpolarized light and the observed standard deviation Δg of the values of g obtained for each aerosol material. The ratio f of aerosol to molecular scattering coefficients inferred from the polar nephelometer data for each material is also indicated.*

3. *Soot*

The shape of the phase function for this type of aerosol (five scans on 31 July and four on 4 August) appeared to be clearly different, as shown in Fig. 6. The significant differences observed between g values in the two polarization planes are apparent. In this case, the asymmetry factors g_1, g_2, g, and f assume the characteristic values listed in Table 1.

4. *Ammonium Sulfate and Soot*

For clarity, Figs. 7(a) and 7(b) separately compare the two phase functions for each polarization P_1 and P_2 with those of the pure ammonium sulfate particles (four cases). It was observed that phase functions for the perpendicular plane of polariztion are similar even though significant differences are exhibited for the data obtained in the parallel plane. Figure 8 shows the values assumed by g_1 and g_2 as a function of time on 4 August. The distinctive characteristic values assumed by g_1 and g_2 for each kind of aerosol generated during that day are clearly recognizable in the figure.

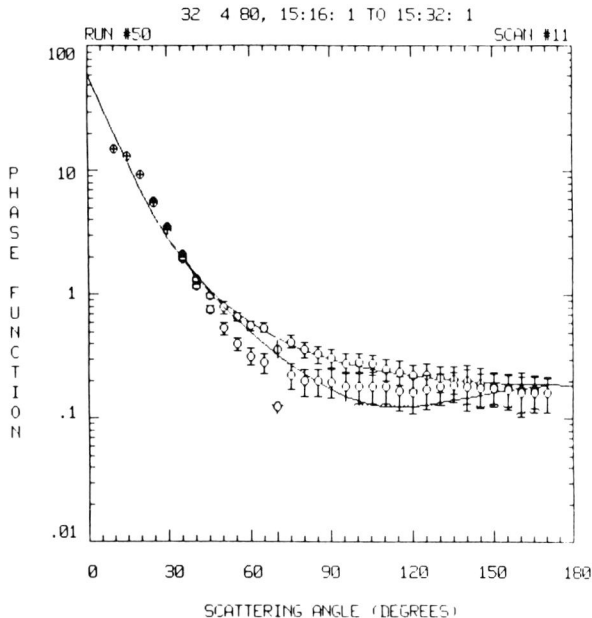

FIGURE 5. Sample phase functions P_1 and P_2 measured for methylene blue during 1 August 1980. The curves and data points have the same meaning as in Figs. 1 to 3.

5. Ammonium Sulfate and MB

Figures 9(a) and 9(b) compare the phase functions in the two polarization planes for (a) ammonium sulfate mixed with 0.3% MB and (b) pure ammonium sulfate. Both of the polarization planes exhibit small but significant differences; it is interesting to note that, for this aerosol with very small concentration of MB, the P_2 values match those for pure ammonium sulfate in the backward direction. This was also true for the case of 4% MB. This may be due to the combined effects of concentration and variations in the size distributions.

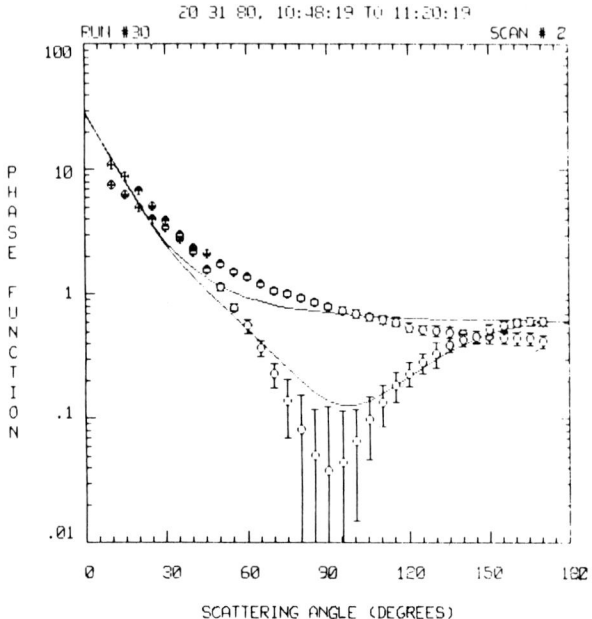

FIGURE 6. Sample phase functions P_1 and P_2 measured for soot on 31 July 1980.

6. Ambient Air

Figure 10 shows typical results for 12 measurements obtained during 4 August. Although significant variations in shape occurred during that day, the phase function curves for each polarization agree in the near-forward directions but they deviate from each other by increasingly larger amounts in the angular interval from 60° to 150°. The Henyey-Greenstein phase function fits were in good agreement with the measurements in all of the 12 different cases.

ANALYSIS OF POLAR NEPHELOMETER DATA

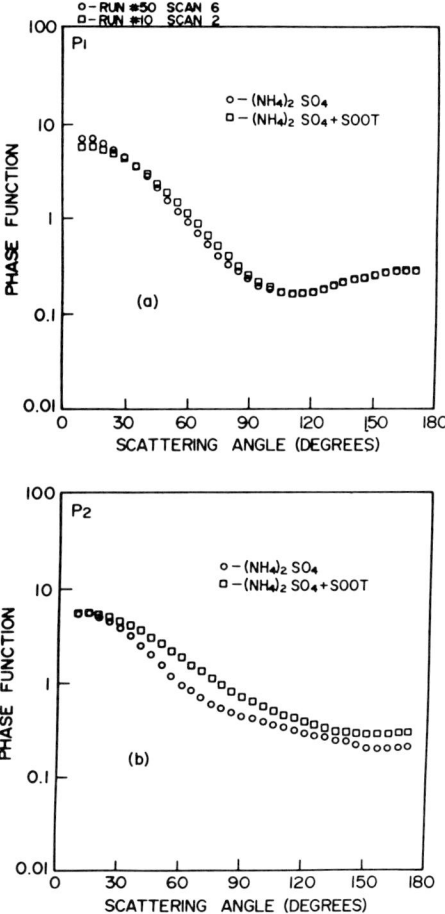

FIGURE 7. Comparison of phase functions measured for pure ammonium sulfate and those measured for ammonium sulfate and soot. (a) Comparison for the perpendicular polarization plane. (b) Comparison for the parallel polarization plane.

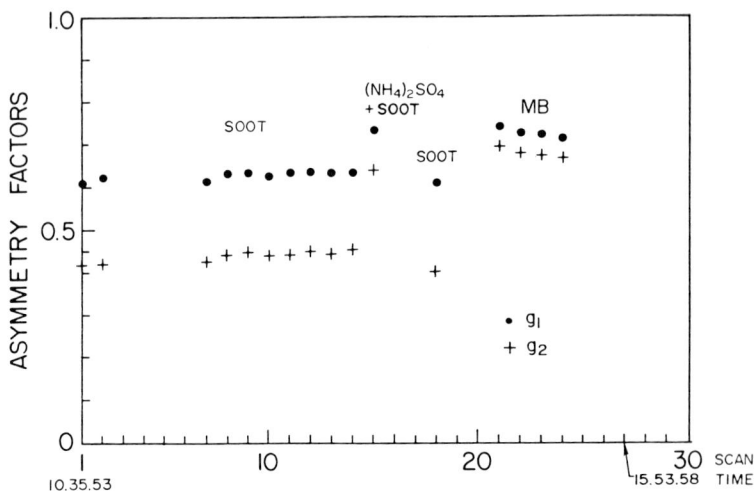

FIGURE 8. Time sequence of the measurements of the asymmetry factors g_1 and g_2 for each polarization plane on 4 August 1980. The sudden changes in the values are related to changes in the aerosol composition as indicated in the figure.

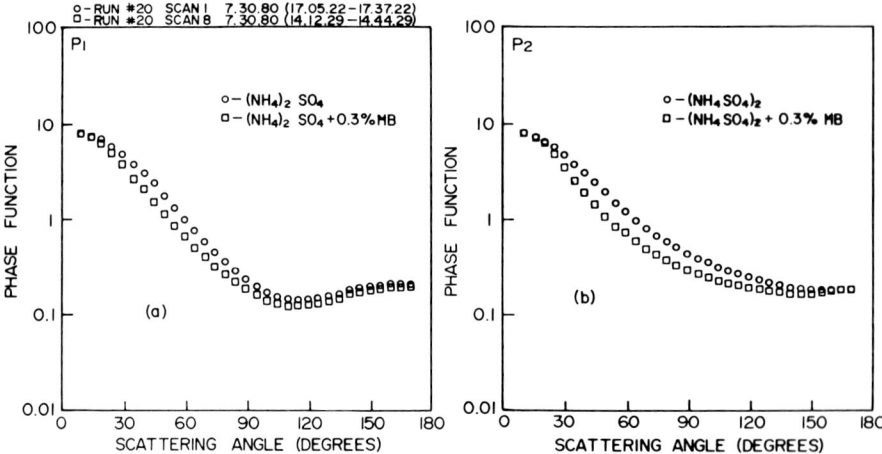

FIGURE 9. Comparison of the phase functions measured for pure ammonium sulfate and ammonium sulfate mixed with 0.3% MB. (a) Comparison for the perpendicular polarization plane. (b) Comparison for the parallel polarization plane.

ANALYSIS OF POLAR NEPHELOMETER DATA 263

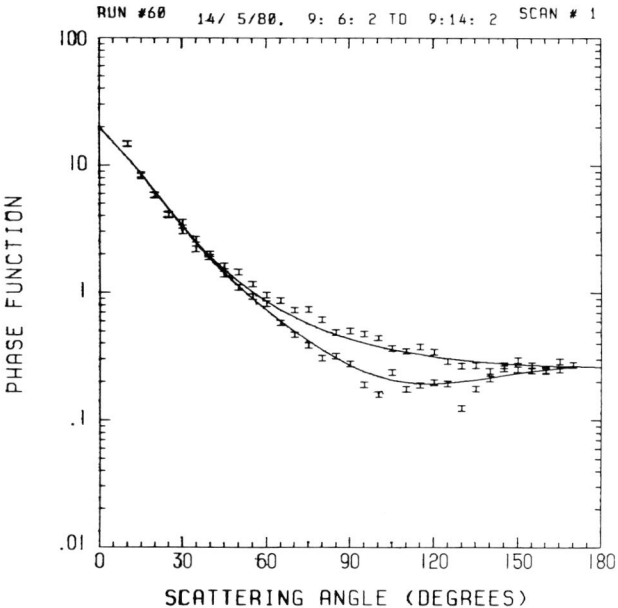

FIGURE 10. Representative phase functions P_1 and P_2 measured for ambient air during 4 August 1980.

III. DISCUSSION

It is planned to extend the present analysis to include several studies on the use of the polar scattering diagrams to obtain data on particle size and refractive index. In particular, there is interest in evaluating a procedure that is based on an empirical theory to relate the scattering due to polydispersions of nonspherical particles to the scattering calculations for a particle size distribution defined in terms of equivalent spherical particles (7). Particles with Mie size parameters ($x = 2\pi r/\lambda$) smaller than 2 or 3 are assumed to scatter in accordance with the predictions of Mie theory; the phase function for larger particles is assumed to have three components due to diffraction, reflection, and transmission. The complete expression for Pollack and Cuzzi's (7) phase function for a polydisperse aerosol will thereby be of the form

$$P(\theta) = \int_0^{X_o} \alpha(x,n)P_{MIE}(x,n,\theta)N(x)dx$$
$$+ \int_{X_0}^{\infty} \beta(x)P_D(x,\theta)N(x)dx + \gamma(n)P_R(n,\theta)$$
$$+ \delta(b)P_T(b,\theta) \qquad (7)$$

where P_{MIE}, P_D, P_R, and P_T are the polar scattering functions for the small-particle Mie component and for the diffracted, reflected, and transmitted components for larger particle sizes. The coefficients α, β, γ, and δ represent the relative scattering contributions for each component, and they must satisfy the condition $\alpha + \beta + \gamma + \delta = 1$. In this formulation, two terms (those for the Mie and the diffracted components) depend on the size distribution, the reflected component depends on the complex reflective index n, $P_T(b,\theta)$ is determined from experimental data by selecting a best-fit constant b to improve the agreement between the observations and the formula, and b is regarded to be a particle shape parameter.

In the Pollack and Cuzzi formulation (7), the polarization of the light is not taken into account. Figure 11 shows the normalized radar cross section for

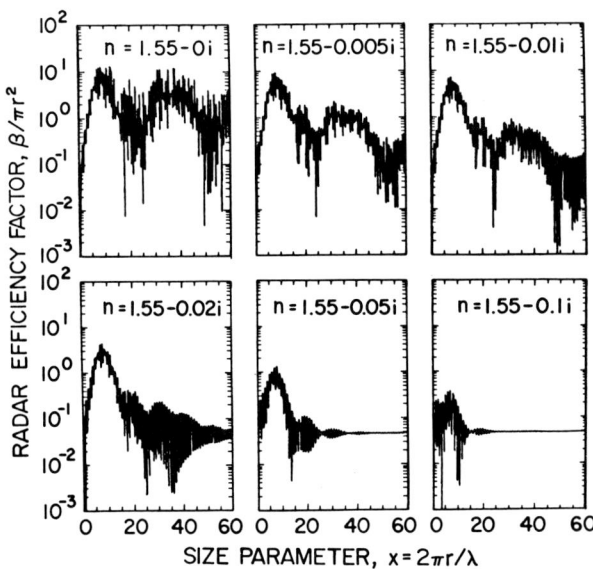

FIGURE 11. Normalized radar cross sections for spherical particles having the indicated complex refractive indices, evaluated and plotted at Mie size parameter increments $\Delta_x = 0.2$.

ANALYSIS OF POLAR NEPHELOMETER DATA

homogeneous spherical particles with $n_1 = 1.55$ and selected values of n_2. In particular, normalized cross sections are shown for a case of no absorption, i.e., $n_2 = 0$, and for cases in which $n_2 = 0.005, 0.01, 0.02, 0.05$, and 0.1, bracketing the expected region for the imaginary part of the index. Note that, for the larger values of n_2, the backscattering cross section seems to approach a constant value at the larger Mie size parameters. This constant, as might be expected, is the Fresnel reflection coefficient for a surface having the given complex refractive index. The formulas for the reflection coefficients are different for each scattering plane and, since the nephelometer used measures separately the light in the two polarization planes, it is planned to modify Pollack and Cuzzi's original formulas to include both polarizations in the curve-fitting procedures.

Hansen and Travis (8) pointed out that aerosol scattering phase functions of polydispersions of Mie scatters are very sensitive to the area-weighted mean particle radius. They also showed, however, that once the effective particle radius was established, the phase functions were extremely insensitive to the detailed shape of the size distribution functions. Figure 12 compares the phase functions of three obviously different size distributions having the same value of the effective radius. It is believed that the close agreement among the

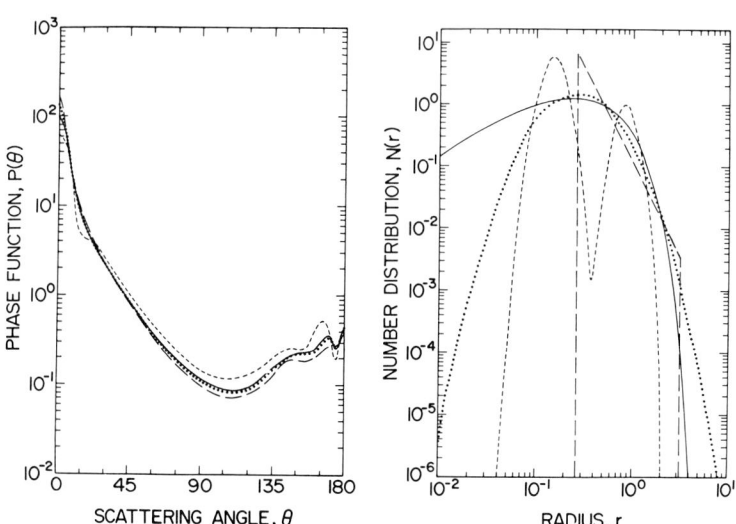

FIGURE 12. Phase function for single scattering of unpolarized incident light. Results are shown for the four size distributions illustrated on the right for which each size distribution has the same effective radius (1 μm) and the same effective variance (0.25). Calculations are for real refractive index 1.5 and wavelength 0.55 μm [after Hansen and Travis (8)].

various phase functions suggests that the mean particle radius is the major contribution to the shape of the phase function, and that the integral over particle size in Pollack and Cuzzi's formula for the diffraction component is not necessary when using their formulas to fit experimental data. It is thereby planned, for future studies, to assume that the aerosol contains only molecules, which act as Rayleigh scatters, and larger particles, for which the scattering can be characterized by the modified Pollack and Cuzzi phase function for some effective particle parameter $\bar{x}$. Then the least-square inversion will assume a phase function of the form

$$P(\theta) = \alpha P_{air}(\theta) + \beta P_D(\bar{x},\theta) + \gamma P_R(n,\theta) + \delta P_T(b,\theta) \qquad (8)$$

This approach will be studied along with the previous approach of using Mie theory and observed size distribution functions to estimate the complex refractive index of the aerosol particles [as, for example, in Grams (3)].

REFERENCES

1. Deirmendjian, D., *Electromagnetic Scattering on Spherical Polydispersions*, Elsevier, New York (1969).
2. Henyey, L. G., and Greenstein, J. L., Diffuse radiation in the galaxy, *Astrophys. J. 93,* 70 (1941).
3. Grams, G. W., Dascher, A. J., and Wyman, C. M., Laser polar nephelometer for airborne measurements of aerosol optical properties, *Opt. Eng. 14,* 85-90 (1975).
4. Van de Hulst, H. C., *Light Scattering by Small Particles*, Wiley, New York (1957).
5. Hansen, J. E., Exact and approximate solutions for multiple scattering by cloudy and hazy atmospheres, *J. Atmos. Sci. 26,* 478-487 (1969).
6. Grams, G. W., In-situ measurements of scattering phase functions of stratospheric aerosol particles in Alaska during July 1979, *Geophys. Res. Lett. 8,* 13-14 (1981).
7. Pollack, J. B., and Cuzzi, J., Scattering by nonspherical particles of size comparable to a wavelength: A new semi-empirical theory and its application to tropospheric aerosol, *J. Atmos. Sci. 37,* 868-881 (1980).
8. Hansen, J. E., and Travis, L. D., Light scattering in planetary atmospheres, *Space Sci. Rev. 16,* 527-610 (1974).

WORKSHOP MEASUREMENTS OF THE AEROSOL ABSORPTION COEFFICIENT WITH AN INTEGRATING PLATE METHOD AND AN INTEGRATING SPHERE PHOTOMETER

Jost Heintzenberg

Department of Meteorology
University of Stockholm

Two measurements of the aerosol absorption coefficient are obtained simultaneously from photometer measurements of light transmitted through a particle-coated filter and light scattered by particles resuspended as a liquid suspension placed in an integrating sphere. Data are presented of measurements with those techniques on workshop aerosols.

I. DESCRIPTION OF INSTRUMENTATION

1. *Filter Sampling*

By drawing sample air through a filter, aerosol particles are collected on the front side and within the filter. The filter has a certain size-dependent particle retention which at present is estimated to be 90% for all sizes (no particle loss corrections are applied to final results of this work). The results are related to a standard volume of 1 m^3 of air by the total volume V_s of the air that has passed through the filter and the net filter area used in the filtration. Data are presented in Table 1.

The filters have an outer diameter of 47 mm. Presently they are used in Nuclepore filter holders. The outer diameter of the exposed filter area is about 41 mm. However, because of the base support grid a significant part of the exposed area remains inactive because no air flow can pass through. For the fiber filters used in this study the net filtration area A_{net} is 8 x 10^{-4} m^2. Because of the support grid, the area reduction can be determined only rather inaccurately for a fiber filter. A ± 20% inaccuracy is estimated for A_{net}, which directly influences the final absolute results.[1]

The sampling length l is given by

$$l = V_s/A_{net} \tag{1}$$

where V_s is the sample volume and l is the length of the air column swept by the filter. With l the optical densities τ determined in the light absorption measurements on the samples can be reduced to light absorption coefficients per unit length.

2. *Sample Analysis*

For the analysis of these samples a new double-beam, single-detector integrating-sphere photometer has been developed for optical measurements on aerosol samples (3). The light measurements are completely digital (photon counting) and controlled by a desk computer. The resolution of the digital photometer gives a 10^{-5} difference between the sample and the reference signals. Taking a typical background aerosol sample (10^5-m length of air column swept by the filter) results in a detection limit of 10^{-10} m^{-1} for the light absorption coefficient of the aerosol particles at about 550-nm wavelength. This sensitivity is sufficient for the detection of elemental carbon in size-segregated samples at ground-level background locations or in stratospheric air.

---

[1]*Subsequent to the preparation of this report a precise determination was made of A_{net}, which was found to be 2.91 x 10^{-4} m^2. This new area affects the values of σ_a in Table 1, which are corrected by multiplying by a factor 2.91/8.0.*

TABLE I. Workshop Aerosol Data

Date and sample no.		Time	Aerosol	M_v^a μg m^{-3}	M_v^b μg m^{-3}	σ_s^c m^{-1}	σ_a^d m^{-1}	σ_a^e m^{-1}	M_v^f μg m^{-3}	Remarks
7/29	1	1529-1600	$(NH_4)_2SO_4$	~2.7 x 20^2	1.9 x 10^2	1.09 x 10^{-3}	3.4 x 10^{-4}	3.4 x 10^{-5}		Four large dark fibers left on filter
7/29	2	1635-1733	$(NH_4)_2SO_4$ + 1% MB	2.2 x 10^2	5.4 x 10^2	7.97 x 10^{-4}	g	4.3 x 10^{-5}		MB = methylene blue^h
7/30	1	1327-1527	$(NH_4)_2SO_4$ + 0.3% MB	5.8 x 10^1	7.0 x 10^1	2.76 x 10^{-4}	g	1.0 x 10^{-5}		
7/30	2	1636-1741	$(NH_4)_2SO_4$ + 4% MB	6.9 x 10^1	7.0 x 10^1	4.18 x 10^{-4}	g	1.0 x 10^{-4}		
7/31	1	1015-1025	Soot		9.5 x 10^1	2.41 x 10^{-4}	4.0 x 10^{-3}	3.8 x 10^{-3}	1.2 x 10^2	
7/31	2	1035-1040	Soot		9.5 x 10^1	2.41 x 10^{-4}	4.9 x 10^{-3}	3.5 x 10^{-3}	1.1 x 10^2	1 0.5-mm plate, 1 large fiber
7/31	3	1048-1113	Soot		9.5 x 10^1	2.41 x 10^{-4}	2.8 x 10^{-3}	2.9 x 10^{-3}	1.1 x 10^2	
7/31	4	1124-1126	Soot		9.5 x 10^1	2.41 x 10^{-4}	g	4.7 x 10^{-3}	1.4 x 10^2	2 2-mm fibers
7/31	5	1133-1136	Soot		9.5 x 10^1	2.41 x 10^{-4}	5.7 x 10^{-3}	4.1 x 10^{-3}	1.3 x 10^2	1 0.5-mm fiber
7/31	6	1143-1145	Soot		9.5 x 10^1	2.41 x 10^{-4}	g	4.6 x 10^{-3}	1.4 x 10^2	1 0.5-mm particle
7/31	7	1157-1205	Soot		9.5 x 10^1	2.41 x 10^{-4}	1.83 x 10^{-3}	4.1 x 10^{-3}	1.3 x 10^2	5 1-mm fibers left on filter
7/31	8	1215-?	Soot							No finishing time recorded
7/31	9	1238-1345	Soot		9.5 x 10^1	2.41 x 10^{-4}	2.1 x 10^{-3}	2.1 x 10^{-3}	2.4 x 10^2	
7/31	10	1442-1452	Soot + $(NH_4)_2SO_4$	1.0 x 10^2	1.4 x 10^2	5.39 x 10^{-4}	1.0 x 10^{-3}	1.3 x 10^{-3}	3.9 x 10^1	
7/31	11	1502-1522	Soot + $(NH_4)_2SO_4$	1.6 x 10^2	1.4 x 10^2	5.39 x 10^{-4}	g	1.0 x 10^{-3}	3.2 x 10^1	
7/31	12	1533-1640	Soot + $(NH_4)_2SO_4$	1.62 x 10^2	1.4 x 10^2	5.39 x 10^{-4}	9.5 x 10^{-4}	9.0 x 10^{-4}	3.3 x 10^1	Small sample loss before analysis
8/1	1	1130-1203	MB			3.86 x 10^{-4}	4.3 x 10^{-4}	2.1 x 10^{-3}		
8/1	2	1212-1218	MB							Filter misplaced during sampling
8/1	3	1226-1239	MB			3.86 x 10^{-4}	6.7 x 10^{-4}	3.0 x 10^{-3}		Support filter missing after sampling
8/1	4	1247-1325	MB	1.7 x 10^2		3.86 x 10^{-4}	7.9 x 10^{-4}	2.1 x 10^{-3}		
8/1	5	1435-1458	Soot		6.1 x 10^1	1.50 x 10^{-4}	1.6 x 10^{-3}	2.5 x 10^{-3}	9.0 x 10^1	1 2-mm brown fiber

TABLE I. Continued

Date and sample no.		Time	Aerosol	M_V^a μg m^{-3}	M_V^b μg m^{-3}	σ_s^c m^{-1}	σ_a^d m^{-1}	σ_a^e m^{-1}	M_V^f μg m^{-3}	Remarks
8/1	6	1505-1510	Soot		6.1×10^1	1.50×10^{-4}	2.4×10^{-3}	2.9×10^{-3}	8.9×10^1	
8/1	7	1517-1527	Soot		6.1×10^1	1.50×10^{-4}	2.1×10^{-3}	2.8×10^{-3}	9.5×10^1	
8/4	1	1053-1113	Soot							Filter misplaced during sampling
8/4	2	1140-1208	50% soot + 50% $(NH_4)_2SO_4$	2.3×10^2	2.4×10^2	1.29×10^{-3}	2.1×10^{-3}	2.5×10^{-3}	9.1×10^1	
8/4	3	1214-1219	50% soot + 50% $(NH_4)_2SO_4$	2.3×10^2	2.4×10^2	1.29×10^{-3}	7.1×10^{-3}	3.3×10^{-3}	1.0×10^2	
8/4	4	1226-1315	50% soot + 50% $(NH_4)_2SO_4$	2.3×10^2	2.4×10^2	1.29×10^{-3}	1.9×10^{-3}	2.0×10^{-3}	8.6×10^1	
8/4	5	1322-1332	50% soot + 50% $(NH_4)_2SO_4$	2.3×10^2	2.4×10^2	1.29×10^{-3}	2.0×10^{-3}	3.0×10^{-3}	9.6×10^1	
8/4	6	1337-1352	50% soot + 50% $(NH_4)_2SO_4$	2.3×10^2	2.4×10^2	1.29×10^{-3}	1.2×10^{-3}	2.9×10^{-3}	9.6×10^1	
8/5	1	1027-?								Incomplete sampling information
8/6	2	1455-1659	96% $(NH_4)_2SO_4$ + 4% soot	5.0×10^2	2.1×10^3	3.73×10^{-3}	5.1×10^{-4}	5.9×10^{-4}	1.7×10^1	

a. TSI 3200 according to size distribution given by Hindman et al. (1).
b. Integral over size distribution according to Hindman et al. (1).
c. Channel 2 of Bodhaine's integrating nephelometer for approximately same time period (2).
d. According to integrating plate method in own integrating sphere photometer.
e. In suspended form in own integrating sphere photometer.
f. According to calibration of integrating sphere photometer with Cabot Monarch 71 soot.
g. Higher transmission than blank average.
h. The solvent used to dissolve the filters can potentially affect the chemistry of methylene blue, hence the listed coefficients for MB are unreliable.

INTEGRATING SPHERE PHOTOMETER

Two independent light absorption parameters can be determined by the instrument: (a) light absorption coefficients according to the integrating plate method (4), and (b) the pure light absorption of the particles independent of the filter on which they are sampled and independent of their scattering properties.

In the second case, the supporting filter is dissolved, the sample is dispersed by ultrasound, and the particles are analyzed in a liquid suspension in an integrating sphere which brings all the light scattered by the soot particles to the optical detector. Figure 1 shows a schematic drawing of the soot photometer plus the data recording equipment.

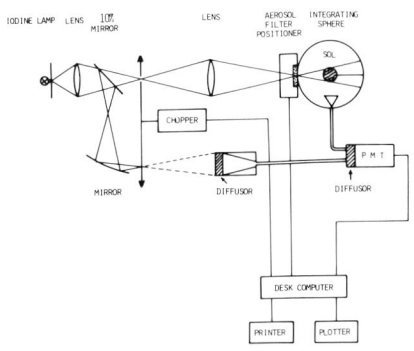

FIGURE 1. Double-beam single-detector integrating-sphere photometer.

3. Calibration in Terms of Equivalent Amounts of Soot

Inferring soot contents from the results of light absorption measurements requires a calibration of the instrument with known amounts of soot plus intercomparisons with measurements which are specific to graphitic carbon, e.g., Raman scattering. A calibration with artificial soot has been performed. A channel-type soot (Monarch M71 of the Cabot Corp.) was chosen which exhibits a narrow size range of about 80 Å radius and consists of about 95% carbon plus 5% volatile material according to the manufacturer. Its specific light absorption coefficient of 9.68 m^2/g is very similar to values found for atmospheric soot aerosols (5). With a microbalance, samples from 100- to 1000-μg of soot were weighed and volume-diluted in suspended form down to about 500 ng.

Figure 2 gives the resulting calibration curve relating the mass of soot in the suspension to the optical density relative to that of the sample flask filled with a dissolved blank filter.

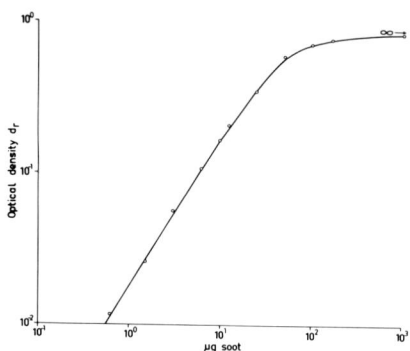

FIGURE 2. *Calibration curve of the soot photometer.*

From this calibration a lower detection limit of about 500 ng and an upper limit of about 100 μg were determined, because of the highly nonlinear response to large amounts of soot. The signal for a sample flask filled completely with soot is marked with an infinity symbol in Fig. 2 to show that the optical method can only see the shell of such a large volume of soot. From a small number of tests, 10% is given as accuracy in terms of M71 soot within the usable mass range.

II. EVALUATION OF PHOTOMETER MEASUREMENTS

The photometer results, evaluated in terms of light absorption coefficients of an aerosol sample, make use of the Bouguer-Lambert law

$$I_\lambda = I_{o\lambda}\, e^{\tau\lambda} \qquad (2)$$

where Eq. (2) is defined strictly for monochromatic radiation in singly scattering media. A rough approximation of Eq. (2) applies to the geometry of the integrating sphere. However, Zuev et al. (6) have demonstrated its validity for optical densities τ up to 30 in dense fogs and smoke. In the present study, Eq. (2) is applied to the absorption of soot and atmospheric aerosols at about 550 nm wavelength. These substances exhibit quite nonselective absorption features in the visible (7), thus justifying the use of Eq. (2).

Several components contribute to the reduction of the incident intensity I_o to the intensity I when passing through the sample

$$\tau = \tau_c + \tau_s + \tau_f + \tau_p \tag{3}$$

where
- τ_c = optical density of the sample flask for the hydrosol
- τ_s = optical density of the solvent
- τ_f = optical density of th blank filter
- τ_p = optical density of the particulates.

In blank measurements, unexposed filters are dissolved in the solvent for the latex fiber filter and measured in the photometer. Thus the sum $\tau_c + \tau_s + \tau_f$ is determined. In separate runs with empty flasks τ_c is determined for each flask used in the data evaluation. As a final result the light absorption coefficient is calculated with

$$\sigma_a = \tau_p/l \tag{4}$$

The comparison of the optical density τ_p of the particles in the suspended state with the calibration curve in Fig. 2 yields absorption equivalent amounts of soot in the sample which are converted to $\mu g\ m^{-3}$ with the sample volume V_s.

REFERENCES

1. Hindman, E., Horn, R. D., and Finnegan, W. G., Generation and characterization of aerosol particles, *Light Absorption by Aerosol Particles* (H. Gerber and E. Hindman, eds.), Spectrum Press, Hampton, Virginia (1982).
2. Bodhaine, B. A., The GMCC four-wavelength nephelometer, *Light Absorption by Aerosol Particles* (H. Gerber and E. Hindman, eds.), Spectrum Press, Hampton, Virginia (1982).
3. Heintzenberg, J., Measurement of light absorption and elemental carbon in atmospheric aerosol samples from remote locations. *Proceedings of the International Symposium Particulate Carbon: Atmospheric Life Cycle*, General Motors Research Laboratrories, Warren, Michigan (1980).
4. Lin, C.-I., Baker, M., and Charlson, R. J., Absorption coefficient of atmospheric aerosol: A method for measurement, *Appl. Opt. 12*, 1356-1363 (1973).
5. Donoian, H. C., and Medalia, A. I., Scattering and absorption of light by carbon black *J. Paint Techn. 39*, 716-727 (1967).
6. Zuev, V. E., Kabanov, M. V., and Sauelev, B. A., Propagation of laser beams in scattering media, *Appl. Opt. 8*, 137-141 (1969).
7. Fischer, K., Measurement of the absorption of visible radiation by aerosol particles, *Beitr. Phys. Atmos. 43*, 244-254 (1970).

SPECTROPHONE MEASUREMENTS OF DIESEL VEHICLE PARTICULATE MATERIAL

S. M. Japar
J. Moore
D. K. Killinger
A. C. Szkarlat

Engineering and Research Staff
Ford Motor Company
Dearborn, Michigan

The design of a spectrophone for the measurement of visible light absorption by airborne carbonaceous particulate is briefly discussed. This has been used successfully to determine the light absorption coefficients of airborne particulate exhaust material from a number of diesel vehicles. A summary of the results from this laboratory is presented.

I. INTRODUCTION

The study of smokes from combustion processes has been hindered by the difficulties inherent in measuring the optical properties of airborne particulate material. However, in recent years the photoacoustic effect (1,2) with its high sensitivity has been used to study the light absorption properties of airborne aerosols (3,4) produced from a number of combustion processes (3,5-11). In the course of these investigations it has been shown that the absorption and photoacoustic spectra of carbonaceous particulates from a propane flame are identical (6) in the wavelength region near 600 nm, that precise optical measurements of carbonaceous particulates from hydrocarbon flames can be made both in the visible and the infrared (IR) regions of the spectrum (6,11), and that measurements on diesel vehicle exhaust can be carried out (8,10).

II. DESCRIPTION OF THE SPECTROPHONE

The light source, a 1.0-W argon ion laser operated in the visible ($\lambda = 0.5145$ µm), was square-wave modulated at about 4 kHz by a variable speed chopper. The photoacoustic cell was a brass cylinder 9.5 cm in length by 5 cm inside diameter. The cell has a longitudinal acoustic resonance mode near 4 kHz (room temperature, atmospheric pressure). The high sensitivity microphone (B & K model no. 4144, -47.6 dB re 1 V/µbar) was placed in the center of the cylinder wall, flush with the cavity wall. The apertures for the laser beam were close to the wall opposite the microphone and the sample was pulled into the cell through one end plate with a vacuum pump. To ensure good mixing in the cell the sample was removed from the cell through the laser apertures in the end plates. The cell sat in a box lined with Conoflex, a commercial multilayered acoustic shielding material. This box sat within a second box similarly lined. Glass windows were placed only on the outer box in order to minimize interferences from minute absorptions of the laser radiation by the window materials. The sample line before and after the box was equipped with mufflers to eliminate noise originating in the dilution tube or the sampling pump. The mufflers were 50-cm cylinders (25-cm outside diameter) filled with the acoustic insulation. The sample line within the mufflers was made from 0.076-mm (0.003 in) FEP Teflon (30-mm outside diameter). This configuration allowed noise in the sample line to escape into the mufflers. The acoustic insulation decreased the broad-band noise at the microphone by a factor of 10^5 to 10^6 during normal operation, thus preventing the overloading of the microphone. The microphone signal was processed by a preamplifier (Gen Rad model no. 1560-P42 operated with a Gen Rad model no. 1560-P62 power supply), and a narrow prefilter was used to reject background noise. The output signals from the microphone/prefilter were detected by a lock-in amplifier and displayed on a strip-chart recorder.

The instrument (spectrophone) response S is proportional to the amount of light absorbed (11) at a given sensitivity

$$S = RW [1 - \exp(-\sigma_a x)] \tag{1}$$

where R is the cell response (dependent upon cell design and modulation frequency), W is the incident optical power, σ_a is the light absorption coefficient, and x is the pathlength. For low-light absorption (a few percent or less) Eq. (1) reduces to

$$S = RW\sigma_a x \tag{2}$$

If the mass concentration, M_v in g m^{-3}, of the light absorbing material is known, then the mass-specific absorption coefficient, $B_a(\sigma_a/M_v$, m^2 g$^{-1})$, of the material can be determined

$$B_a = S(RW\,M_v\,x)^{-1} \qquad (3)$$

Conversely, if the mass-specific absorption coefficient is known, then the concentration of the aerosol can be determined directly.

The cell response factor, R in Eq. (1), was determined using NO$_2$ (32 parts per million in N$_2$). This calibration can be used with Eq. (1) to calculate either the light absorption coefficient for the diesel vehicle exhaust particulates, or, if the latter is known, to determine the particulate mass loading in the exhaust plume.

III. SUMMARY OF RESULTS

The initial studies of airborne particulate material involved propane flames (6). Using a tunable dye laser, it was demonstrated that the photoacoustic and absorption spectra of the carbonaceous particulates formed in such flames is identical in the wavelength region near 600 nm. It was also shown that light scattering was a negligible contribution to total light attenuation under the experimental conditions employed.

Subsequent studies of particulate material have centered on the particulate exhaust emissions from diesel vehicles. With the use of an argon ion laser ($\sim$ 1.0 W at 514.5 nm), the absorption coefficients for the airborne exhaust particulate material have been determined for a number of diesel vehicles. For the total mass of the particulate emissions (predominantly "elemental" carbon with a nonlight-absorbing adsorbed organic component), the absorption coefficients for the various vehicles range from 2.4 to 7.6 m^2 g^{-1}, (fleet-averaged value is 5.0 m^2 g^{-1}), while for the light-absorbing "elemental" carbon fraction of the particulate mass, the absorption coefficients range from 8.2 to 9.4 m^2 g^{-1} (fleet-averaged value is 8.3 m^2 g^{-1}) (8,12).

The spectrophone has also been compared to the integrating plate method for determining aerosol light absorption (13), using diesel vehicle exhaust particulates (9). An excellent correlation between the two techniques was found. However, the regression coefficient was found to depend on the orientation of the Teflon-backed membrane filter used for the integrating plate analysis. With the collected particulate material between the filter backing and the integrating plate (IP), the IP response is 1.85 times that for the filter reversed. In both cases the response ratio of the IP method to the photoacoustic method is greater than 1.0, i.e., 2.43 and 1.30.

ACKNOWLEDGMENTS

The authors would like to thank R. W. Terhune for many helpful discussions and W. R. Pierson for his interest and support.

REFERENCES

1. Pao, Y.-H., ed., *Optoacoustic Spectroscopy and Detection*, Academic Press, New York (1977).
2. Rosencwaig, A., *Photoacoustics and Photoacoustic Spectroscopy*, Wiley & Sons, New York (1980).
3. Terhune, R. W., and Anderson, J. E., Spectrophone measurements of the absorption of visible light by aerosols in the atmosphere, *Opt. Lett 1*, 70-72 (1977).
4. Bruce, C. W., and Pinnick, R. G., In situ measurements of aerosol absorption with a resonant cw laser spectrophone, *Appl. Opt. 16*, 1762-1765 (1977).
5. Truex, T. J., and Anderson, J. E., Mass monitoring of carbonaceous aerosols with a spectrophone, *Atmos. Environ. 13*, 507-509 (1979).
6. Japar, S. M., and Killinger, D. K., Photoacoustic and absorption spectrum of airborne carbon particulate using a tunable dye laser, *Chem. Phys. Lett. 66*, 207-210 (1979).
7. Killinger, D. K., Moore, J., and Japar, S. M., The use of photoacoustic spectroscopy to characterize and monitor soot in combustion processes, *ACS Symposium Series, 134*, 457-462 (1980).
8. Japar, S. M., and Szkarlat, A. C., Measurement of diesel vehicle exhaust particulate using photoacoustic spectroscopy, *Comb. Sci. Tech. 24*, 215-219 (1981).
9. Szkarlat, A. C., and Japar, S. M., Light absorption by airborne aerosols: Comparison of integrating plate and spectrophone techniques, *Appl. Opt. 20*, 1151-1155 (1981).
10. Faxvog, F. R., and Roessler, D. M., Optoacoustic measurements of diesel particulate emissions, *J. Appl. Phys. 50*, 7880-7882 (1979).
11. Roessler, D. M., and Faxvog, F. R., Optoacoustic measurement of optical absorption in acetylene smoke, *J. Opt. Soc. Amer. 69*, 1669-1704 (1979).
12. Japar, S. M., and Szkarlat, A. C., Real-time measurement of diesel vehicle exhaust particulate using photoacoustic spectroscopy and total light extinction, *Society of Automotive Engineers Paper No. 811184* (1981).
13. Lin, C.-I., Baker, M. B., and Charlson, R. J., Absorption coefficient of atmospheric aerosol: A method for measurement, *Appl. Opt. 12*, 1356-1363 (1973).

DIFFUSE REFLECTANCE AND TRANSMISSION MEASUREMENTS OF AEROSOL ABSORPTION: REPORT ON RESULTS OF FIRST INTERNATIONAL WORKSHOP ON LIGHT ABSORPTION BY AEROSOLS[1]

E. M. Patterson
B. T. Marshall

School of Geophysical Sciences
Georgia Institute of Technology
Atlanta, Georgia

Measurements of aerosol absorption using bulk-sample techniques were made as part of the First International Workshop on Light Absorption by Aerosols. Both diffuse reflectance and filter transmission measurements were made on ammonium sulfate, methylene blue, soot, and crustal aerosols. The data suggest systematic differences between these two methods. They also indicate that the absorption measured using the filter transmission technique will depend on filter composition and orientation; there will also be a possible dependence on the nature of the absorbing aerosol.

[1]*This work was supported in part by NSF Grant ATM-8005356.*

I. INTRODUCTION

The Georgia Institute of Technology measurement program at the First International Workshop on Light Absorption by Aerosols held at Fort Collins (1980) included a series of filter collections of the aerosols for subsequent laboratory analysis by two techniques. The diffuse reflectance techniques are described by Lindberg and Laude (1) and Patterson et al. (2). The integrating plate method is described by Lin et al. (3) and the laser transmission method is reported in Rosen et al. (4).

The results of these measurements are reported in terms of the volume absorption coefficient for the aerosol suspension, σ_a, and of absorption parameters characteristic of the aerosol material itself, k, B_a, and n_2. The volume absorption coefficient, σ_a, is defined analogously to the extinction coefficient σ_e, and is the total energy removed by absorption from the incident beam per unit path length by the individual aerosol particles. For the aerosol (particles and suspending medium)

$$\sigma_a + \sigma_s = \sigma_e \tag{1}$$

and the single scattering albedo is given by the relation

$$\tilde{\omega} = \frac{\sigma_s}{\sigma_s + \sigma_a} \tag{2}$$

The basic parameter describing the absorption of light in a bulk material is the absorption coefficient k which is defined by the equation

$$\frac{I}{I_o} = e^{-kx} \tag{3}$$

with I_o equal to the incident intensity, I the emerging intensity, and x the path length in the material. The definition assumes that scattering processes may be neglected. The mass absorption coefficient or specific absorption coefficient is defined by the equation

$$B_a = \frac{\delta_A}{M} \tag{4}$$

for transmission spectroscopy, with M the mass of material in the beam, A the cross sectional area of the beam subtended by the absorbing material, and δ the measured optical depth. The definition in Eq. (4) is equivalent to

$$B_a = \frac{k}{\varrho} \tag{5}$$

or to

$$\frac{\sigma_a}{M_v} \tag{6}$$

where M_v is the mass concentration of the aerosol suspension. The mass absorption coefficient is a directly measurable quantity that is independent of the density σ of the absorbing material.

Another widely used absorption parameter for a bulk material is the imaginary part of the complex index of refraction n_2 which is related to k by the equation

$$n_2 = \frac{k\lambda}{4\pi} \tag{7}$$

with λ the wavelength of the incident light; n_2, for example, is the absorption parameter that is used directly in Mie calculations.

II. DIFFUSE REFLECTANCE MEASUREMENTS

Diffuse reflectance measurements were made on representative samples of each of the separate aerosol types generated during the absorption workshop, as well as on bulk material that was representative of the generated aerosol. The standard technique was the measurement of the total diffuse reflection, with analysis using the Kubelka-Munk theory described by Kortum (5).

The samples were collected on Delbag polystyrene filters which were then dissolved in organic solvents (benzene or toluene). The aerosol was separated from the dissolved filter material by sedimentation. Following repeated washing with solvent, the excess solvent was removed by evaporation to obtain the pure aerosol material. In general, the dilution method described by Patterson et al. (2) was used, in which the total diffuse reflectance of a mixture of aerosol and white standard was measured using an integrating sphere. The absorption properties of the aerosol are then determined from the known reflectance of the standard, the previously measured scattering properties of the standard, and the known concentration of aerosol in the dilution. This dilution method is used for most ambient aerosols because of the relatively high absorption of the aerosol for which direct measurement of an undiluted sample is impractical. Because of the extremely low absorption of pure ammonium sulfate, it was possible to determine its absorption directly, following procedures described in Ref. 2.

The Kubelka-Munk scattering and absorption coefficients were determined from reflectance measurements of two thicknesses of the ammonium sulfate. The measured packing density of the ammonium sulfate was used to determine a mass absorption coefficient, which was then converted to a bulk absorption coefficient k and an imaginary index of refraction n_2 using the measured bulk density of ammonium sulfate.

A listing of the analyses made using these diffuse reflectance techniques, together with the values of B_a and n_2 determined for these samples for $\lambda = 633$ nm, is given in Table I.

TABLE 1. Diffuse Reflectance Measurements

Sample type	Time of sample	Sample ID	$B_a(m^2/g)$ $\lambda = 633$ nm	$k\ (cm^{-1})$ $\lambda = 633$ nm	n_2 $\lambda = 633$
$(NH_4)_2SO_4$	7/29/80 14:41-16:09	ASA	1.8×10^{-4}	3.2	1.6×10^{-5}
$(NH_4)_2SO_4$	Bulk sample	ASB	8×10^{-7}	1.4×10^{-2}	7×10^{-8}
Methylene blue	8/4/80 15:04-15:44	MBA	3.23	4.2×10^4	2.1×10^{-1}
Methylene blue*	Bulk sample	MBB	20.3	2.6×10^5	1.3×10^0
Methylene blue + salt (1%)	7/29/80 16:26-17:30	MBC	3.47×10^{-2}	6.1×10^2	3.07×10^{-3}
Soot	7/31/80 10:13-11:12	SB	2.49	5.0×10^4	2.5×10^{-1}
Soot	7/31/80 11:38-13:46	SA	5.24	1.05×10^5	5.3×10^{-1}
Soot	7/31/80 10:37-11:28	SC	3.37	6.7×10^4	3.4×10^{-1}
$(NH_4)_2SO_4$ + soot	7/31/80 15:28-15:49	SD	0.62	1.10×10^4	5.53×10^{-2}
$(NH_4)_2SO_4$ + soot	8/6/80 14:57-17:00	SE	0.143	2.53×10^3	1.28×10^{-2}
Arizona road dust	8/6/80 10:42-13:33	AA	0.0021	4.2×10^1	2.1×10^{-4}
Arizona road dust	Bulk sample	AB	0.0048	9.6×10^1	4.8×10^{-4}

*Aqueous solution.

III. AMMONIUM SULFATE MEASUREMENTS

Measured B_a at $\lambda = 633$ nm for the ammonium sulfate aerosol sample was $\sim 2 \times 10^{-4}$ m²/g, with a corresponding n_2 of 1.6×10^{-5}. These values, although quite low, are considerably higher than the values of n_2 for single crystals of ammonium sulfate ($< 10^{-7}$ throughout the visible wavelength range) reported by Toon et al. (6). It is possible that some trace contamination could have been introduced into the ammonium sulfate material during the sample preparation procedures, or during the aerosol generation or transport procedures. To compare these measurements with the literature values of n_2 for ammonium sulfate, the direct Kubelka-Munk measurements described were made on a highly purified ammonium sulfate (Fisher-certified standard). For this sample the average B_a value between 500 and 700 nm was $\sim 7 \times 10^{-8}$ m^2/g. It is estimated that the uncertainty in these direct measurements of B_a and n_2 for the ammonium sulfate is approximately a factor of two. This value of n_2 for the bulk material is in agreement with the Toon et al. data.

IV. METHYLENE BLUE MEASUREMENTS

Diffuse reflectance measurements were made on two methylene blue (MB) aerosol samples, one consisting of methylene blue only and the other consisting of 1% by weight of methylene blue in ammonium sulfate. A density of 1.3 was used to convert mass absorption coefficients to imaginary index of refraction values. The measured absorption values for $\lambda = 633$ nm are shown in Table 1. In addition, the mass absorption values as a function of wavelength are shown in Fig. 1.

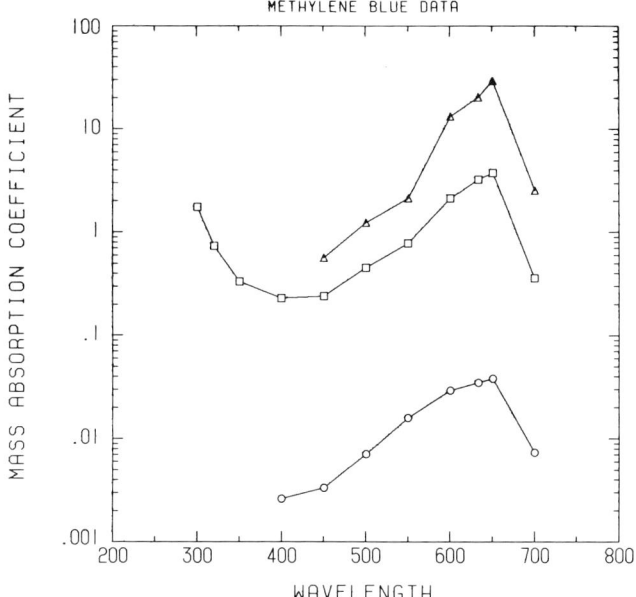

FIGURE 1. Mass absorption coefficients for methylene blue (MB) aerosol (☐), aqueous solution of MB (∆), and 1% mixture of MB and $(NH_4)_2SO_4$ (○).

The two sets of the MB diffuse reflectance measurements (pure MB and $(NH_4)_2SO_4$ + 1% MB) appear to be quite consistent. At λ = 633 nm, for example, the 1% MB mixture has a mass absorption that is 1.08% of the mass absorption for the pure MB solution. This consistency over approximately two orders of magnitude of absorption suggests that even though the MB is highly absorbing, saturation effects discussed by Jennings et al. (7) are not significant for these measurements.

The absorption of a dilute aqueous solution of methylene blue was also measured using transmission spectroscopy. The transmission measurements were on the same lot of MB used to generate the aerosol. Absorption values for these measurements are also shown in Table 1 and Fig. 1.

The absorption properties of the pure methylene blue in an aqueous solution appear to be quite different from the absorption properties determined for the solid MB. The absorption of the aqueous solution is considerably higher throughout the region of measurement. The absorption at 650 nm is approximately a factor of eight higher for the aqueous solution than for the powder. In addition, the relative absorption determined for the two forms varies with wavelength.

The measured aqueous-solution MB absorption is consistent with values of MB singlet absorption presented by Ohno et al. (8). The peak value for the measured solid MB absorption is consistent with that measured for excited triplet absorption, although the wavelength dependence is more characteristic of the singlet absorption. Similar differences between transmission and reflectance measurements for the other organic dyes have been discussed by Kortum (5) and have been attributed to chemical differences between the dyes in solution and in powdered form or when adsorbed on some other matrix.

V. CARBONACEOUS AEROSOLS

Diffuse reflectance carbonaceous aerosol data are shown in Figs. 2 and 3 as well as in Table 1. The values of B_a at 633 nm for pure soot range from 2.5 to 5.2; the soot and salt mixtures have lower values as expected. The wavelength dependent B_a values from the five carbonaceous aerosol samples are shown in Fig. 2; each sample appears to have the same wavelength dependence. Values of n_2 determined using an assumed density of $2g/cm^3$ for samples SA (pure soot) and SE [$(NH_4)_2SO_4$ + soot mixture] are shown in Fig. 3. Each of these samples shows the approximately constant value of n_2 with the wavelength that is associated with elemental carbon absorption. Values for B_a for pure soot are somewhat lower than those measured for "graphitic" carbon (9). There is no apparent increase of n_2 with decreasing wavelength that is characteristic of the organic hydrocarbon fraction (10), although the data are not sufficient to rule out the presence of small amounts of organic material.

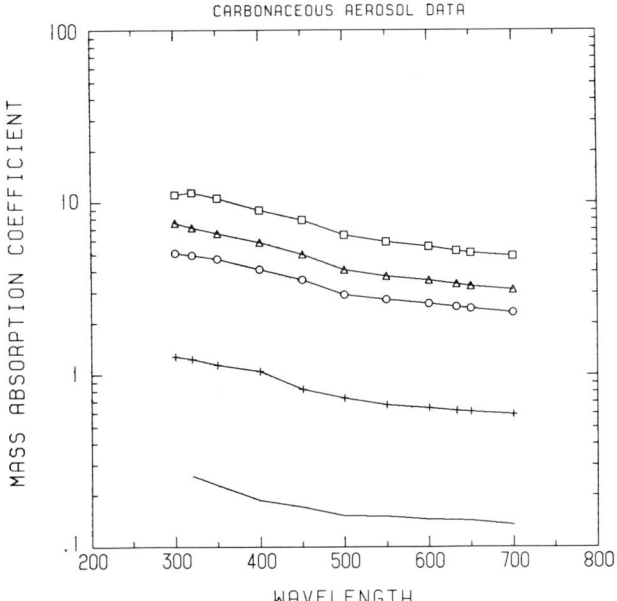

FIGURE 2. Mass absorption coefficients for three soot samples (SA, □; SB, ○; SC, △) and two mixtures of soot and ammonium sulfate (SD, +; SE, −).

VI. ARIZONA ROAD DUST

Two diffuse reflectance measurements of this silicate material were made; one for the aerosol sample generated on 6 August and one for a bulk sample of the Arizona road dust from which the larger-size fractions had been removed by sedimentation. The values measured for both the aerosol fraction of the Arizona road dust (AA) and the small-size fraction of the bulk material (AB) are shown in Figs. 4 and 5. Each of these samples has the wavelength dependence for absorption that is expected for a crustal silicate material. The absolute values are approximately an order of magnitude less than those measured for crustal aerosol far from the source regions for the aerosol, but are generally consistent with data of Lindberg and Gillespie (11) for their larger size fractions. Due to questions concerning the calibration of the particle counters, no reliable size distribution data are available.

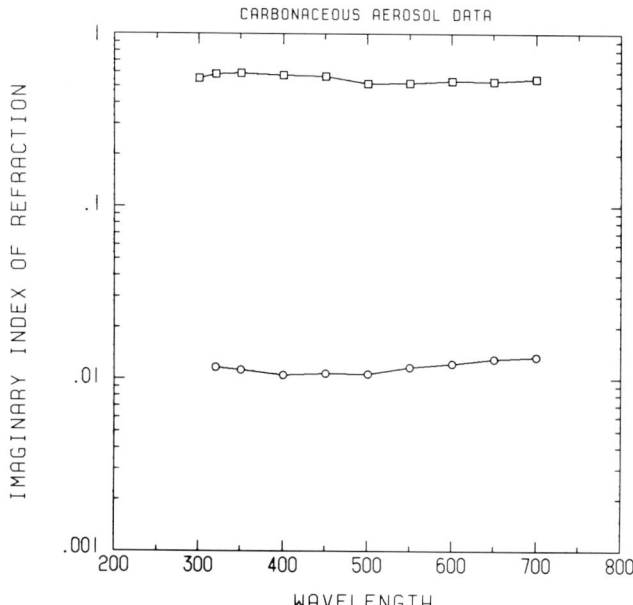

FIGURE 3. Values of n_2 for soot sample SA (□) and soot + salt sample SE (○).

VII. TRANSMISSION MEASUREMENTS

Transmission measurements of aerosol absorption were designed to determine σ_a and B_a for comparison with other measured values, and to assess the effect of differing filter material and filter orientation on the measured response for the experimental geometry. In these procedures the transmission of a filter is measured with and without aerosol loading. An optical depth is calculated from these two transmission measurements, and is attributed to the aerosol absorption only. The volume absorption coefficient σ_a is calculated from the measured optical depth and the optical path length that is determined from the filter area and air volume sampled. If the mass of material on the filter is known, B_a may be calculated as described in the introduction. Both 25-mm and 47-mm filters were used in the transmission measurements. For the filter media comparisons, a filter holder was used that allowed collection of simultaneous samples with a common inlet line from a sampling port on the aerosol transport tube. The transmission was measured at $\lambda = 633$ nm. The experimental geometry included an HeNe laser, the filter whose absorption is to be measured, an acrylic plastic diffuser, and a photomultiplier detector.

VIII. EFFECT OF DIFFERING FILTER MATERIAL AND ORIENTATION

In the original discussion of the integrating plate method, Nuclepore polycarbonate filters were used to collect the particles, and the experimental geometry was such that the particles (collected on the surface of the Nuclepore filters) were away from the light source and were between the filter and the opal glass diffuser. The geometry described by Rosen et al. (4) was such that the particles were collected on quartz fiber filters (a depth rather than a surface filter), and the filter was oriented so that the particles were on the source side of the filter. Sadler et al. (12) discussed the relative response of the two experimental setups. Szkarlat and Japar (9) discussed the effects of filter orientation on measurements for Teflon membrane filters.

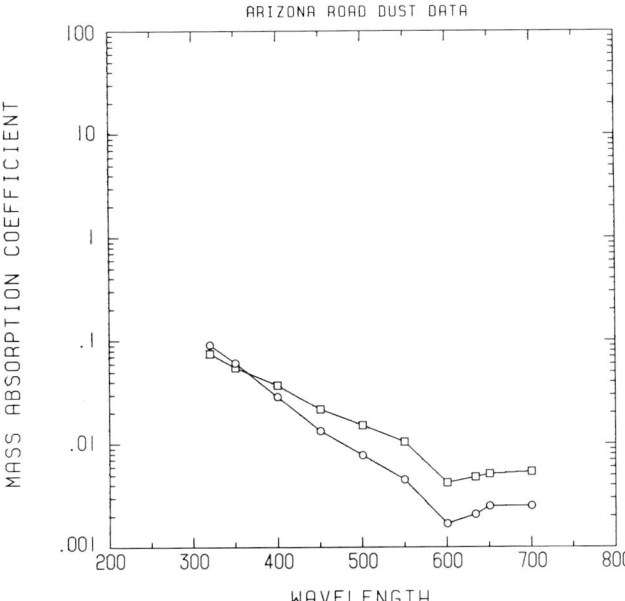

FIGURE 4. Mass absorption coefficients for Arizona road dust aerosol (○) and small-size fraction of bulk material (□).

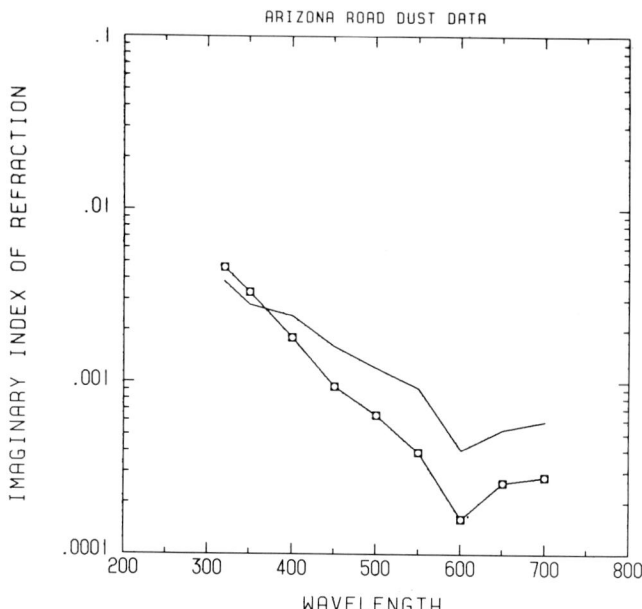

FIGURE 5. Values of n_2 for Arizona road dust aerosol (□) and small-size fraction of bulk material (−).

The effects of filter orientation and media were measured for three types of filters, 0.4-μm-pore-size Nuclepore polycarbonate filters, Gelman glass fiber filters, and Whatman 41 cellulose filters. The Nuclepore filters are surface collectors, while the Whatman and fiberglass filters are fibrous depth filters.

The effects of differing orientation for the three types of filters measured are shown in Figs. 6 to 8. There was an orientation effect in each case, with the measured optical depths consistently higher when the particles were on the side away from the incident beam (toward the detector) than when the particles were facing the beam. The measured values of the slope of the best fit line describing the data are 1.23 for the Nuclepore filters, 1.14 for the glass fiber filters, and 1.10 for the Whatman filters. Although there do not appear to be any major differences with respect to orientation effects among the filter materials tested, the differences that are seen are consistent with what might be expected on the basis of differing penetration depths in the filters. The Nuclepore filter, the suface filter, has the highest measured relative response ratio, while the glass fiber and Whatman filters, both depth filters, have lower relative response ratios. The penetration depth in the Whatman filter is greater than in the glass fiber filter; and the measured relative response for the Whatman is, in turn, less than that for the glass fiber filter.

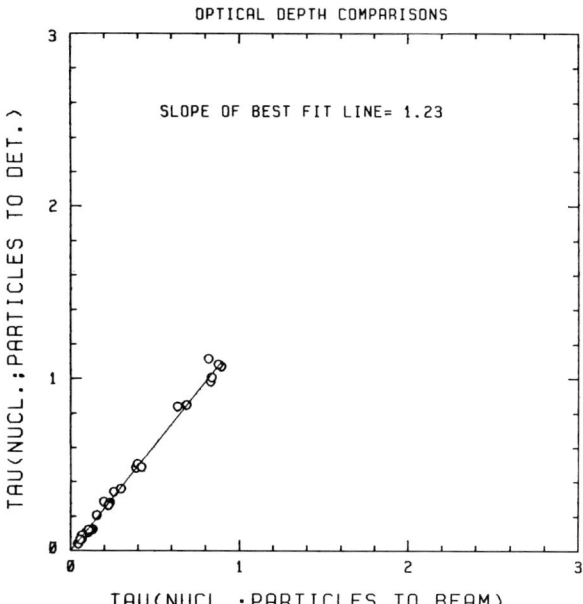

FIGURE 6. Effect of differing filter orientation on measured absorption response of Nuclepore filters.

By comparison, Szkarlat and Japar (9) have found a relative response ratio of 1.85 for Zefluor membrane filters, which are also surface filters. The comparison of the measured optical depths for two sets of filter samples (Whatman vs. Nuclepore and glass fiber vs. Whatman), which included the various types of aerosol generated during the absorption workshop, are shown in Figs. 9 and 10. The data for the Whatman and Nuclepore filters are shown in Fig. 9. Each of the measurements was made with the filters oriented so that the particles faced the incident beam. Although the data do not cover a wide range of optical depths, the fitted slope is 1.50. This figure also includes some data determined following the workshop.

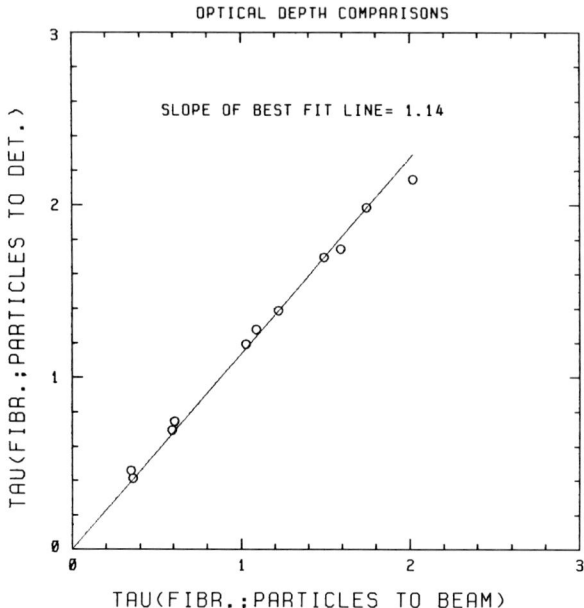

FIGURE 7. *Effect of differing filter orientation on measured absorption response for glass fiber filters.*

Measurements of the optical depths measured on the simultaneous samples collected on the glass fiber and Whatman filters, both oriented with the particles toward the incident light, are shown in Fig. 10. In this figure, data points for the soot-containing particles are indicated with a circle and data points for the methylene blue particles are indicated by triangles. There is an obvious difference in the relative response for the two types of particles. Fitted lines for each of the data sets show a slope of 2.2 for the soot-containing particles and 0.6 for the methylene blue particles. Although the reasons for this difference are not well understood, the particles have similar absorption coefficients at λ = 633 nm. The major difference in the particles appears to be their different aerodynamic size, which possibly results in different penetration depths in the filter and causes differences in the ambient light field within the filter at the absorbing particles.

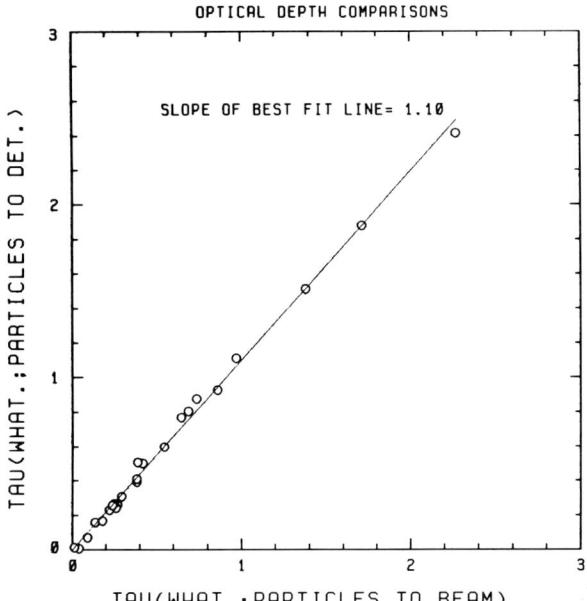

FIGURE 8. Effect of differing filter orientation on measured absorption response for Whatman filters.

In general these transmission results confirm the measurements that show orientation effects and filter effects. In addition, these measurements suggest that particle size and/or composition may have a significant effect on the response of the method to absorbing particles. A direct comparison of these response results with other such measurements has not been made. The glass fiber filters may not have the same characteristics as the quartz fiber filters used by Rosen et al. (4). If it is assumed that these filters do have the same characteristics, and if these data are used to calculate a relative response for the differing filters and geometries of Lin et al. (3) and Rosen et al. (4), a relative response of 0.37 is obtained for the Nuclepore filters relative to the quartz fiber filters for carbonaceous particles. Measurements of Sadler et al. (12), by comparison, show a relative response between the two methods of 0.35 for carbonaceous particles.

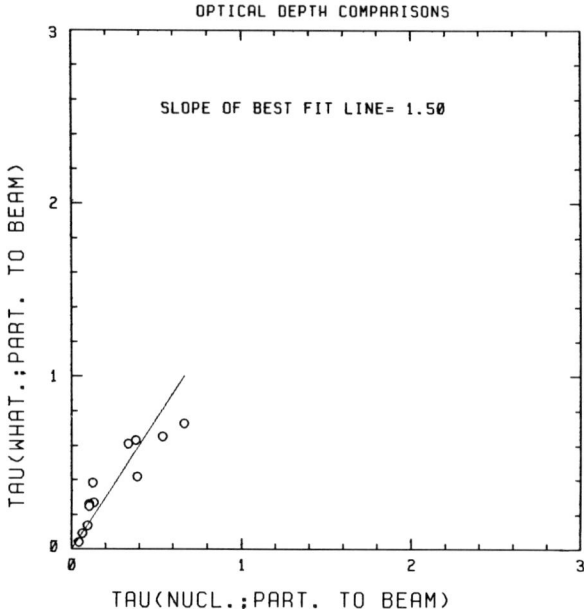

FIGURE 9. *Measured relative response characteristics for Whatman and Nuclepore filters.*

IX. MEASUREMENTS OF σ_a and B_a

Measurements of σ_a and B_a using Nuclepore filters with the particles oriented toward the incident beam are given in Table 2. The basic measurements are of σ_a, the volume absorption coefficients. The Colorado State University mass concentrations have also been used to determine mass absorption coefficients for these samples. Since there was some uncertainty in the measured values of mass concentration, at the present time these B_a values must be considered to be tentative results. Although more complete discussion of these measurements must await detailed comparisons with the other simultaneous measurements of absorption, an initial comparison of B_a values obtained by the two techniques shows that the filter transmission measurements result in significantly higher B_a values than those obtained by

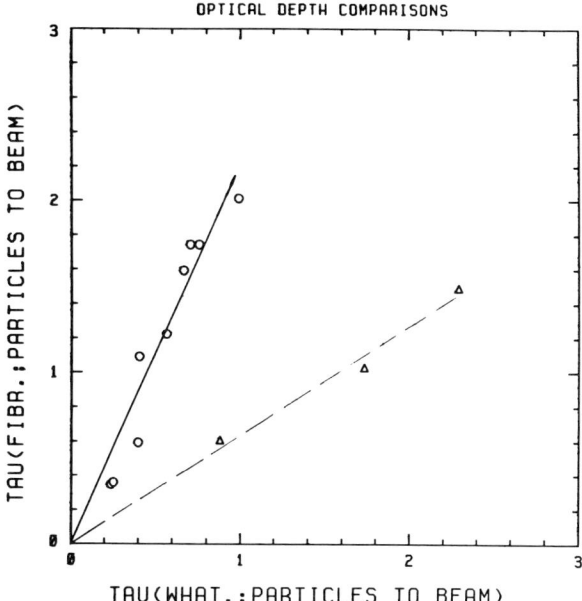

FIGURE 10. *Measured relative response characteristics for glass fiber and Whatman filters for soot (O) and methylene blue (Δ).*

the diffuse reflectance measurements for the relatively nonabsorbing ammonium sulfate and Arizona road dust. The MB B_a values determined by the two methods appear to be similar; the soot values determined by the transmission technique appear to be somewhat higher than those determined using diffuse reflectance.

X. CONCLUSIONS

The major results of these absorption measurements may be summarized as follows:

1. The diffuse reflectance measurements are internally consistent. The 1% mixture of methylene blue in ammonium sulfate has, as expected, a mass absorption coefficient which is 1% of that measured for the pure methylene blue. Absorption measurements of methylene blue in an aqueous solution show significant differences from those measured on the solid material, differences that may be related to chemical differences between the powder and the aqueous solution.

TABLE 2. Transmission Measurements Data

Sample type	Time of Sample	Sample ID	$\sigma_a (m^{-1})$	$M_v, \mu g/m^3$	B_a
$(NH_4)_2SO_4$	7/29/80 14:38-15:41	N111	1.5×10^{-5}	308(f)	7.8×10^{-2}
$(NH_4)_2SO_4$	8/8/80 11:44-13:37	N236	3.2×10^{-6}	30(s)	1.1×10^{-1}
$(NH_4)_2SO_4$	8/8/80 11:44-13:37	N420*	9.9×10^{-7}	489(f)	2.0×10^{-3}
Methylene blue	8/1/80 11:31-13:11	N130	5.9×10^{-4}	171(m)	3.5
Methylene blue	8/1/80 13:17-13:45	N131	6.2×10^{-4}	171(m)	3.6
$(NH_4)_2SO_4$ + 0.3% MB	7/30/80 13:33-15:26	N121	1.8×10^{-5}	58(m)	0.31
$(NH_4)_2SO_4$ + 0.3% MB	7/30/80 13:29-15:27	N401*	1.4×10^{-4}	--	--
$(NH_4)_2SO_4$ + 1% MB	7/29/80 16:33-16:48	N112	9.3×10^{-5}	224(m)	0.42
$(NH_4)_2SO_4$ + 1% MB	7/29/80 17:03-17:33	N113	9.2×10^{-5}	224(m)	0.41
$(NH_4)_2SO_4$ + 4% MB	7/30/80 16:33-16:36	N122	9.7×10^{-5}	68(m)	1.4
$NH_4)_2SO_4$ + 4% MB	7/30/80 17:12-17:53	N403*	9.8×10^{-4}	2506(f)	0.32
Soot	7/31/80 10:13-11:08	N124	7.7×10^{-4}	95(s)	8.1
Soot	7/31/80 11:40-12:02	N125	8.7×10^{-4}	95(s)	9.1
Soot	7/31/80 14:06-14:20	N227	1.1×10^{-3}	95(s)	10.9
Soot	8/6/80 17:42-17:44	N414	1.1×10^{-3}	85(s)	13
Soot	8/6/80 17:46-17:50	N415	1.1×10^{-3}	85(s)	14
Soot	8/6/80 17:52-18:03	N416	9.8×10^{-4}	85(s)	12
$(NH_4)_2SO_4$ + soot	7/31/80 15:28-15:49	N128	4.1×10^{-4}	145(f)	2.7
$(NH_4)_2SO_4$ + soot	7/31/80 16:17-17:00	N129	4.1×10^{-4}	145(f)	2.7
$(NH_4)_2SO_4$ + soot	8/4/80 13:52-14:02	N133	9.4×10^{-4}	590(f)	1.6
$(NH_4)_2SO_4$ + soot	8/6/80 14:58-17:00	N137	2.7×10^{-4}	516(f)	0.52
$(NH_4)_2SO_4$ + soot	8/7/80 10:51-13:38	N138	2.9×10^{-4}	355(f)	0.81
Arizona road dust	8/6/80 10:39-13:34	N136	2.8×10^{-5}	132(f)	0.21
Ambient aerosol	8/5/80 10:25-16:15	N134	1.2×10^{-5}	18(f)	0.67
Ambient aerosol	8/5/80 17:00- 8/6/80 08:33	N135	2.4×10^{-6}	17(f)	0.14
Ambient aerosol	8/6/80 10:57-17:15	N334	1.2×10^{-5}	--	--

*High concentration line; all others, low concentration line.
f = CSU filter; s = CSU size distribution; m = CSU mass monitor.

2. The wavelength-dependent measurements of the imaginary index of refraction for the soots do not show a contribution from organic hydrocarbons, although their presence cannot be ruled out. Measurements of B_a for the soot by the diffuse reflectance technique are somewhat lower than those previously measured for elemental or graphitic carbon.

3. Values of n_2 measured for a bulk sample of ammonium sulfate are consistent with literature values for this quantity. The ammonium sulfate aerosol sample, by comparison, had a significantly higher absorption.

4. The transmission measurements show that filter orientation can be significant for each of the measured filter media, and that the measured response will depend upon the filter material. Measured response ratios appear to be related to penetration depth in the filter. These data also suggest that the measured response may be dependent on the nature of the absorbing aerosol as well.

5. Due to the uncertainties in the mass concentration measurements, a detailed comparison of the B_a values determined from these transmission measurements with other measurements must be done at some future time. There do appear to be differences between the transmission measurements and the reflectance measurements for the relatively nonabsorbing materials, as well as for the soots.

REFERENCES

1. Lindberg, J. D., and Laude, L. S., Measurements of the absorption coefficient of atmospheric dust, *Appl. Opt. 13,* 1923-1927 (1974).
2. Patterson, E. M., Gillette, D. A., and Stockton, B. H., Complex index of refraction between 300 and 700 nm for Saharan aerosols, *J. Geophy. Res. 82,* 3153-3160 (1977).
3. Lin, C. I., Baker, M., and Charlson, R. J., Absorption coefficient of atmospheric aerosol: A method of measurement, *Appl. Opt. 12,* 1356-1363 (1973).
4. Rosen, H., Hansen, A. D. A., Gundel, L., and Novakov, T., Identification of the graphitic carbon component of source and ambient particulates by Raman spectroscopy and an optical attenuation technique, *Appl. Opt. 17,* 3859-3861 (1978).
5. Kortum, G., *Reflectance Spectroscopy*, Springer-Verlag, New York (1969).
6. Toon, O. B., Pollack, J. B., and Khare, B. N., The optical constants of several atmospheric aerosol species: Ammonium sulfate, aluminum oxide, and sodium chloride, *J. Geophys. Res. 81,* 5733-5748 (1976).

7. Jennings, S. G., Pinnick, R. G., and Gillespie, J. B., Relation between absorption coefficient and imaginary index of atmospheric aerosol constituents, *Appl. Opt. 18,* 1368-1371 (1979).
8. Ohno, T., Osif, T. F., and Lichtin, N. N., A previously unreported intense absorption band and the PKA of protonated triplet methylene blue, *Photochem. Photobiol. 30,* 541-546 (1979).
9. Szkarlat, A. C., and Japar, S. M., Light absorption by airborne aerosols: Comparison of integrating plate and spectrophone techniques, *Appl. Opt. 20,* 1151-1155 (1981).
10. Patterson, E. M., Optical properties of urban aerosols containing carbonaceous material, In *Carbonaceous Particles in the Atmosphere* (T. Novakov, ed.), pp. 247-251, LBL-9037, CONF-7803101, UC-11. [Available from NTIS, Springfield, Virginia.]
11. Lindberg, J. D., and Gillespie, J. B., Relationship between particle size and imaginary refractive index in atmospheric dust, *Appl. Opt. 16,* 2628-2630 (1977).
12. Sadler, M., Charlson, R. J., Rosen, H., and Novakov, T., An intercomparison of the integrating plate and the laser transmission methods for determination of aerosol absorption coefficients, *Atm. Environ, 15,* 1265-1268 (1981).

UNIVERSITY OF ARIZONA AEROSOL ABSORPTION MEASUREMENTS

J. A. Reagan

Department of Electrical Engineering
University of Arizona
Tucson, Arizona

B. M. Herman
R. M. Schotland

Institute of Atmospheric Physics
University of Arizona
Tucson, Arizona

Three remote sensing techniques for determining the aerosol imaginary index are described. The techniques involve a passive method and two active methods. The passive method utilizes the ratio of the direct to diffuse transmitted solar radiation. The two active methods utilize a bistatic and monostatic lidar to determine aerosol extinction to backscatter ratios and angularly dependent polarization ratios. These measurements may be interpreted to yield the imaginary part of the aerosol refractive index.

I. INTRODUCTION

During the past decade the University of Arizona Atmospheric Optics Group has developed a wide array of remote sensing techniques to monitor atmospheric aerosols and variable gases to determine their effects on solar radiative transfer. A considerable portion of this effort has been directed toward determining the absorptive properties of atmospheric aerosols and subsequent effects on the radiative field. To this end three remote sensing methods have been developed to infer an "effective" imaginary component of the index of refraction of the aerosols. The term "effective" here is defined as that one value of imaginary index which satisfies the measured parameters. It is to be understood that individual aerosol particles may have indices quite different from this effective value.

The three techniques which have been developed all require that the particle size distribution be known. This size distribution function is also determined from remotely sensed values of the wavelength dependent aerosol optical depth determined from the University's solar photometer. These optical depths are then inverted to yield the particle size distribution function (1). The three techniques are summarized in the following section.

II. METHODS

1. *Direct-Diffuse Technique*

The direct-diffuse technique utilizes the extreme sensitivity of the imaginary part of the index of refraction to the ratio of directly transmitted to diffusely transmitted solar radiation (2,3). The method requires a knowledge of the aerosol optical depth and columnar size distribution function, both of which are determined from solar photometry data. It is also useful to know the ground albedo, but as pointed out in the two previous references this latter parameter may also be solved for by this technique. The technique, fortunately, is not sensitive to the value of the real part of the index of refraction, and therefore its precise value need not be known.

The direct-to-diffuse ratios are measured with an eight-channel narrowband (10.0-nm) filter wheel spectral pyranometer developed at the University of Arizona (4). The cosine response of this instrument is polarization independent and lies within 2% of the true cosine function. The direct and diffuse fluxes are determined from measurements made of the total flux incident on the pyranometer and from the incident flux with the sun occulted. The accuracy of the measurement over the spectral range 400 nm to 800 nm is on the order of 1.5%.

2. Extinction to Backscatter Ratio Technique

The ratio of attenuated light to backscattered light, referred to here as the S ratio, is sensitive to both aerosol size distribution and refractive index. Given the size distribution function and the real part of the index, the imaginary part may then be inferred. If the real part of the index is not known, it is still possible to set limits on the imaginary part for a reasonable range of real index values (5).

The S ratio may conveniently be measured by means of a monostatic lidar system utilizing the slant path technique also described in Ref. 5. This technique requires reasonable atmospheric homogeneity, and the resulting S ratio applies to an atmospheric column (typically the mixing layer) as opposed to a point measurement. This technique has been successfully employed in Tucson on numerous occasions utilizing a ruby lidar ($\lambda = 694.3$ nm) and yields results which compare favorably with the other techniques described herein. Results to date show that the imaginary index in the Tucson area lies in the range 0.00 to 0.01 and is typically less than 0.005.

3. S Ratio Combined with Angular Scatter Measurements

This technique combines determinations of the S ratio with measurements of the polarization by scattered light. Since the polarization is sensitive to both the real and imaginary parts of the index, the technique yields estimates of both parts of the index. The particle size distribution again must be known, as all parameters are also sensitive to it. The specific measure of polarization used in this method is the polarization ratio, defined as the ratio of the parallel component to the perpendicular component arising from scatter by the aerosols only. An added feature of this method is that the angular scattered intensities combined with the index determined from this method may be inverted to determine the aerosol size distribution function. This function may then be compared to that determined from the solar photometry data, thus providing a consistency check (6).

Both the S ratio and polarization ratios are measured with the University of Arizona combination bistatic, monostatic ruby lidar system (7). Again, reasonable horizontal homogeneity is required. Results for Tucson (6) indicate a real index near 1.45 (typical for H_2SO_4-H_2O solutions) and an imaginary index less than 0.01.

REFERENCES

1. King, M. D., Byrne, D. M., Herman, B. M., and Reagan, J. A., Aerosol size distributions obtained by inversion of spectral optical depth measurements, *J. Atmos. Sci. 35,* 2153-2167 (1978).
2. Herman, B. M., Browning, R. S., and DeLuisi, J., Determination of the effective imaginary term of the complex index of refraction of atmospheric dust by remote sensing: The diffuse-direct radiation method, *J. Atmos. Sci. 32,* 918-925 (1975).
3. King, M. D., and Herman, B. M., Determination of the ground albedo and the index of absorption of atmospheric particulates by remote sensing—Part I: Theory, *J. Atmos. Sci. 36,* 163-173 (1979).
4. Schotland, R. M., and Copp, J. D., A narrow band spectral pyranometer. IRS 1980 International Radiation Symposium, *Vol. Extended Abstracts,* pp. 562-564, Intl. Assoc. Meteor. and Phys., 11-16 August 1980, Fort Collins, Colorado (1980).
5. Spinhirne, J. D., Reagan, J. A., and Herman, B. M., Vertical distribution of aerosol extinction cross section and inference of aerosol imaginary index in the troposphere by lidar technique, *J. Appl. Meteor. 19,* 426-438 (1980).
6. Reagan, J. A., Byrne, D. M., King, M. D., Spinhirne, J. D., and Herman, B. M., Determination of the complex refractive index and size distribution of atmospheric particulates from bistatic-monostatic lidar and solar radiometer measurements, *J. Geophys. Res. 85,* 1591-1599 (1980).
7. Reagan, J. A., and Herman, B. M., Bistatic lidar investigations of atmospheric aerosols, 14th Radar Meteorology Conference, *Conf. Reprints,* pp. 275-280, 17-20 November 1970, Tucson, Arizona (1970).

PHOTOACOUSTIC STUDIES OF AEROSOL OPTICAL PROPERTIES

D. M. Roessler

Physics Department
General Motors Research Laboratory
Warren, Michigan

Photoacoustic techniques for measuring light absorption by carbonaceous materials are described, and measurements on cigarette smoke, acetylene smoke, and diesel exhaust are summarized.

I. INTRODUCTION

The photoacoustic effect has been used to study the fundamental optical properties of absorbing aerosols. Instrumentation has been developed to focus on this aspect rather than the ultimate sensitivity of the photoacoustic technique itself. In particular, measurements have been made on the single-scattering albedo $\tilde{\omega}$ and on the specific absorption coefficient B_a for a variety of aerosols but with emphasis on carbonaceous materials. This report summarizes these investigations, which are described more fully elsewhere (1-3).

II. DETERMINATION OF FRACTIONAL ABSORPTION σ_a/σ_e

By combining simultaneous transmission and photoacoustic measurements in aerosols, the fraction of the attenuated light which is due to absorption has

been determined. A schematic of the apparatus is given in Fig. 1.
The transmitted light intensity is

$$I = I_o \exp(-\sigma_e L) \tag{1}$$

where σ_e is the extinction coefficient and L is the path length. The microphone signal S generated from the photoacoustic effect can be written as

$$S = R \frac{\sigma_a}{\sigma_e} [1 - \exp(-\sigma_e L)] \frac{W}{L} \tag{2}$$

where σ_a is the absorption coefficient, W is the time-averaged incident optical power, and R is the spectrophone responsivity. The latter depends on the cell geometry, the microphone sensitivity, and the frequency of amplitude modulation of the incident light. Combining these equations gives the fractional absorption as

$$\frac{\sigma_a}{\sigma_e} = \frac{SL}{R\left(1 - \frac{I}{I_o}\right)W} \tag{3}$$

The cell response R is determined by introducing a gas for which $\sigma_a \approx \sigma_e$ at the incident light wavelength, e.g., nitrogen dioxide for the argon laser at $\lambda = 0.5145$ μm. For routine, nonresonant operation, a cell 100 mm long and 10 mm in diameter was used which contained a small hearing-aid microphone (Knowles BT 1753) with a sensitivity of about 10 mV·Pa^{-1} at the modulation frequency (500 Hz). The value of R was found to be 12.4 ± 0.9 mV/(m^{-1}·W).

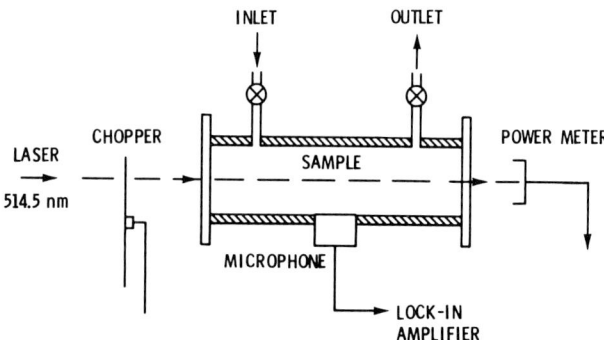

FIGURE 1. Schematic of spectrophone used to determine factional absorption, σ_a/σ_e, and specific absorption coefficient, B_a.

For aged cigarette smoke having a refractive index of n = 1.45 - 0.00133 i at λ = 0.5145 µm the Mie theory predicts σ_a/σ_e to be in the range 0.5% to 2.0%. The experimentally determined value of 1.0 ± 0.3% was in excellent agreement.

In the case of carbon smoke particles, the applicability of the Mie theory is not obvious because of clustering effects (4). For single carbon spheres, the Mie theory predicts σ_a/σ_e to be in the range 70% to 100%, depending on the refractive index and size distribution chosen. The measurements reported here gave σ_a/σ_e = 84 ± 9% at λ = 0.5145 µm for acetylene smoke particles. At 10.6 µm, σ_a/σ_e was found to be 90 ± 10%, the large uncertainty stemming from the low sensitivity in the transmission measurement.

III. DETERMINATION OF SPECIFIC ABSORPTION COEFFICIENT B_a

The value of σ_a can be determined itself, using the apparatus described, simply by eliminating the extinction coefficient σ_e between Eqs. (1) and (2). By measuring the mass concentration M_v of the aerosol, the specific absorption coefficient $B_a = \sigma_a/M_v$ can thus be derived.

For these measurements the aerosol is pumped continuously through the spectrophone at flow rates between 0.1 and 10 l/min^{-1}. The choice of flow rate is dictated by a compromise between the required response time and the need to keep the flow noise below the level of the true photoacoustic signal. Further, the apparatus was modified to include Brewster windows and purge air ports to minimize background signals. Measurements were made at both λ = 0.5145 µm (argon ion laser) and λ = 10.6 µm (CO_2), the windows being changed from quartz to zinc selenide appropriately. The cell responsivity R was calibrated with nitrogen dioxide for the visible region and with trichloroethylene for the infrared.

For acetylene smoke particles produced by a laboratory burner, B_a was determined to be 8.2 ± 0.9 m^2 g^{-1} at λ = 0.5145 µm and 0.85 ± 0.2 m^2 g^{-1} at λ = 10.6 µm. The large uncertainty in the infrared is thought to indicate variations in B_a for different combustion conditions rather than solely experimental imprecision. For diesel smoke particles, very substantial variations in B_a have been found (and also in B, the total specific extinction coefficient). These are discussed more fully elsewhere and have been attributed to variations in the physical morphology and chemical composition of the particles (5). At λ = 0.5145 µm, B was found from conventional transmission measurements to vary in the range 2 to 12 m^2 g^{-1}, a value of about 5 m^2 g^{-1} being more typical of the exhaust from passenger cars at average road load driving conditions. At λ = 10.6 µm, B_a was about an order of magnitude lower, but again varied strongly with engine operation.

IV. PHOTOACOUSTIC MEASUREMENT OF DIESEL PARTICULATE MASS CONCENTRATION

The genesis of this interest in aerosol optical properties was in the extent to which optical measurements could be used to determine diesel particulate mass concentrations. Figure 2 shows a schematic of the system used to monitor diesel emissions photoacoustically. The detectability limits are determined primarily by exhaust noise and by interference from absorbing species (such as gases) other than those of immediate interest.

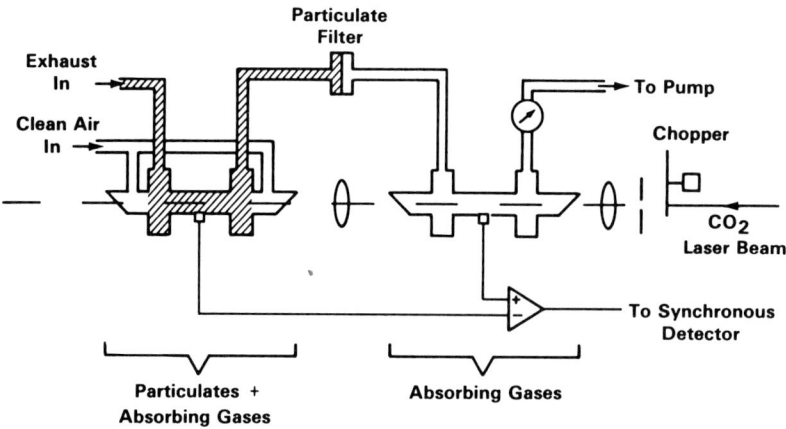

FIGURE 2. *Schematic of photoacoustic system used to monitor diesel exhaust particle mass concentrations. A dual-cell arrangement permits subtraction of the signal from interfering exhaust gases.*

Measurements were made at 9.6 μm with a 3-W CO_2 laser operating at 2.3 kHz. The system incorporates zinc selenide and barium fluoride optics, and a double-cell approach is used to subtract out the signal due to interfering exhaust gases. The cell bores are 5 mm, the main cavity being 50 mm long. Resonant operation occurs at about 3.4 kHz but better stability is obtained by choosing modulation frequencies lower than this.

Under these conditions, values of M_v could be measured down to about 100 μg m^{-3}, this limit being due to flow noise and background absorption as indicated. Although the measurements can be made with a precision of about 15%, the absolute values of mass concentration are uncertain by about a factor of two because of the variability of B_a with exhaust particle variations (5). Table 1 summarizes the basic optical properties determined by these techniques.

TABLE 1. Summary of Optical Data Derived from Photoacoustic Measurements.

Aerosol	σ_a/σ_e $\lambda = 0.5145$ μm	σ_a/σ_e $\lambda = 10.6$ μm	$B_a (m^2 g^{-1})$ $\lambda = 0.5145$ μm	$B_a (m^2 g^{-1})$ $\lambda = 10.6$ μm
Cigarette smoke	0.01 ± 0.003	--	0.04 ± 0.01[a]	--
Acetylene smoke	0.84 ± 0.090	0.9 ± 0.1	8.30 ± 0.90	0.85 ± 0.2
Diesel particles[b]	--	--	2-10 ± 15%	0.20-1.2 ± 30%

[a]Calculated from Mie theory, with $n = 1.45$ to 0.00133 i.
[b]The ranges shown for B_a indicate widely different engine operating conditions.

REFERENCES

1. Roessler, D. M., and Faxvog, F. R., Photoacoustic determination of optical absorption to extinction ratio in aerosols, *Appl. Opt.* **19**, 578-581 (1980).
2. Faxvog, F. R., and Roessler, D. M., Optoacoustic measurements of diesel particulate emissions, *J. Appl. Phys.* **50**, 7880-7882 (1979).
3. Roessler, D. M., and Faxvog, F. R., Optoacoustic measurement of optical absorption in acetylene smoke, *J. Opt. Soc. Am.* **69**, 1699-1704 (1979).
4. Roessler, D. M., and Faxvog, F. R., Optical properties of agglomerated acetylene smoke particles at 0.5145-μm and 10.6-μm wavelengths, *J. Opt. Soc. Am.* **70**, 230-235 (1980).
5. Roessler, D. M., Faxvog, F. R., Stevenson, R., and Smith, G. W., Optical properties and morphology of particulate carbon: Variation with air/fuel ratio, *Proceedings of GM Research Labs Symposium on Particulate Carbon*, (1980).

PHOTOACOUSTIC DETERMINATION OF LIGHT ABSORPTION BY AEROSOLS[1]

R. Röhl[2]
R. A. Palmer

Paul M. Gross Chemistry Laboratory
Duke University
Durham, North Carolina

W. A. McClenny

ESRL, EPA, Research Triangle Park
Durham, North Carolina

The absorptivity of soot and methylene blue particles collected on Teflon filters is derived from photoacoustic measurements by least-squares fitting a simple expression based on Beer's law to the experimental data. Refinements of the expression take into account the diffuse reflection of light by the filter substrate, yielding a base-10 absorptivity at 600 nm for soot of 3.00 ± 0.37 m^2/g. This value is in close agreement with the result of transmission measurements performed on the same samples (3.08 ± 0.05 m^2/g).

[1]*The research was partially supported by EPA Grant 807407 and NSF Grant ATM 8005356.*
[2]*Present address: Tennenloher Street 14, 8521 Utteureuth, FRG.*

I. INTRODUCTION

Aerosols were collected at the Colorado State University aerosol generation facility and subsequently analyzed at Duke University in Durham, North Carolina. Sampling was accomplished by placing 1-μm pore size, type 501 Teflon filters (Ghia Corp.) into ERC manual dichotomous samplers and pulling aerosols from a part of the downstairs manifold onto the Teflon filter surfaces at approximately 16.7 l/min. Analysis was done by placing the intact loaded filter in a specially designed photoacoustic cell and recording the photoacoustic response of the PAR Model 6001 photoacoustic spectrometer as a function of wavelength. The resulting spectra were then interpreted as explained in the text.

II. THEORY

In transmission spectroscopy, the absorptivity of a substance is defined as $a = A/bc$ (2.3026 $a = B_a$), where A is the absorbance, b is the pathlength of light through a sample, and c is the concentration of absorbing material in the sample cell (1). The absorbance is defined by

$$A = \log I_o/(I_o - I_a) \tag{1}$$

with I_o representing the intensity of the unattenuated light beam and I_a the the intensity of the absorbed light. In photoacoustic spectroscopy (PAS), I_a corresponds to the signal from a given sample S, and I_o corresponds to the signal from a blackbody reference R. The PAS analog to absorbance, A_{pas}, can therefore be written as

$$A_{pas} = \log R/(R - S) = \log 1/(1 - S/R) \tag{2}$$

Equation (2) assumes that all absorbed light contributes to the photoacoustic (PA) signal. This assumption is justified if a low chopping frequency is used in the PA experiment so that the samples are thermally thin. This approach to relating absorbance to PA response is similar to that taken by Vidrine (2). Identical results can be obtained by solving the idealized one-dimensional, boundary value problem (3) and invoking simplifying assumptions (4) to obtain Eq. (2). Using the definition of absorptivity and a transformation of Eq. (2), one can express the PA signal S as follows

$$S = R(1 - 10^{-abc}) \tag{3}$$

With the goal of determining absorptivities, the application of Eq. (3) to the PA analysis of solid materials collected on a filter substrate also requires knowledge of the product bc. A simple derivation shows that this product is given by the loading density of the filter (e.g., in g/m²).

Assuming the absorption "cell" has a thickness b and the same cross-sectional area A' as the light beam, the volume V of the cell is $A'b$. The mass of material within the cell, m', is the total mass on the filter, M, divided by the total filter area A'' and multiplied by A'. The concentration c is then given by $c = m'/V = MA'/A''A'b = M/A''b$. Therefore, the product bc is equal to the loading density M/A''. This parameter will henceforth be denoted by the symbol L.

III. EXPERIMENTAL

For photoacoustic analyses, samples were mounted in a specially designed cell which accepts whole dichotomous virtual-impactor filters in their slide frames. Spectra extending from 200 nm to 2000 nm were obtained with a Princeton Applied Research 6001 photoacoustic spectrometer. The selected chopping frequency was 15 Hz, the monochromator slit width was 2.0 mm, and the scan rate was 100 nm/min. PA signals were corrected for variations in the xenon arc lamp spectrum by dividing the measured response by the output of a pyroelectric detector illuminated by a small (< 10%) but constant fraction of the total beam intensity. A Teflon filter loaded with 2 mg of carbon black from a propane soot generator was used to generate the blackbody reference signal R. Carbon and methylene blue particles were collected on Teflon filters with a dichotomous sampler. The particles are divided by this type of sampler into two fractions, fine (< 2.5-μm aerodynamic diameter) and coarse (2.5- to 15-μm aerodynamic diameter). Two identical samplers were used, one by researchers at Duke University and one by researchers at North Carolina State University. The values of absorption parameters inferred from analysis of the fine particle fraction of one of the samplers are reported here. The total set of carbon samples consisted of EPA filters 122227, 122228, and 122230 to 122233 taken on 7 August 1980 and EPA filters 222103 and 222107 taken on 31 July during the late morning and afternoon. Methylene blue absorptivities were obtained using EPA filters 222130 to 222134 loaded near midday of 1 August. To achieve different loadings for a given aerosol generation rate, sampling times were varied over a range of 1 to 10 minutes and the constant sampling rate of 16.7 l/min.

IV. RESULTS AND DISCUSSION

1. *Soot Samples*

Spectra of Teflon filters loaded with soot during the First International Workshop on Light Absorption by Aerosol Particles (Fort Collins, Colorado, August 1980) exhibited a general decrease of PA response with increasing wavelength (Fig. 1). More detailed analyses were limited to inspection of signals at 600, 780, 1200, and 1600 nm respectively; the 780-nm wavelength was selected because the spectrometer changes gratings at 800 nm.

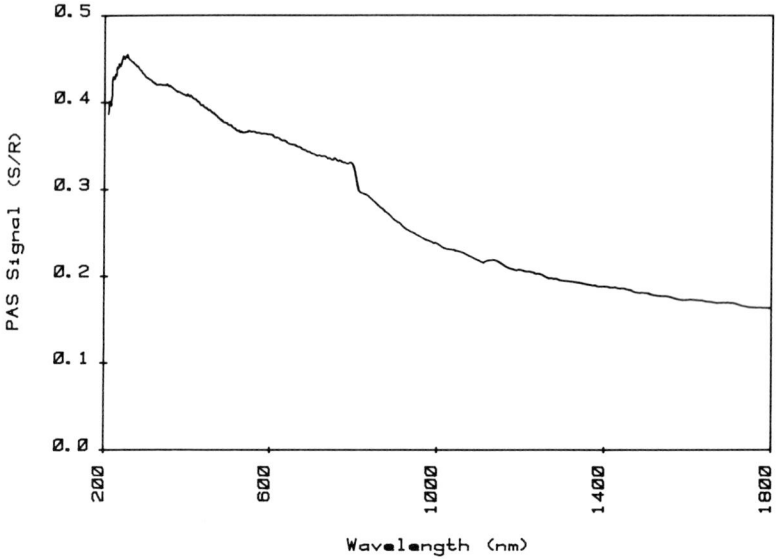

FIGURE 1. *Photoacoustic spectrum of soot on a Teflon filter. The break at 800 nm is caused by a grating change in the spectrometer. Reference: filter loaded with 2 mg of carbon black.*

Comparison of curves obtained at different soot loadings showed that the PA signal approached saturation at levels considerably below the value obtained with a 2-mg reference filter (Figs. 2-5). After considering several possible explanations, this observation was attributed to different physical (thermal) properties of the soot on the sample and reference filters. A less dense packing of carbon particles on the reference filter could have been caused by the decrease in flow velocity at higher loading and resulted in an enhanced PA signal.

PHOTOACOUSTIC MEASUREMENTS 311

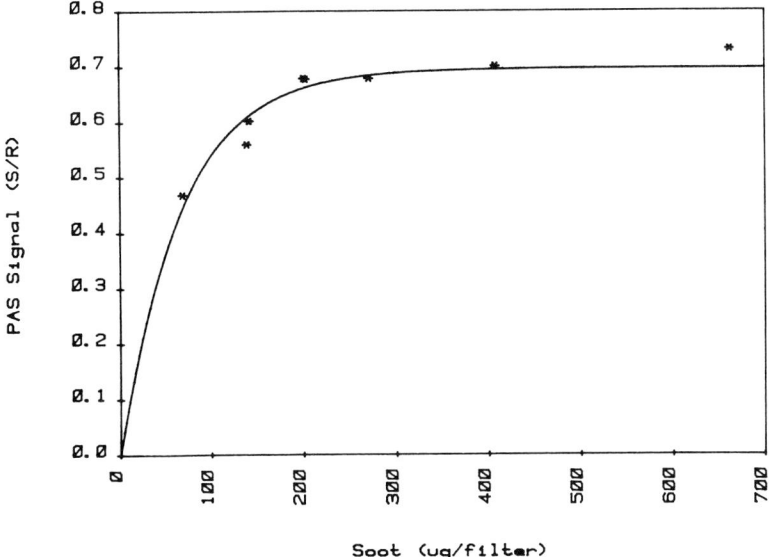

FIGURE 2. Photoacoustic signal at 600 nm as a function of soot loading.

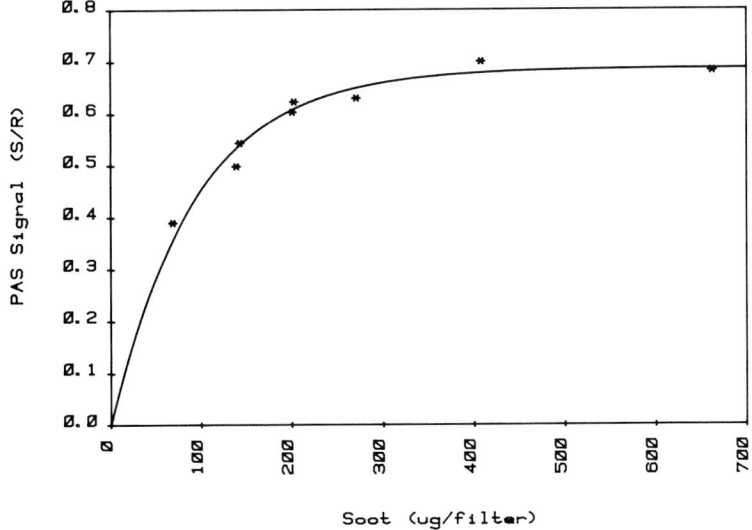

FIGURE 3. Photoacoustic signal at 780 nm as a function of soot loading.

Since very thick soot samples from the workshop were not available for comparison, appropriate reference levels were estimated from the measured data.

It was initially assumed that the PAS data followed the expression

$$S = R(1 - 10^{-aL}) \tag{4}$$

The two unknowns, R and a, were found by iterative least-squares fitting of Eq. (4) to the experimental data points, using R and a as fitting variables. A graphical two-parameter fitting procedure was used to ensure the localization of a unique solution, i.e., an absolute minimum of χ^2.

The uncertainty or standard deviation of the fitted parameters was calculated by the method described by Bevington (5). Briefly, the shape of the function $\chi^2 = f(R,a)$ near the minimum was assumed to be parabolic and the uncertainty of each parameter was determined by parabolic extrapolation of three points on the $\chi^2(R,a)$ surface near the minimum. The experimental uncertainty at a given wavelength was assumed to be the same for all data points and was estimated from two pairs of near-replicate samples with 140 ± 1 µg and 200 ± 2 µg soot respectively.

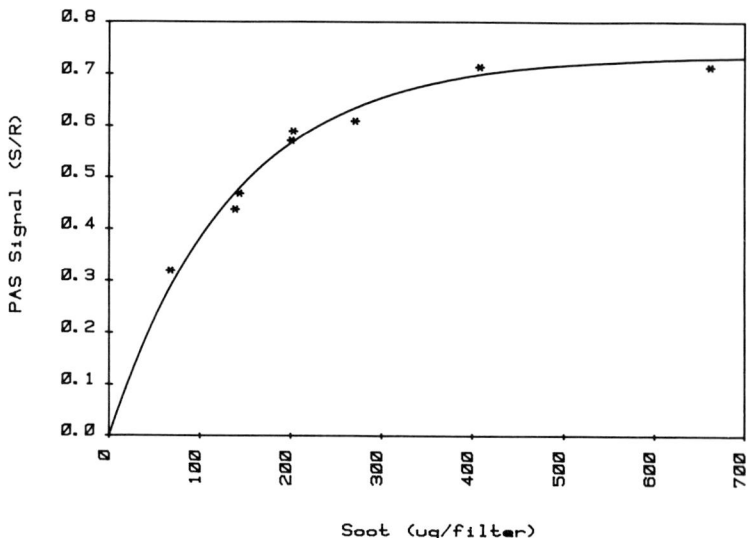

FIGURE 4. *Photoacoustic signal at 1200 nm as a function of soot loading.*

PHOTOACOUSTIC MEASUREMENTS

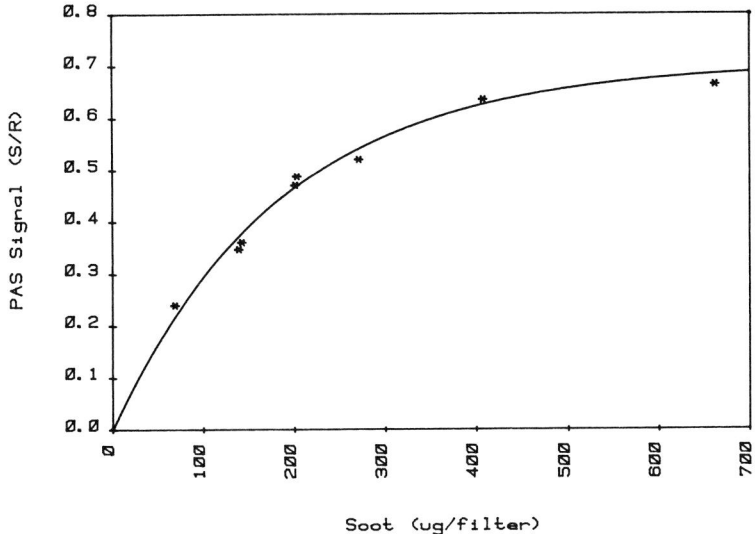

FIGURE 5. Photoacoustic signal at 1600 nm as a function of soot loading.

Results of this analysis for the PA measurements taken at four different wavelengths are listed in Table 1. The corresponding best-fitting curves are plotted along with the experimental data points in Figs. 2 to 5. Base e absorptivities (B_a) varied from 4.31 ± 0.39 at 1600 nm to 11.9 ± 3.27 at 600 nm. Values for the imaginary part of the refractive index were calculated using the formula $n_2 = B_a \varrho \lambda / 4\pi$ and are given for base e only, in keeping with standard usage. The relative standard deviation of the fitted absorptivities was higher at shorter wavelengths because most samples were close to the saturation level (Fig. 2).

2. Methylene Blue Samples

Photoacoustic spectra of methylene blue collected on Teflon filters showed a signal maximum at 710 nm (Fig. 6). The analysis described in the previous section was applied to PAS measurements for five filters loaded at this workshop with 170 µg to 659 µg of methylene blue. Again, a close fit of Eq. (4)

TABLE 1. Summary of Results for Carbon Particles.

Wavelength (μm)	PAS reference $R \pm 2s^a$	Absorptivity $a \pm 2s^a (m^2/g)$	Imaginary index of refraction, $n_2{}^c$
0.600	0.696 ± 0.016	5.18 ± 1.42	
		11.90 ± 3.27[b]	1.14 ± 0.31[b]
0.780	0.687 ± 0.018	3.73 ± 0.72	
		8.59 ± 1.66[b]	1.07 ± 0.21[b]
1.200	0.736 ± 0.008	2.54 ± 0.17	
		5.85 ± 0.39[b]	1.12 ± 0.07[b]
1.600	0.703 ± 0.008	1.87 ± 0.17	
		4.31 ± 0.37[b]	1.10 ± 0.10[b]

[a] s = sample standard deviation.
[b] Base e.
[c] Assuming a particle density of 2.0 grams/cm^3 and using the formula $n_2 = a\varrho\lambda/4\pi$.

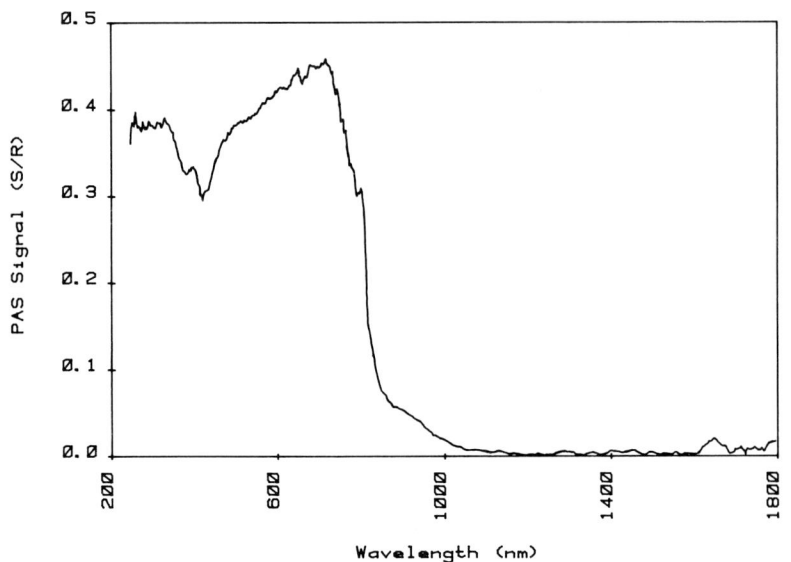

FIGURE 6. Photoacoustic spectrum of methylene blue on a Teflon filter.

to the experimental data at 710 nm was possible (Fig. 7). The derived absorptivity was 0.773 m²/g using base 10 and 1.780 using base e. Since no replicate samples were available, the uncertainity of this value could not be calculated. However, judging from the good fit illustrated by Fig. 7, the relative standard deviation should be less than 10%.

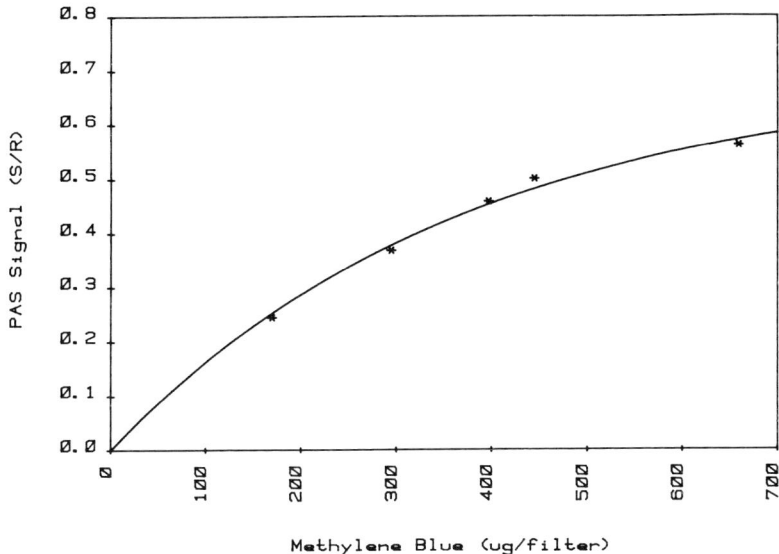

FIGURE 7. Photoacoustic signal at 710 nm as a function of methylene blue loading.

3. Error Sources and Refinement of Fitting Model

Only a portion of the filters loaded during this workshop were analyzed, mainly due to the difficulties in obtaining accurate sample weights. In addition to the errors in weight determinations and PAS measurements (which were assumed to be random), the calculated absorptivities may be subject to several types of systematic errors. Since soot tends to form larger particles by aggregation, the surface coverage at low loadings is heterogeneous. Nonuniform surface coverage in turn leads to reduced light absorption compared to the uniform deposit assumed in the derivation of Eq. (4). Transmission measurements, which are briefly discussed below, showed that this effect was negligible for the range of loadings investigated in this study. Another type of systematic error is also expected to be most prominent at low loading levels, but with an opposing effect on apparent absorptivities: The analogy to transmission spectroscopy assumes that light passes the sample only once, i.e., that light not absorbed during the first pass is lost by transmission through the filter. However, since Teflon filters reflect a significant portion of the light impinging on their surface, the absorption length is increased, thus increasing the apparent absorptivity.

In order to refine the estimate of the absorptivity of soot at 600 nm, the reflectivity of blank Teflon filters was measured relative to a freshly prepared magnesium oxide standard. Using the PAR 6001 xenon arc lamp and monochromator as a light source and a pyroelectric detector in a $_0R_{45}$ geometry (6), a relative reflectivity F of 0.55 ± 0.02 was found for unsupported filters and a value of 0.58 ± 0.02 for the same filters backed by the quartz window of the photoacoustic cell. Thus, the first refinement applied to the calculations was to assume that 58% of the light reaching the filter surface was reflected back through the sample, again in a straight line. Equation (5) describes the PAS signal as a function of loading L when these assumptions apply.

$$S = R(1 - 10^{-aL}) + R F 10^{-aL} (1 - 10^{-aL}) \tag{5}$$

This model was further refined by taking into account that Teflon filters reflect light diffusely and not straight back through the sample. This leads to an increased effective pathlength which depends on the angular intensity distribution $I_r(\theta)$ of the reflected light. For a perfect diffuse reflector, this distribution is described by Lambert's cosine law (6)

$$I_r(\theta) = B \cos \theta \tag{6}$$

In Eq. (6), θ is the angle between the incident light beam (perpendicular to the planar sample surface) and the direction of observation, and B is a constant proportional to the intensity of the illuminating light beam. The effective pathlength of reflected light through a layer of thickness b is given by $b/\cos \theta$. Weighting this pathlength with the angular intensity distribution given by Eq. (6) and integrating over all possible angles (0° to 90°) yields the average effective pathlength b'.

$$b' = \int_0^{\pi/2} \cos \theta (b/\cos \theta) d\theta = \int_0^{\pi/2} b \, d\theta = b\pi/2 \tag{7}$$

With this second refinement, the PAS response as a function of loading is given by

$$S = R(1 - 10^{-aL}) + R F 10^{-aL}(1 - 10^{-aL\pi/2}) \tag{8}$$

PHOTOACOUSTIC MEASUREMENTS 317

For reference purposes the three different models fitted to the experimental PAS data are summarized in Fig. 8. Results obtained with the three models for the PAS measurement at 600 nm are compared in Table 2. Since the refinements of the initially used simple model increase the assumed pathlength of light through the sample, they decrease the calculated absorptivity of soot. The calculated absorptivities from model 1 to model 3 (Fig. 8) are in the ratios 1.0:0.67:0.58. Very similar ratios (1.0:0.67:0.57) were obtained when a test data set was calculated from model 3 and fitted with the equations of models 1

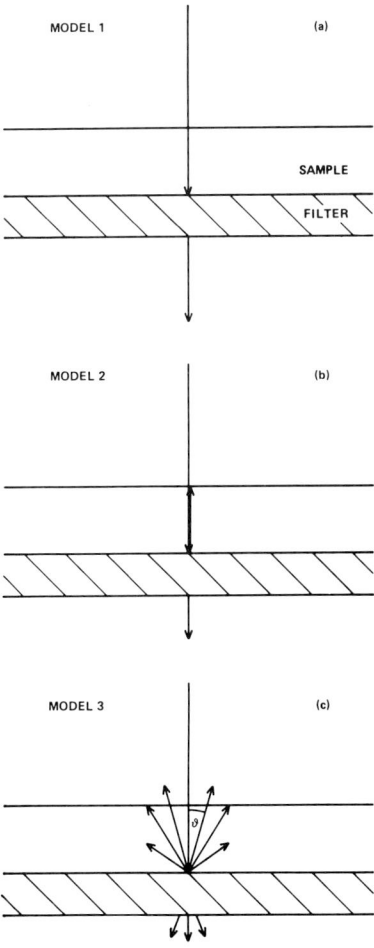

FIGURE 8. Illustration of the three models used for fitting PAS data.

TABLE 2. *Absorptivity of Soot at 600 nm as Determined by Three Different Methods of Fitting the Experimental PAS Data (note that χ^2 decreases with increasing refinement of the fitting model).*

Model	Equation	$\chi^2 \times 10^5$	$a_{10} \pm 2s^*$	$a_e \pm 2s^{**}$
1	4	513	5.18 ± 1.41	11.9 ± 3.27
2	5	462	3.45 ± 0.82	7.95 ± 1.88
3	8	445	3.00 ± 0.74	6.91 ± 1.70

* $a_{10} B_a/2.3026$
** $a_e = B_a$

and 2. The values in Table 2 may also be compared to the absorptivity calculated from transmission measurements performed on five of the soot samples analyzed by PAS. The transmission measurements were made using a modified form of the integrating plate method (7). A base-10 absorptivity of 3.55 ± 0.06 m²/g was obtained at 520 nm, which corresponds to 3.08 ± 0.05 m²/g at 600 nm if the imaginary part of the refractive index of soot is assumed to be constant between 520 nm and 600 nm (8). The latter value is in close agreement with the absorptivity of soot derived from PAS data using a model which takes into account the diffuse reflection of light at the surface of the sample substrate, i.e., the Teflon filter (model 3).

ACKNOWLEDGMENTS

The authors wish to thank C. A. Bennett, Jr., for providing optical transmission measurements, C. C. Owen for making filter loading measurements, and R. K. Stevens of the Environmental Protection Agency (EPA) for loan of particle collectors.

REFERENCES

1. Spectrometry nomenclature, *Anal. Chem.* 53, 141-142 (1981).
2. Vidrine, D. W., Photoacoustic Fourier transform infrared spectroscopy of solid samples, *Appl. Spec.* 34, 314-319 (1980).
3. Rosencwaig, A., and Gersho, A., Theory of photoacoustic effect with solids, *J. Appl. Phys.* 47, 64-69 (1976).

4. Yasa, Z., Amer, N. M., Rosen, H., Hansen, A. D. A., and Novakov, T., Photoacoustic investigation of urban aerosol particles, *Appl. Opt. 18,* 2528-2530 (1979).
5. Bevington, P. R., *Data Reduction and Error Analysis for the Physical Sciences,* McGraw-Hill, New York (1969).
6. Kortum, G., *Reflexionsspektroskopie,* Springer-Verlag, Berlin (1969).
7. Lin, C. I., Baker, M. B., and Charlson, R. J., Absorption coefficient for atmospheric aerosol: A method of measurement, *Appl. Opt. 12,* 1356-1363 (1973).
8. Rosen, H., Hansen, A. D. A., Gundel, L., and Novakov, T., Identification of the graphitic carbon component of source and ambient particulates by Raman spectroscopy and an optical attenuation technique, In *Proceedings, Conference on Carbonaceous Particles in the Atmosphere,* pp. 49-55, Lawrence Berkeley Laboratory, March 1978 (LBL-9037, Conf. 7803101, UC-11) (1979).

LAWRENCE BERKELEY LABORATORY LASER TRANSMISSION METHOD[1]

H. Rosen
T. Novakov

Lawrence Berkeley Laboratory
University of California
Berkeley, California

The Lawrence Berkeley Laboratory laser transmission method has been used to determine the absorption coefficients of aerosol particles generated at the First International Workshop on Light Absorption of Aerosol Particles (Fort Collins, Colorado, August 1980). Analysis of the results confirms that the optical attenuation measurement is insensitive to the scattering properties of the aerosol. A simple model calculation is presented which explains these observations and points out the critical role of the filter substrate as an almost perfect diffuse reflector in the technique.

---

[1] This work was supported by the Department of Energy under Contract W-7405-ENG-48 and by the National Science Foundation under Grants ATM 80-13707 and ATM 80-05356.

I. INTRODUCTION

An optical attenuation technique has been developed at Lawrence Berkeley Laboratory (LBL) to determine the absorption coefficient of aerosol particles (1). This technique has been applied on a routine basis to the analysis of filter samples collected in many urban (2) and rural areas in the United States, several sites in Europe, and in the Arctic region (3).

The LBL technique compares the transmission of a 633-nm He-Ne laser beam through a loaded filter (Millipore, quartz, and certain other filter media) relative to that of a blank filter (Fig. 1). The loaded filters are placed in the beam with the loaded side toward the laser; after multiple scattering through the filter substrate, the light is collected by an f/1 lens and focused on a photomultiplier tube. The absorption coefficient σ_a is determined from the Beer-Lambert law in a way similar to that outlined for the integrating plate method (4). This method measures the absorbing component of aerosol particles and is apparently insensitive to its scattering properties, as demonstrated by a wide range of experiments, including solvent-extraction/thermal-analysis studies (5) and photoacoustic spectroscopy (6). (Also see results on $(NH_4)_2SO_4$ + mixtures of soot and $(NH_4)_2SO_4$ presented in this report.) Furthermore, Raman scattering (1) and thermal analysis experiments (7) indicate that the optical attenuation measurement is proportional to the graphitic content of aerosols collected directly from combustion sources or urban air.

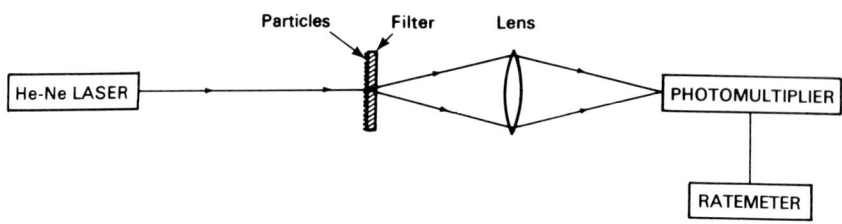

FIGURE 1. Schematic of LBL laser transmission apparatus.

The most direct substantiation that the optical attenuation technique selectively measures the absorbing component of aerosol particles is provided by a set of photoacoustic measurements (5) done in collaboration with the Applied Laser Spectroscopy group headed by Dr. N. M. Amer at LBL. Unlike conventional optical absorption techniques, photoacoustic spectroscopy measures the energy deposited in a sample due to absorption. Since questions have been raised regarding whether the LBL laser transmission method exclusively measures the absorbing rather than the scattering component of the aerosol, a comparison between photoacoustic and optical attenuation measurements made on the same aerosol sample should help resolve this ambiguity.

The photoacoustic measurements were made in an acoustically nonresonant detector with cylindrical geometry (Fig. 2). A Knowles microphone (Model BT-1759) was used, and the cell was 2.1 cm in diameter and 0.3 cm in length. The gas in the detector cell was air at atmospheric pressure. A He-Ne laser operating at 632.8 nm with 0.5 mW of power was used as the light source, and the experiments were performed at a modulation frequency of 20 Hz. The aerosol particles, collected on 1.2-μm Millipore filter substrates, were mounted on a 1.5-μm-thick Pyrex backing with the particles facing the incident light beam. Experiments were also performed with the laser beam first incident on the filter substrate.

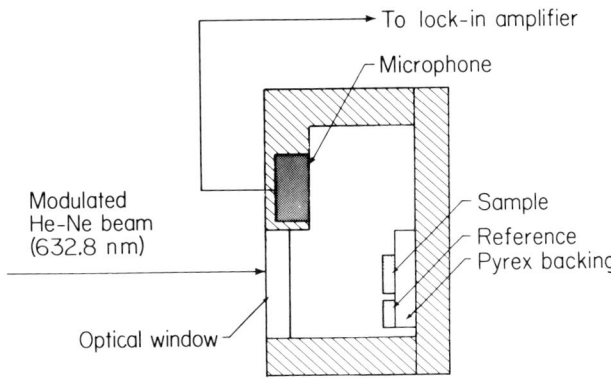

FIGURE 2. *Photoacoustic experimental arrangement.*

It is easy to show in the limit of low frequency light modulation (8) (≤ 100 Hz) that the ratio of the photoacoustic signal (V) to a reference sample for which the signal is saturated (V_{sat}) is given by

$$S_{ph} = V/V_{sat} = 1 - \exp(-\alpha l) \tag{1}$$

where α is the absorption coefficient and l is the effective pathlength.
This saturable behavior was observed for highly absorbing samples, and the sample which yielded the largest photoacoustic signal was used as the reference V_{sat}. Note that such samples yield values of $\alpha l \geq 3$, as deduced from the optical attenuation measurements; hence the highest signal obtained from available samples is close to the actual saturation value.

The experimental setup for the optical attenuation measurements is described above. In this technique the signal S_{op} is defined as $1 - \exp(-x)$, where x is the optical attenuation of the sample and is given by $-\ln I/I_o$, where I is the transmitted intensity of a loaded filter and I_o is the transmitted intensity of a blank filter.

Figure 3 presents a plot of the normalized photoacoustic signal S_{ph} versus S_{op} for a wide range of ambient samples and samples collected directly from combustion sources. The samples include urban particulates collected over a 24-hour period in Fremont and Anaheim, California, Denver, Colorado, and New York, New York, and particles collected in a highway tunnel and from an acetylene torch. The least squares fit of the experimental points yields a correlation coefficient r of 0.98 and a slope of 1.03, which would be expected if both techniques measure the same optical property of the aerosol particles. Since the photoacoustic signal is proportional to the heat generated by absorption, it is concluded that the optical attenuation method measures the light absorbing component of the aerosol particles.

From a theoretical point of view this result is somewhat surprising, since aerosol particles have a large scattering coefficient which would be expected to contribute to the optical attenuation measurement and not to the photoacoustic signal. This is especially true where the absorbing component represents only a small fraction of the aerosol mass. A simple model calculation will be presented which explains these observations and points out the critical role of the filter substrate as an almost perfect diffuse reflector in the technique. Similar considerations may also apply to the opal glass method of Lin et al. (4).

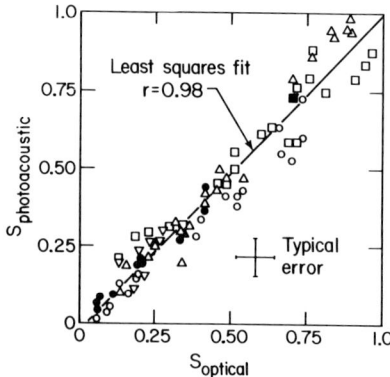

FIGURE 3. Plot of S_{ph} versus S_{op} for various samples: ▽—Fremont; □—Anaheim; ○—Denver; △—New York City; ■—highway tunnel; ●—acetylene torch. The solid line is a least squares fit of the data.

II. INFLUENCE OF AEROSOL SCATTERING

For this model calculation it will be assumed that the particles and the filter media can be treated independently and the geometry shown in Fig. 4 will be considered. A similar treatment, in which the light beam is first incident on the particles, gives identical results. After the light beam passes through the filter medium, it is incident on the particles with an intensity I_o. The particles forward-scatter a fraction of the incident light, backward-scatter a fraction, and absorb a fraction. These components in the low loading limit are given by $n_s\sigma_F I_o$, $n_A\sigma_B I_o$, and $n_A\sigma_A I_o$, where n_s is the number of scattering aerosol particles per unit area, n_A is the number of absorbing aerosols per unit area, σ_F is the forward scattering cross section, σ_B is the backward scattering cross section, and σ_A is the absorption cross section.

Since the optical attenuation technique measures only the forward-scattering light, it would seem as if the backscattered light would be lost to the system and would contribute to the attenuation. However, in this method, the filter is an almost perfect reflector. Under these circumstances, the backscattered light will be reflected in the forward direction and will again be incident on the particles. This process will continue until almost all the backscattered radiation is collected by the optics and therefore does not contribute to the optical attenuation. This result can be put in mathematical form, where I is the light detected by the collection optics and R_F is the reflectivity of the substrate.

$$\begin{aligned}I &= I_o(1 - n_s\sigma_B - n_A\sigma_A) + I_o(1 - n_s\sigma_B - n_A\sigma_A)n_s\sigma_B R_F \\ &\quad + I_o(1 - n_s\sigma_B - n_A\sigma_A)n_s^2\sigma_B^2 R_F^2 \\ &\quad + \ldots I_o(1 - n_s\sigma_B - n_A\sigma_A)n_s^n\sigma_B^n R_F^n \\ &= I_o \left(\frac{1 - n_s\sigma_B - n_A\sigma_A}{1 - n_s\sigma_B R_F} \right)\end{aligned} \quad (2)$$

Consider several limits. Where $R_F \cong 0$, which normally would be considered an ideal substrate, Eq. (2) reduces to

$$I = I_o(1 - n_s\sigma_B - n_A\sigma_A) \quad (3)$$

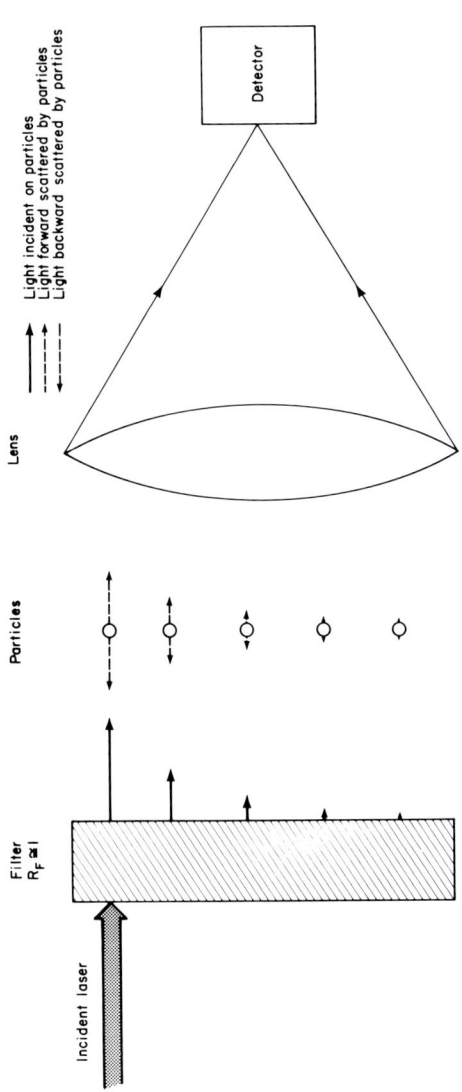

FIGURE 4. Schematic of experimental arrangement used for model calculation.

Under these conditions the backscattered radiation will contribute significantly to the optical attenuation and make the technique unsuitable for exclusively measuring the absorbing properties of the aerosol. In this method, however, $R_F \cong 1$ and Eq. (2) becomes

$$I = I_o \left(1 - \frac{n_A \sigma_A}{1 - n_s \sigma_B} \right) \quad (4)$$

Or, the optical attenuation in the low loading limit is

$$-\ln \frac{I}{I_o} \cong I_o - I = \frac{I_o n_A \sigma_A}{1 - n_s \sigma_b} \quad (5)$$

From this expression it is clear that a purely scattering aerosol will make no contribution to the optical attenuation; this is consistent with experimental results. However, for partially absorbing aerosol the optical attenuation is somewhat dependent on the scattering properties of the aerosol, although in the low loading limit this effect is small. For example, if the substrate has 50% coverage, and if the scattering cross section of the particles σ_s is twice the particle area, then $n_s \sigma_s \cong 1$. If σ_B is about 20% of σ_s [the maximum value measured by Charlson and coworkers (9)], then

$$1 - n_s \sigma_B \cong 0.8 \quad (6)$$

so that even for this rather high loading, the error in the absorption measurement due to scattering of the aerosol is only about 20%. This treatment should only be viewed as giving physical insight into the LBL method for determining absorption coefficients and clearly is approximate since it assumes that the scattering properties of the particles are not affected by the filter substrate and neglects the penetration of the particles into the substrate. Future analysis will try to evaluate the significance of these effects.

III. RESULTS OF THE INTERCOMPARISON

For the purpose of this intercomparison, parallel filter samples were collected on 47-mm Millipore (1.2-μm nominal pore size, type RATF) and prefired quartz fiber filters (Pallflex type 2500) at a nominal flow rate of 2.4×10^{-4} m³/sec. The flow rate through the filters was monitored continuously by a pressure sensor at the inlet of the pumps, and absolute flow rate calibrations were done on a daily basis using a Rockwell Model 5110 gas meter. (No altitude or temperature corrections were made in the gas meter flow rate calibration.) Optical attenuation measurements at a wavelength of 0.63 μm were done on both the Millipore and quartz fiber substrates. The optical attenuation is defined as ATN = $-100 \ln I/I_o$, where I_o is the intensity of the light transmitted through a blank filter substrate and I is the intensity through a loaded filter. The quartz filters were also used to determine the total carbon content of the aerosols by a combustion method (10). Typical loading of dynamic blanks was 1 ± 0.3 μg/cm^2 of total carbon.

The optical attenuation measurements and total carbon loadings are reported in Tables 1 and 2 for the various aerosols investigated in the intercomparison. The optical attenuation measurements on quartz and Millipore substrates show a high degree of correlation as shown in Fig. 5. The least squares fit of the data has a correlation coefficient r = 0.99 and a slope of 1.15. The differences in the optical attenuation measurement on these two rather different collection media appear to be minimal. The optical attenuation measurements also show a high degree of correlation with the total carbon content of the combustion-generated aerosols (soot and soot + sulfate) as shown in Fig. 6. Such a correlation would be expected when graphitic carbon represents a fixed fraction of the total carbon content of the aerosol. No systematic differences are found in the slope of this line, even when considerable amounts of $(NH_4)_2SO_4$ are present. Although this data set is limited, the result is a further confirmation of the fact that the optical attenuation measurement is insensitive to the scattering properties of the aerosol. It should also be noted that the optical attenuation of a pure $(NH_4)_2SO_4$ aerosol was below the detection limit of the measurement (see Table 1).

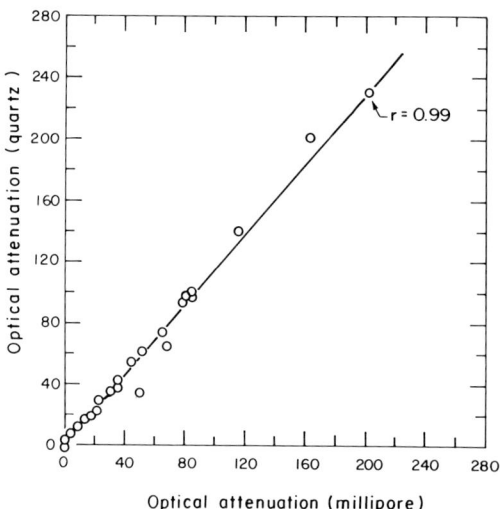

FIGURE 5. *Comparison of optical attenuation determined on quartz and Millipore substrates.*

TABLE 1. Optical Attenuation Measurements, Absorption Coefficients, and Specific Absorption Coefficients Determined from Millipore Substrates.

Sample	Date	Time	Aerosol	Flow (m^3/cm^2)	ATN	σ_a (m^{-1})	µg C/m^{3a}	Specific absorption coefficient (m^2/g)
7/29/1	7/29	1529-1600	$(NH_4)_2SO_4$	5.0×10^{-2}		0		
7/29/2	7/29	1635-1733	$(NH_4)_2SO_4$ + 1% MB					
7/30/1	7/30	1120-1232	$(NH_4)_2SO_4$	9.4×10^{-2}	52	3.4×10^{-4}		
7/30/2	7/30	1327-1527	$(NH_4)_2SO_4$ + 0.3% MB	1.08×10^{-1}	-4.8	b		
7/30/3	7/30	1636-1741	$(NH_4)_2SO_4$ + 4% MB	1.8×10^{-1}	8.6	3.0×10^{-5}		
7/31/1	7/31	1015-1025	Soot	9.8×10^{-2}	85	5.4×10^{-4}		
7/31/2	7/31	1035-1040	Soot	1.4×10^{-2}	44	1.9×10^{-3}	193	9.8
7/31/3	7/31	1048-1113	Soot	6.8×10^{-3}	21	1.9×10^{-3}		
7/31/4	7/31	1124-1126	Soot	3.5×10^{-2}	79	1.4×10^{-3}	180	7.8
7/31/5	7/31	1133-1136	Soot	Abort				
7/31/6	7/31	1143-1145	Soot	4.2×10^{-3}	14	2.1×10^{-3}		
7/31/7	7/31	1157-1204	Soot	2.4×10^{-3}	3.4	---		
7/31/8	7/31	1215-1230	Soot	1.0×10^{-2}	35	2.2×10^{-3}	177	10.7
7/31/9	7/31	1238-1345	Soot	2.1×10^{-2}	65	1.9×10^{-3}	159	6.9
7/31/10	7/31	1442-1452	Soot + $(NH_4)_2SO_4$	9.5×10^{-2}	163	1.1×10^{-2}		
7/31/11	7/31	1502-1522	Soot + $(NH_4)_2SO_4$	1.4×10^{-2}	18	8.0×10^{-4}		
7/31/12	7/31	1533-1640	Soot + $(NH_4)_2SO_4$	2.8×10^{-2}	35	7.8×10^{-4}		
8/1/1	8/1	1130-1203	MB	9.5×10^{-2}	86	5.6×10^{-4}	48	11.7
8/1/2	8/1	1212-1218	MB	3.9×10^{-2}	118	1.9×10^{-3}		
				7.8×10^{-3}	36	2.9×10^{-3}		

TABLE 1. Continued.

Sample	Date	Time	Aerosol	Flow (m^3/cm^2)	ATN	σ_a (m^{-1})	$\mu g\ C/m^{3a}$	Specific absorption coefficient (m^2/g)
8/1/3	8/1	1226-1239	MB	1.7×10^{-2}	85	3.1×10^{-3}		
8/1/4	8/1	1247-1325	MB	4.9×10^{-2}	133	1.7×10^{-3}		
8/1/5	8/1	1435-1458	Soot	2.8×10^{-2}	68	1.5×10^{-3}		
8/1/6	8/1	1505-1510	Soot	6.6×10^{-3}	18	1.7×10^{-3}		
8/1/7	8/1	1517-1527	Soot	1.3×10^{-2}	31	1.5×10^{-3}		
8/4/1	8/4	1053-1113	Soot	2.8×10^{-2}	69	1.5×10^{-3}		
8/4/2	8/4	1140-1208	Soot + $(NH_4)_2SO_4$	3.9×10^{-2}	85	1.4×10^{-3}		
8/4/3	8/4	1214-1219	Soot + $(NH_4)_2SO_4$	6.6×10^{-3}	28	2.6×10^{-3}		
8/4/4	8/4	1226-1315	Soot + $(NH_4)_2SO_4$	6.5×10^{-2}	99	9.4×10^{-4}		
8/4/5	8/4	1322-1332	Soot + $(NH_4)_2SO_4$	1.3×10^{-2}	37	1.8×10^{-3}		
8/4/6	8/4	1337-1352	Soot + $(NH_4)_2SO_4$	2.0×10^{-2}	47	1.5×10^{-3}		
8/5/1	8/5-8/6	1027-0827	Ambient	1.7	22	8×10^{-6}		
8/6/1	8/6	1044-1331	Arizona	2.1×10^{-1}	4.1	b		
8/6/2	8/6	1455-1659	Soot + $(NH_4)_2SO_4$	1.6×10^{-1}	113	4.4×10^{-4}		
8/6/3	8/6	1741-1802	Soot	2.7×10^{-2}	83	1.9×10^{-3}	190	10.0
8/7/1	8/7	1145-1224	Soot + $(NH_4)_2SO_4$	4.9×10^{-2}	50	6.3×10^{-4}		
8/8/1	8/8	0952-1051	MB	7.5×10^{-2}	219	1.8×10^{-3}		Average specific absorption = $9.5 \pm 1.8\ m^2/g$

[a] Only total carbon loadings with an error of less than ± 25% are reported. [b] Below detection limit.

TABLE 2. Optical Attenuation Measurements and Total Carbon Loadings Determined from Quartz Substrates.

Sample	Date	Time	Aerosol	Flow (m^3/cm^2)	ATN	ATN^a	$\mu g\ C\ cm^2$	$\mu g\ C\ m^3$
7/29/1	7/29	1529-1600	$(NH_4)_2SO_4$	4.8×10^{-2}	1.7	b		
7/29/2	7/29	1635-1733	$(NH_4)_2SO_4$ + 1% MB	8.7×10^{-2}	57	62		
7/30/1	7/30	1120-1232	$(NH_4)_2SO_4$	1.0×10^{-1}	-2	b		
7/30/2	7/30	1327-1526	$(NH_4)_2SO_4$ + 0.3% MB	1.7×10^{-1}	12	13		
7/30/3	7/30	1636-1741	$(NH_4)_2SO_4$ + 4% MB	9.4×10^{-2}	93	97		
7/31/1	7/31	1015-1025	Soot	1.4×10^{-2}	57	57	2.7	193
7/31/2	7/31	1035-1040	Soot	7.3×10^{-3}	32	30		
7/31/3	7/31	1048-1113	Soot	3.6×10^{-2}	97	94	6.5	180
7/31/4	7/31	1124-1126	Soot	2.9×10^{-3}	9.3			
7/31/5	7/31	1133-1136	Soot	4.4×10^{-3}	18	17		
7/31/6	7/31	1143-1145	Soot	2.5×10^{-3}	6.4	6.1		
7/31/7	7/31	1157-1204	Soot	1.1×10^{-2}	46	42		
7/31/8	7/31	1215-1230	Soot	2.2×10^{-2}	76	73	3.9	177
7/31/9	7/31	1238-1345	Soot	9.7×10^{-2}	206	202	15.4	159
7/31/10	7/31	1442-1452	Soot + $(NH_4)_2SO_4$	1.3×10^{-2}	18	19		
7/31/11	7/31	1502-1522	Soot + $(NH_4)_2SO_4$	2.9×10^{-2}	41	39		
7/31/12	7/31	1533-1640	Soot + $(NH_4)_2SO_4$	9.7×10^{-2}	100	98	4.7	48
8/1/1	8/1	1130-1203	MB	4.1×10^{-2}	185	176		
8/1/2	8/1	1212-1218	MB	9.0×10^{-3}	114	99		
8/1/3	8/1	1226-1238	MB	1.8×10^{-2}	153	145		
8/1/4	8/1	1247-1325	MB	5.8×10^{-2}	169	143		
8/1/5	8/1	1435-1458	Soot	3.4×10^{-2}	79	65		
8/1/6	8/1	1505-1510	Soot	7.5×10^{-3}	19	17		
8/1/7	8/1	1517-1527	Soot	1.5×10^{-2}	41	36		
8/6/3	8/6	1741-1802	Soot + $(NH_4)_2SO_4$	3.2×10^{-2}	116	98	6.1	190
8/7/1	8/7	1145-1224	MB	5.9×10^{-2}	41	34		
8/8/1	8/8	0952-1051	Soot + $(NH_4)_2SO_4$	8.9×10^{-2}	310	261		
8/6/2(?)	8/6	1455-1659		0.19	140	118		

^{a}Normalized to Millipore flow rate. ^{b}Below detection limit.

The absorption coefficients reported in Table 1 were obtained from optical attenuation measurements on the Milipore substrates in the following manner

$$b_{ab}(m^{-1}) = \frac{(ATN)(10^{-6})C}{F} \tag{7}$$

where F is the flow rate in m^3/cm^2 and C is an experimentally determined correction factor which takes into account the penetration of particles in the filter substrate. For the data reported here, a correction factor of 0.62 has been used. This correction term has been determined in a preliminary way by comparing the optical attenuation of parallel filter samples of urban air collected on Millipore and very thin, almost transparent Teflon substrates where enhanced absorption due to multiple scattering is expected to be small (Fig. 7). The optical attenuation measurement on the Teflon filters was made with the geometry shown in Fig. 4, using a blank Millipore filter as the diffuse reflector. Also included in Table 1 are values of the specific absorption coefficient for the soot aerosols. These values are given whenever the carbon loading is determined with an accuracy of ± 25% or better. The absorption coefficients reported here are consistent with the optical constants of graphitic carbon and are expected to have an accuracy of better than a factor of two.

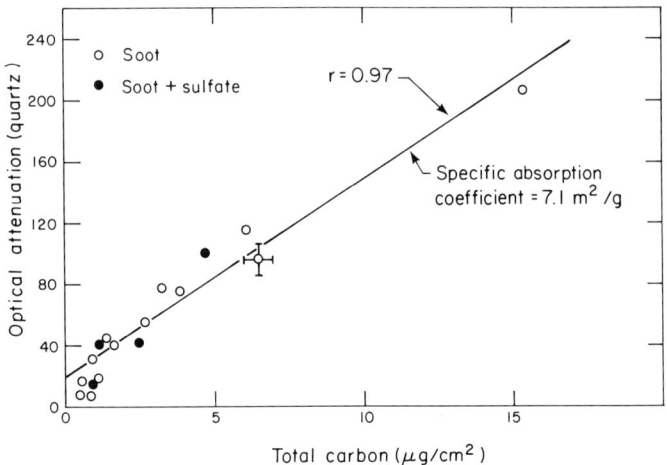

FIGURE 6. *Comparison of optical attenuation determined on quartz substrates with total carbon loading for combustion-generated aerosols.*

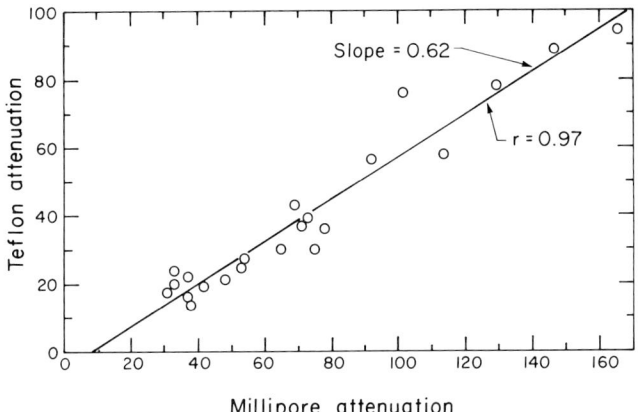

FIGURE 7. Comparison between optical attenuation measurements on Millipore and Teflon substrates.

REFERENCES

1. Rosen, H., Hansen, A. D. A., Gundel, L., and Novakov, T., Identification of the optically absorbing component in urban aerosols, *Appl. Opt. 17*, 3859 (1978).
2. Rosen, H., Hansen, A. D. A., Dod, R. L., and Novakov, T., Soot in urban atmospheres: Determination by an optical absorption technique, *Science, 208*, 741 (1980).
3. Rosen, H. Novakov, T., and Bodhaine, B., Soot in the Arctic, *Atmos. Environ. 15*, 1371 (1981).
4. Lin, C. I., Baker, M., and Charlson, R. J., Absorption coefficient of atmospheric aerosols: A method for measurement, *Appl. Opt. 12*, 1356 (1973).
5. Rosen, H., Hansen, A. D. A., Gundel, L., and Novakov, T., Identification of the graphitic carbon component of source and ambient particulates by Raman spectroscopy and an optical attenuation technique, *Proceedings Conference on Carbonaceous Particles in the Atmosphere*, p. 49, Lawrence Berkeley Laboratory Report LBL-9037 (1979).
6. Yasa, Z., Amer, N. M., Rosen, H., Hansen, A. D. A., and Novakov, T., Photoacoustic investigation of urban aerosol particles, *Appl. Opt. 18*, 2528 (1979).
7. Gundel, L., Dod, R., and Novakov, T., Determination of black carbon by thermal analysis, *Environmental Pollutant Studies*, p. 5, Lawrence Berkeley Laboratory Report LBL-11986 (1981).

8. Rosencwaig, A., and Gersho, A., Theory of the photoacoustic effect with solids, *J. Appl. Phys. 47,* 64 (1976).
9. Weiss, R., Charlson, R. J., Waggoner, A. P., Baker, M. B., Covert, D., Thorsell, D., and Yuen, S., Application of directly measured aerosol radiative properties to climate models, *Atmospheric Aerosols: Their Optical Properties and Effects,* NASA CP-2004 (1976)
10. Mueller, P. K., Mosley, R. W., and Pierce, L. B., Carbonate and non-carbonate carbon in atmospheric particulates, *Proceedings, Second International Clean Air Congress,* Academic Press, New York (1971).

REMOTE SENSING OF AEROSOL IN THE FREE ATMOSPHERE BY PASSIVE OPTICAL TECHNIQUES[1]

Glenn E. Shaw

Geophysical Institute
University of Alaska
Fairbanks, Alaska

Passive, monochromatic measurements were made at Fort Collins, Colorado, of the sun's irradiance, the solar aureole intensity, and the diffuse downwelling radiant flux from the sky. The objective of the measurements was to determine, by inversion, physical properties of the aerosol suspended in the air column. The mathematical inversion algorithms employed constrained linear smoothing; the input data to the inversion routine consisted of a combination of aureole intensity and wavelength dependence of atmospheric optical thickness. The aerosol optical thickness (at 500 nm wavelength) above Fort Collins averaged 0.07 and underwent a strong diurnal variation, generally increasing throughout the day and reaching maximum values just prior to the onset of cumulus cloudiness in the afternoons. Though the size distribution spectrum of the aerosol particles also changed systematically throughout the day, these variations were relatively modest. A bimodal aerosol size distribution was found at Fort Collins, with particles centered around a mean radius of 3 μm consisting of crustal material, and particles centered around 0.2-μm radius consisting of hygroscopic sulfur-rich particles. The mean-column mass loadings of the crustal and sulfur-bearing particle components were 14×10^{-7} and 5×10^{-7} g cm^{-2}, respectively. It was inferred, by matching the observed downwelling flux of radiation from the sky to theoretical predictions of a two-stream radiative transfer model, that the particles above Fort Collins were lightly absorbing; the single scattering albedo $\tilde{\omega}$ was estimated to be equal to 0.95 ± 0.05.

[1] This research was sponsored by the National Science Foundation under Grant DPP 79-19816 and by the Office of Naval Research under Contract N-00014-76C-0435.

I. INTRODUCTION

The spectral composition of light from the sun and from the sky is strongly affected by the presence of even extremely minute amounts of submicron-size particles suspended in the atmosphere. This is because particles with dimensions comparable to the wavelengths of light resonate with photons and cause them to be scattered and, in some instances, absorbed. To appreciate the sensitivity of the particle-wave interactions caused by particles in the atmosphere, consider that the optical transmissivity for light (wavelength 500 nm) passing through a pure oxygen-nitrogen Rayleigh-scattering atmospheric column is about 90% while the Mie optical transmission through the same column mass loading of 0.3-μm-radius particles would be less than 1 part in 10^{10} (Fig. 1).

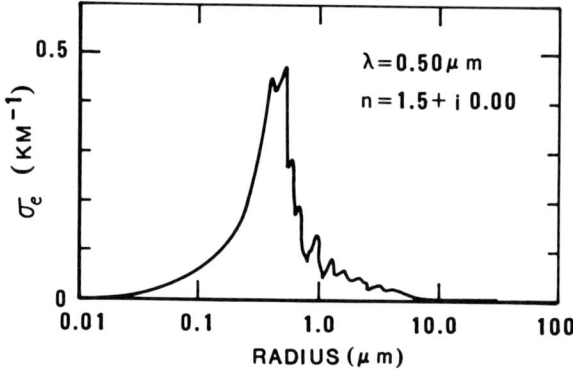

FIGURE 1. *Computed aerosol volume extinction coefficient σ_e arising from 50 μg m^{-3} mass loading of monodisperse particles of radius r. Particles $\sim$ 0.3-μm radius are most effective for a given mass loading of particles at attenuating light.*

The extreme sensitivity of the particle-wave interaction, for submicron particles, formed the basis of the Geophysical Institute's aerosol experiments at the First International Workshop on Light Absorption by Aerosol Particles. There were two scientific objectives for this study, (a) to employ passive sensing spectral optical techniques to study the physics of suspended particles in a relatively nonpolluted semi-arid region of the western United States, and (b) to judge the quality of the indirectly sensed quantities by intercomparing them with independent data obtained by other investigators who used different experimental techniques.

II. THEORY OF OPTICAL INVERSIONS TO RECOVER AEROSOL PROPERTIES.

The aim of making the optical spectral measurements of skylight and sunlight in the natural environment at Fort Collins was to deduce the optical and physical characteristics of suspended particles in the atmosphere. To understand how this was done, one must realize that the angular and wavelength dependence of direct and scattered solar radiation are functions, through linear mathematical relationships (solutions to Maxwell's equations), of the size and composition of the scattering centers which, in this case, are the particles contained in the size range $\sim 10^{-2}$ to 10 μm in diameter. Particles smaller than the specified range coagulate rapidly, while larger ones fall out quickly once they are introduced into the atmosphere.

The functional dependencies of the radiant parameters on particle size and composition make it possible to invert the electromagnetic scattering relationships to recover the aerosol size spectrum. It should be kept in mind, however, that the process of formal inversion to recover the particle size spectrum and the particle's absorptive characteristics does not necessarily generate any unique solution. Since the set of linear equations relating optical variables and particle variables tends to be poorly conditioned, even tiny errors in the measured optical parameter introduce severe instabilities in the solution. Because of such mathematical difficulties, a strategy must be developed to confront the possibility of nonuniqueness in the solutions. Fortunately it turns out that, of the many allowed solution vectors (in this case the solution vector is the particle size distribution spectrum), most contain high frequency components which cannot possibly correspond to real particle distributions in the natural environment, because spikes in an aerosol size distribution function would rapidly smear out due to particle coagulation. Thus, some *a priori* knowledge of the general frequency characteristics of the many allowed solutions that can occur in nature, and in particular the fact that particle size spectra tend to be reasonably well behaved continuous functions, allows one to introduce smoothing constraints into the inversion process. Provided the correct measurements are made, and provided that they are made to high enough accuracy, acceptable estimates of the actual aerosol size distribution function existing in the atmosphere can in fact be obtained. An excellent discussion of the capabilities and limitations of such inversion methods can be found in Parker (1) and Twomey (2).

Making the assumption that the scattering of light from the aerosol particles can be adequately modeled by assuming the individual particles to be spherical, one can, for a heterodisperse cloud of aerosol, express the spectral aerosol optical extinction at wavelength λ in terms of an optical thickness $\tau(\lambda)$

$$\tau(\lambda) = H \int_0^\infty C_e (2\pi r/\lambda) N(r) dr \tag{1}$$

where N(r)dr represents the aerosol number density for particles in the free atmosphere between radius limits r and r + dr, and C_e is the particle extinction cross section (cm^2) computed by classical electromagnetic theory (3). The parameter C_e, as in all diffraction and interference problems, is a function only of the ratio λ/r and of the real and imaginary components of the particle's refractive index. The value H represents the length of an equivalent atmospheric column containing homogeneous aerosol of number density N(r); H is typically about 1 km for natural aerosol particles.

In principle N(r) as a function of r may be recovered from measurements of $\tau(\lambda)$ by inverting the integral in Eq. (1). However, Twomey (2) has shown that the information content from such inversions is quite small; typically there are only about three linearly independent eigenvectors for the case where $\tau(\lambda)$ is measured to an accuracy of 1%. To improve the accuracy of the determination of N(r), and especially to make the inferences more meaningful for "giant" aerosols with r $\gtrsim$ 1 μm, one can profit by simultaneously measuring the sky intensity (sky radiance) at angles close to the sun. Such sky brightening near the sun is due to light from the sun diffracting around the particles, which to first approximation is similar to Fraunhofer diffraction by a disk of radius r. The diffracted light is contained mainly within primary scattering angles θ_o, where θ_o is given approximately by $\theta_o = \sin^{-1}(0.6\lambda/r)$. Thus, for light of wavelength 0.5 μm, particles with r = 10 μm cause a brightening (an aureole) around the sun of angular radius $\theta_o \cong 2°$, while 1-μm particles result in an aureole around 20°. One can, therefore, immediately sense that the size distribution of particles in the range r > 1 μm might be deduced by measuring the solar Fraunhofer pattern out to about 20° from the sun and using inversion theory.

To quantify the theory of the solar aureole brightening, one must solve the problem of plane waves scattering from dielectric spheres. Given in terms of Mie-scattering expressions (3,4), the solar aureole sky intensity is given by

$$I(\lambda,\theta)/F_o(\lambda) = \frac{H}{k^2} \int_0^\infty \left(\frac{M_1 + M_2}{2}\right) N(r)dr \qquad (2)$$

Equation (2) relates the sky radiance I for an overhead sun (erg cm^{-3} steradian^{-1}) at wavelength λ and at primary scattering angle θ to the particle size spectrum N(r). The Mie-scattering terms M_1 and M_2 (4) are functions of λ/r, the particle's refractive index, and θ; k is the wave index $2\pi/\lambda$, and $F_o(\lambda)$ is the incident spectral solar irradiance (erg cm^{-3}).

The kernels of Eqs. (1) and (2) are shown in Fig. 2, from which it can be seen that combined spectral extinction measurements in the wavelength range 400 < λ < 1000 nm and aureole brightness measurements in the angular range 1° < θ < 30°, if considered simultaneously, allow one to recover the particle size spectrum for particles in the approximate size range 0.1 < r < 10 μm.

PASSIVE OPTICAL TECHNIQUES

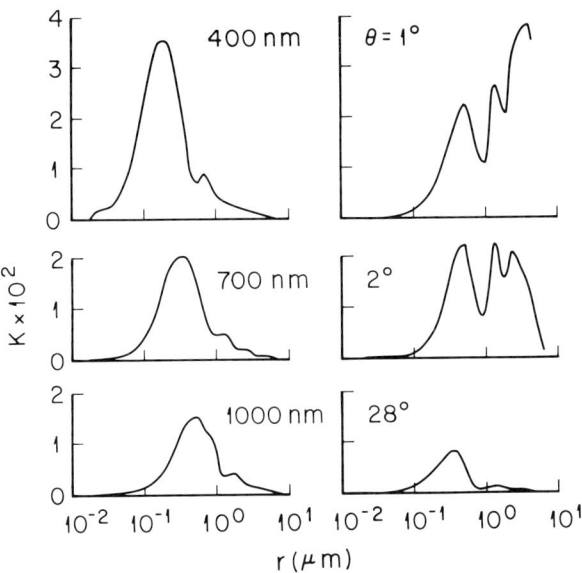

FIGURE 2. Kernel functions in Eqs. (1) and (2) weighted by a standard aerosol of 50 μg m^{-3} in a 1-km column and having a power law size distribution of the form dn/dlnr ∝ r^{-3}. The scattering kernels are evaluated at a wavelength of 700 nm.

The integrals in Eqs. (1) and (2), when broken into sums over discrete values of r by quadrature, form a set of linear equations of the form

$$[g] = [A][F] \tag{3}$$

which by a linear transformation relate g_i, the optical quantities (spectral extinction τ and aureole sky brightness I_i), and particle size characteristics f_j, at radii r_j. The solution to Eq. (3), realizing that $g' = g + \varepsilon$, where ε is a measurement error vector which is subject to the constraint of minimizing the quadratic sum of the solution's second differences

$$Q_{MIN} = \sum_{2}^{n-1} (F_{i-1} - 2f_i + f_{i+1})^2 \tag{4}$$

is given by

$$F = (A^*A + \gamma H)^{-1} A^*g \tag{5}$$

where A^* is the transpose matrix of A, γ is a parameter that determines the amount of smoothing used (which depends on the magnitude of ϵ), and H is a smoothing matrix (5).

It is important to remember in this particular application of inversion theory that the information content, which is related to the degree of independence of the measurements of radiation quantities, has been greatly improved over previously existing remote sensing experiments by incorporating a combination of optical scattering (its wavelength dependence) and optical extinction (its angular dependence).

III. EXPERIMENTAL DETERMINATIONS OF $\tau(\lambda)$ AND $I_{SKY}(\theta)$

1. Optical Extinction

The optical thickness referred to the vertical direction (sometimes called the optical depth) τ of a planar atmosphere is related to the monochromatic transmission $\delta(\lambda)$ of plane waves passing through the atmosphere from the sun at zenith angle z_0 by

$$\delta(\lambda) = F(\lambda)/F_0(\lambda) = \exp(-\tau/\cos z_0) \tag{6}$$

where F and F_0 are the solar irradiance values at the bottom and top of the atmosphere, respectively. Samuel Langley in the last years of the nineteenth century devised a scheme, now named after him, to determine τ by measuring the relative strength of direct monochromatic sunlight F at different solar zenith angles z_i. The method involves fitting a curve through a Cartesian coordinate plot of log $V(\lambda)$ (V is the spectrometer's reading at wavelength λ) against $[\cos(z_0)]^{-1}$, which, as an inspection of Eq. (4) shows, is expected to be linear. This old method of Langley's is still favored by scientists as being the most accurate way to experimentally deduce $\tau(\lambda)$ and it was the underlying method of obtaining the values of $\tau(\lambda)$ being reported.

At Fort Collins, though it is not greatly polluted, there is detectable pollution, which, moreover, changes throughout the day. It was therefore thought best not to employ the Langley method directly at Fort Collins. Instead, a very stable (drift rate 0.5%/year) narrow-band multiwavelength photometer was accurately calibrated by the Langley method before and after the aerosol workshop at Mauna Loa Observatory, where conditions over the day are extremely stable. Values of $\tau(\lambda)$ obtained with this instrument are accurate to $\tau_0 \pm 0.002$.

2. Solar Aureole Radiance and Its Gradient

In the case of the measurements of sky intensity $I_0(\lambda)$ and its angular gradient at angles within a few degrees of the sun, a photoelectric coronameter

PASSIVE OPTICAL TECHNIQUES   341

was employed to ensure that stray light diffracted into the photometer was negligible (6). A drawing of the instrument is shown in Fig. 3. Stray light is a potentially very serious cause of systematic error.

Calibration of the sky radiance in units of incoming solar irradiance at the top of the atmosphere F_o [see Eq. (2)] was obtained by sensing the light from a Lambertian screen (with known diffuse reflectance) which was illuminated sequentially by global radiation (sky + sun) and then by diffuse radiation. Subtraction of the two readings gave the instrument's calibration constant (in terms of direct solar irradiance) after correcting for the transmission loss of the solar beam through the atmosphere. The calibration of the coronameter has been constant to 1% over 3 years.

3. Diffuse Downwelling Spectral Flux Determinations

The diffuse radiant flux from the sky at monochromatic wavelengths was measured for reasons that will be described later; this quantity is defined as

$$F\downarrow = \int I(\mu,\phi)\mu d\omega \tag{7}$$

where ϕ is the azimuth of solid angle $d\omega$ with respect to the solar vertical plane, and the integral is taken over the upward hemisphere. This flux can be measured in principle by viewing the radiance transmitted or reflected by a shaded horizontal diffusing plate. In practice Lambert's cosine law is not obeyed exactly for a diffusing disk, and to improve its response one can devise an arrangement consisting of a diffusing vertical cylinder surrounded by concentric step rings so that the combination of umbral and penumbral shadows cast by the rings brings the system's response into closer accord with Lambert's law. The cross section of the geometry used, which was determined empirically, is shown in Fig. 4, along with the experimentally derived departure from the cosine response law.

Calibration of the spectral flux instrument was done (in terms of the monochromatic solar irradiance, $F\downarrow/F_o$) by exposing the detector sequentially to diffuse and global radiation and subtracting the two to determine the instrument's output to direct solar irradiance. One must, of course, correct for the transmissivity of the atmosphere.

IV. RESULTS

1. Time Variations in the Column-Integrated Aerosol Spectral Extinction

Spectral radiation data were obtained on 3 exceptionally clear days, 4, 6, and 7 August 1980, at the workshop laboratory in Fort Collins. These were hot days, with afternoon temperatures rising to approximately 40 °C. The sky appeared to be quite clear; however, a distinct whitish aureole was apparent

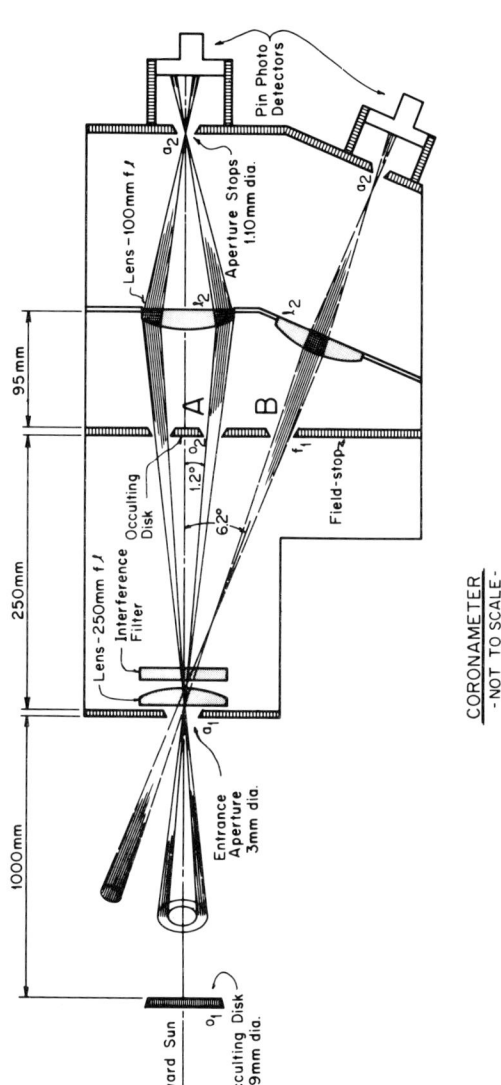

FIGURE 3. Optical diagram of the sky-brightness photoelectric coronameter (not to scale).

PASSIVE OPTICAL TECHNIQUES

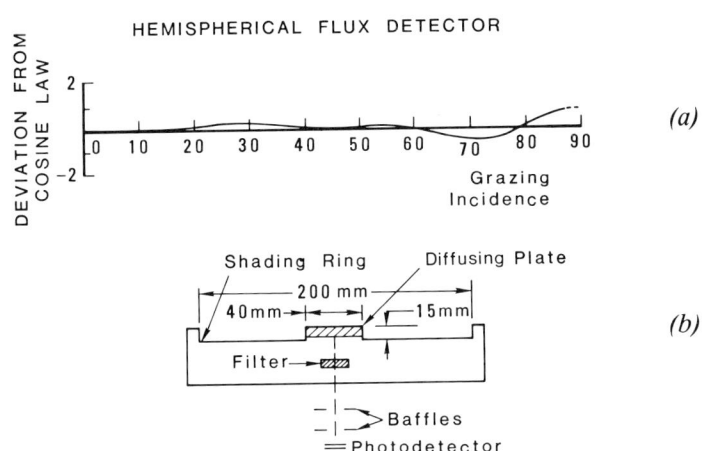

FIGURE 4. *Characteristics of sensor used to determine monochromatic diffuse sky flux. (a) Deviation of angular response from Lambert's cosine law in percent. (b) Cross-sectional diagram showing geometric relationships.*

around the sun, seeming to be about 15° to 20° in diameter. One had the impression when watching the sky that conditions were changing during the day, and indeed the sun photometry measurements, which are shown in Fig. 5, indicated that the total aerosol load underwent a change of about a factor of two during the day, with the most predominant pattern seeming to be an increase with time in the morning hours and fairly abrupt increases an hour or so before the onset of anomalous cloudiness in the late afternoon. These variations are understandable and seem to be caused, respectively, by the advection of pollution from the south and east and by hygroscopic particle growth in the high humidities preceding cloud formation.

2. Spectral Variations in Aerosol Extinction

The size of the mean aerosol in the air column above Fort Collins altered somewhat, but not greatly, throughout the day and also from day to day as indicated qualitatively from variations in the wavelength dependence of optical extinction. A diagram of the aerosol spectral extinction spectrum (aerosol optical thickness against log wavelength) is shown in Fig. 6 for several times each day during which measurements were made. The indication from these data is that the Fort Collins aerosol is not homogeneous but changes its concentration and, to a lesser extent, its size in a period of hours.

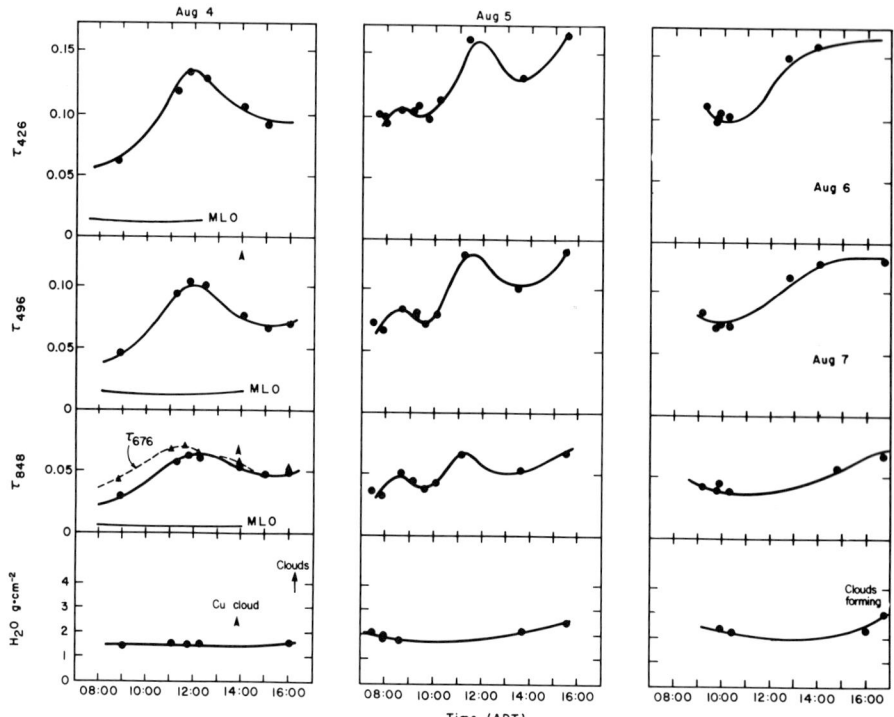

FIGURE 5. The variation in the atmospheric water vapor content is given in the bottom three graphs. Temporal variation of the aerosol optical thickness at three wavelengths of light.

3. The Diffracted Light Surrounding the Sun

It was found at Fort Collins that the aureole brightness 1° away from the sun's limb underwent strong variation, as shown in Fig. 7. Since, for the optically thin approximation, the aureole brightness would be expected to be inversely proportional to $\mu = \cos(z)$, the data shown in Fig. 7 are normalized by dividing the aureole intensity, $I(1°)/F_0$, by μ_0. It can be seen that, as one would expect, the aureole brightness waxes and wanes in synchronism with the optical thickness. Radiation removed by the direct beam by scattering reappears in the sky radiation field. A typical representative mid-morning aerosol aureole diffraction pattern at 1000 MDT, 6 August is illustrated in Fig. 8 for radiance at 700 nm wavelength.

PASSIVE OPTICAL TECHNIQUES

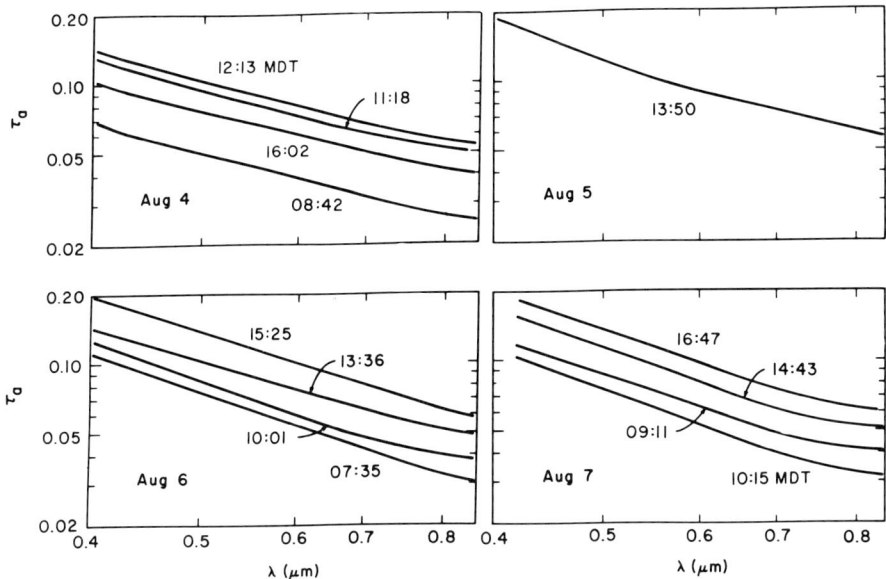

FIGURE 6. The spectral extinction spectrum for aerosols at Fort Collins at various times during the period 4 to 7 August 1980.

4. The Diffuse Monochromatic Radiant Flux from the Sky

In the experiments to derive the wavelength behavior of downwelling diffuse sky flux, the detector was shadowed from direct sunlight with an opaque disk which had an umbral angular diameter of 2° and a penumbral angular diameter of 8°; thus the radiance from the sky that was sensed by the detector did not include the light in the solar aureole out to a mean angular diameter of approximately 4°. However, since the aureole intensity was measured independently, it was possible to estimate the error caused by ignoring the part of the sky blocked out by the occulting disk; the necessary correction was about 2%.

The sky flux, of course, changed throughout the day as the solar elevation changed; to a first approximation the sky flux would be expected to be proportional to $\mu_0 = \cos(z_0)$. Hence, as in the case of the solar aureole measurements, the data are most effectively presented in terms of $(F\downarrow/F_0)\mu_0^{-1}$, since this quantity should be fairly constant if the aerosol loading remains unchanged. In fact, the flux normalized as just mentioned did change, and, as one would expect, this change was in synchronism with the optical thickness, as shown in Fig. 7.

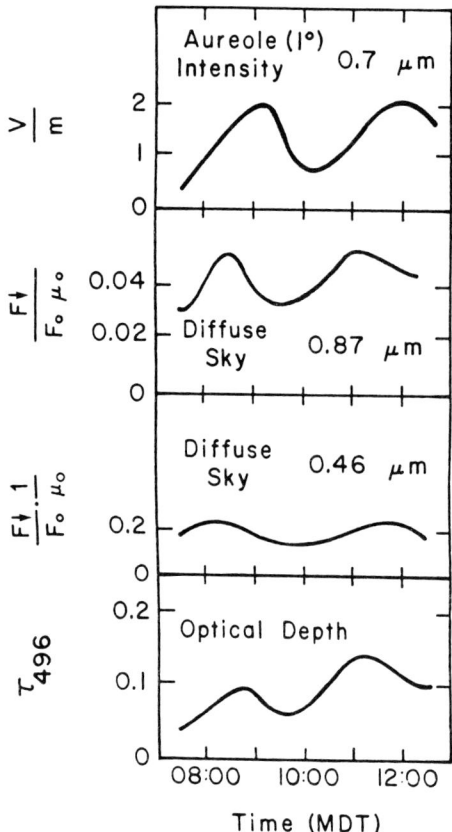

FIGURE 7. Temporal variations of sky brightness 1° from the sun at wavelength 700 nm; spectral downwelling diffuse flux from the sky at 870 nm and 460 nm wavelength; aerosol optical thickness, τ_A at 496 nm at Fort Collins on 6 August 1980.

Figure 9 illustrates the wavelength dependence of the diffuse sky flux at different times on 6 August 1980. The data indicated by circles were obtained by the author, whereas the data indicated by triangles were obtained simultaneously by the University of Arizona team. There is reasonably close agreement between the two sets of data.

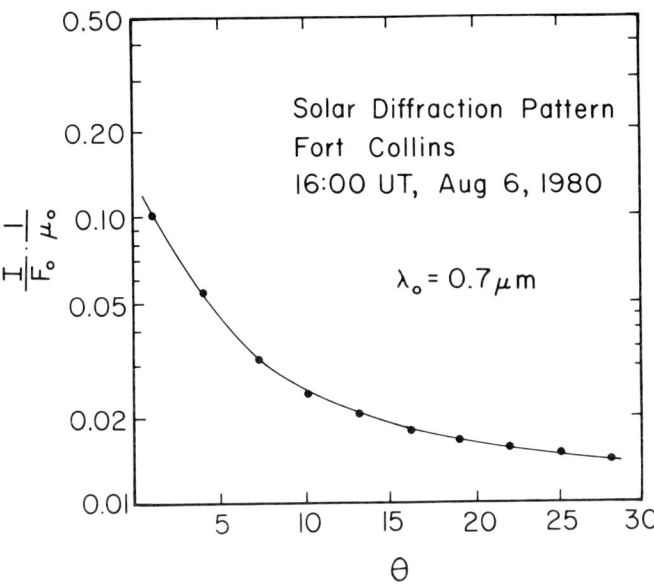

FIGURE 8. *Solar aureole diffraction pattern (sky brightness referred to vertical illumination);* θ *is the distance in degrees from the sun.*

5. *Particle Size Distribution in the Free Atmosphere Derived by Inversion*

Linear constrained inversion was applied to estimate the size spectrum of the atmospheric particles overlying Fort Collins, using as input the wavelength dependence of the aerosol optical thickness and the angular variation of diffracted light within the angular scattering range of 1° to 28° from the sun. The matrix A in Eq. (3) contained in its first eight columns the quadrature elements of the optical extinction integral [Eq. (1)] for eight wavelengths in the region 400 nm to 850 nm, and in columns 9 through 18 it contained the quadrature expansion coefficients of Eq. (2) for scattering angles in 3° increments from 1° to 28°. The matrix A contained in its 20 rows values of r_i equally spaced in increments of lnr, where r ranged from 0.05 μm to 10 μm.

Since the particle size distribution function can more or less be expected to be a rapidly decreasing function of increasing r, the inversion was performed to obtain a scaling function of r, $\xi(r)$, which, when multiplied by a power law size distribution function, $dn/d \ln r \propto r^3$, would provide the estimate for the actual size distribution. Experience with modeling such inversion has shown that one gains improved accuracy

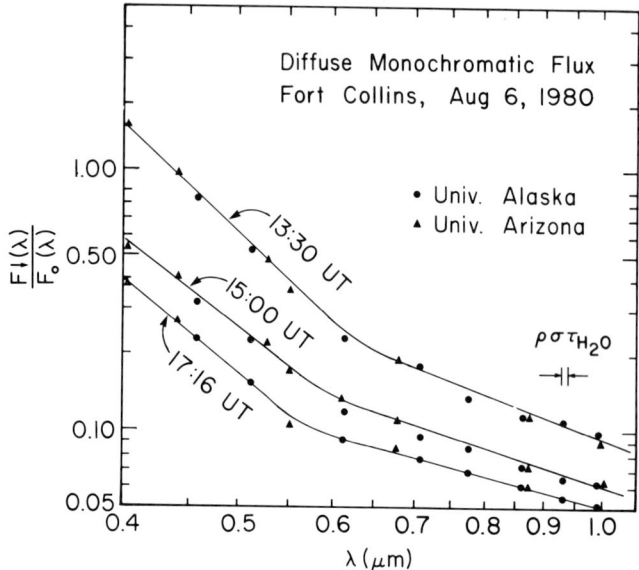

FIGURE 9. *Wavelength dependence of the diffuse monochromatic flux from the sky at Fort Collins 6 August 1980.*

by this technique, and it also has the advantage that $\xi(r)$ is proportional to $dv/d \ln r$ (the size distribution of aerosol volume or mass). It was necessary to provide some scaling in the sky brightness quadrature elements of A to make the vector of row i semicontinuous; in this way there is no problem in mixing extinction and scattered radiation data together to formulate a single matrix A.

The aureole intensity measured at the surface had to be corrected to put it in terms of the brightness for an aerosol layer illuminated by a vertical sun and containing no component due to Rayleigh scattering from the gaseous atmosphere. Implicit, too, is the assumption that the just-proposed hypothetical aerosol layer is optically thin so that primary scattering predominates. The scattering properties for this equivalent aerosol layer were arrived at by making the measurements at long wavelengths where both the Rayleigh and aerosol optical thickness were $\ll 1$.

To a first approximation the sky intensity in the solar almucantar, at wavelengths where $\tau \ll 1$ and where no gaseous absorption occurs, is expressible as

$$I(\phi) = \frac{F_0}{\mu_0} [\tau_a (P_a/4\pi) + \tau_R (P_R/4\pi)] e^{-(\tau_a + \tau_R)/\mu_0} \qquad (8)$$

where the subscripts a and R refer to aerosol and Rayleigh components, P is the scattering phase function normalized so that its integral over 4π steradians is equal to 4π, and ϕ is the scattering angle. One can improve on the estimation of P_a from measured $I(\phi)$ in situations where the condition $\tau < 1$ is not well met by employing a perturbation scheme developed by Deirmendjian (7) which performs a perturbation about an equivalent Rayleigh scattering atmosphere with optical thickness $\tau_R + \tau_a$. This excellent scheme has not received the attention it should from people working in the field of radiative transfer; it is powerful and accurate, even for relatively large optical thicknesses.

Use of the theory just outlined was applied to aureole and optical extinction data measured at Fort Collins at 1000 MDT (0600 UT) 6 August 1980 and provided the aerosol size spectrum shown in Fig. 10. This shows that there are two particle size modes; one is located at ≈ 0.1 μm radius and the other is at ≈ 3.0 μm.

An indication of the cause of the two particle modes in Fig. 10 came from X-ray spectrographic examination of individual particles collected on

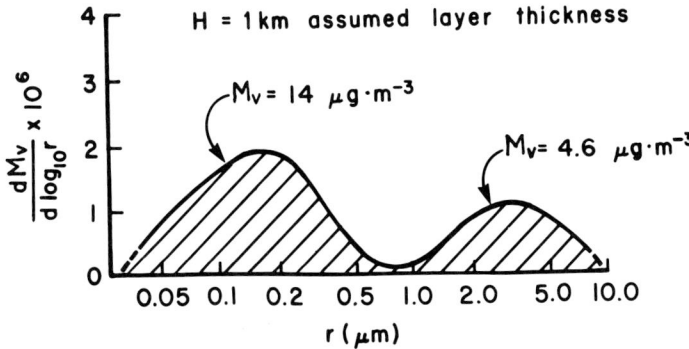

FIGURE 10. *Aerosol mass size distribution function $d M_v/d \log r$ obtained by inverting sky intensity and spectral optical extinction measurements at 1000 MDT 6 August 1980 at Fort Collins.*

Nuclepore filters at the surface during the experiments. Figure 11 shows typical X-ray spectra of collected particles. The X-ray analysis indicated that the large particles were rich in Al, Mg, and Si, and were therefore probably crustal-derived material injected into the unstable airmass, probably from blowing dust in the surrounding arid region. The small particles, which were typically 0.1 μm to 0.5 μm in diameter (in agreement with the smaller mode in Fig. 10) showed sulfur signatures which suggests that this mode is due to a sulfate aerosol. Possible sources for sulfate are the industrial emissions surrounding Fort Collins, the nearby cement plant, and, what seems more probable judging from the fact that it seemed hazier to the south, imported sulfates from the Boulder-Denver area.

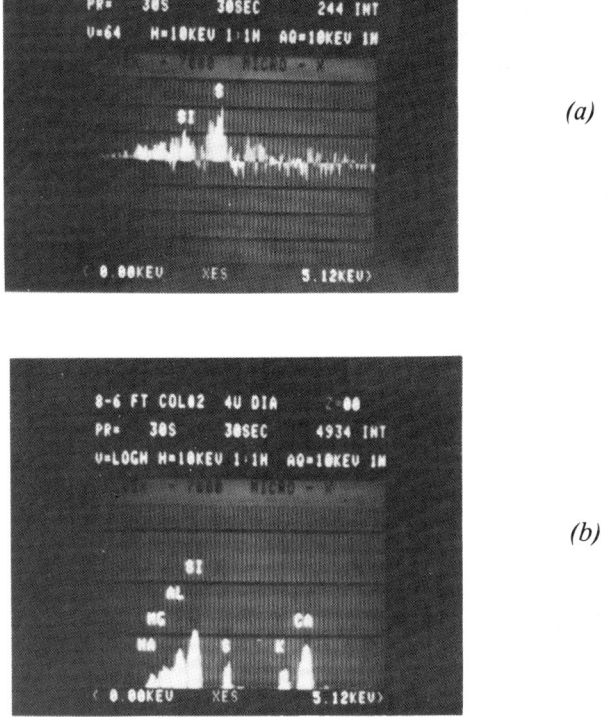

FIGURE 11. Typical X-ray spectra of a small aerosol particle and a large particle collected near the surface at Fort Collins. (a) Small particle. (b) Large particle.

6. Spectroscopic Determinations of Column Water Vapor

One filter in the sun photometer was located at $\lambda = 945$ nm, a wavelength in the $\varrho\sigma\tau$ complex of water vapor bands. A curve of growth relating water vapor amount (in g cm^{-2}) along the solar ray path in the atmosphere to the optical extinction had been derived previously for the instrument. This was done by assuming that the individual vibration-rotation lines comprising the $\varrho\sigma\tau$ complex are distributed according to a Poisson distribution, for which assumption one can apply Goody's theory using the Lorentz line shape to derive an analytic expression for the curve of growth

$$\delta_{\varrho\sigma\tau} = \exp[-C_1 w(1 + C_2 w^2/w^*)^{-1/2}] \tag{9}$$

where δ represents the optical transmission of the sun along the slant path for the filter ($\Delta\lambda = 7.5$ nm) at 950 nm; w is the water vapor (g cm^{-2}) along this path; and $w/w^* = (P/P_o)(T_o/T)^{1/2}$, where P_o and T_o are the environmental pressure and temperature for which the constants C_1 and C_2 are derived. For the filter and detector combination used, the constants C_1 and C_2 were estimated by using the LOWTRAN program, and also by curve-fitting observations of $\delta(w)$ to the water amount obtained by scaling radiosonde humidity profiles.

In applying the above theory to actual measurements of H_2O, one of the most pressing problems is establishing a calibration for the photometer so that a reference voltage, which would correspond to the case of unity transmission (no water vapor along the path), can be determined. It would not be appropriate, of course, to use the Langley method for such a calibration. Instead, a tungsten-filament lamp was used to provide radiation, which was normalized to the sun's extraterrestrial irradiance with the sun photometer at wavelengths close to the $\varrho\sigma\tau$ bands but unaffected by water vapor absorption. Since the lamp source was a quasi-Planckian one and had a smoothly varying irradiance with wavelength, one could then extrapolate to determine its radiance at $\varrho\sigma\tau$ wavelengths and then employ it to calibrate the sun photometer.

In Fig. 5, at the bottom, the temporal variations of water vapor in a vertical column overlying Fort Collins show the form of the diurnal variation. The variation of total w is surprisingly small.

V. USE OF THE RADIATION PARAMETERS TO ESTIMATE AEROSOL ABSORPTION

Aerosol absorption can be estimated, though only roughly, by comparing the expected downward monochromatic diffuse radiation calculated for the aerosol size distribution shown in Fig. 10 to the actual values measured with the hemispherical flux detector (8,9). Atmospheric absorption will obviously lower this downwelling flux.

The point of departure for putting this on a somewhat more quantitative base is to analyze the diffuse radiation field with a simplified approximation to the equation of radiative transfer applied to planar atmospheres. The most commonly used approximation, which turns out to be adequate for the problem being addressed here, is the Schuster-Schwarzschild two-stream approximation

$$\frac{1}{2}\frac{dI^-}{d\tau} = -I^- + \tilde{\omega}(1-\bar{\beta})I^- + \tilde{\omega}\bar{\beta}I^+ + F_o\tilde{\omega}(1-\beta')e^{-\tau/\mu_0} \qquad (10a)$$

$$-\frac{1}{2}\frac{dI^+}{d\tau} = -I^+ + \tilde{\omega}(1-\bar{\beta})I^+ + \tilde{\omega}\bar{\beta}I^- + F_o\tilde{\omega}\beta' e^{-\tau/\mu_0} \qquad (10b)$$

where I^+ is the upward-directed, hemispherically isotropic, diffuse radiant flux, I^- is the downward-directed, hemispherically isotropic, diffuse flux, F_o is the solar irradiance entering the top of the atmosphere, $\tilde{\omega}$ is the albedo of single scattering, β is the backscattered fraction for incident diffuse radiation at optical depth τ, and $\beta'(\mu_0)$ is the backscattered fraction of diffuse radiation for monodirectional radiation incident at the sun's zenith angle $z = \cos^{-1}\mu_0$.

The boundary conditions for solving Eq. (10) are $I^-(0) = 0$ (sky at the top of the atmosphere is black) and $I^+(\tau_T) = A I^-(\tau_T)$, where τ_T is the optical thickness of the entire atmosphere and A is the Lambertian reflectance, or albedo, of the surface.

Intercomparisons made between the two-stream solutions and exact solutions to the equation of radiative transfer carried out with the discrete-ordinate method (10) indicated that the two-stream approximation was accurate to about 2% for $\tau < 0.1$. In the work reported here measurements are analyzed only at red wavelengths where the condition $\tau < 0.1$ was well met.

To remove ambiguity which sometimes arises in the definition of $\bar{\beta}$ and β, the definition of these quantities is explicitly stated in Eq. 11 (11).

$$\beta'(\mu_o) = \frac{1}{2} \int_0^1 \bar{P}(-\mu, \mu_o) d\mu \qquad (11a)$$

$$\bar{\beta} = \int_0^1 \beta(\mu) d\mu \qquad (11b)$$

where $\bar{P}$ is the azimuthally integrated phase function

$$\bar{P}(\mu, \mu') \equiv \frac{1}{\pi} \int_0^\pi P[\mu, \mu' + (1-\mu^2)^{1/2}(1-\mu'^2)^{1/2} \cos\phi] d\phi \qquad (11c)$$

Note that $\bar{\beta}$ and $\beta(\mu_o)$ are triple and double integrals, respectively, over the scattering phase function P.

Equations (10) and (11) were solved for the conditions at Fort Collins using a phase function P calculated for the aerosol size distribution shown in Fig. 10 and for a wavelength of 710 nm. The computed downward flux was calculated for a surface albedo (at 710 nm) of 0.3 (which was estimated by viewing the surrounding terrain with a photometer) and came out to be $F\downarrow/F_o = 0.069$ at 1000 MDT on 6 August, which can be compared to the measured value of 0.0695. The close agreement indicates that the albedo of single scattering $\tilde{\omega}$ is close to unity and probably larger than 0.95.

VI. CONCLUSIONS

The remote sensing experiments conducted at Fort Collins on 4, 6, and 7 August indicated that the general passive optical techniques described can provide certain insights into the behavior of the overlying aerosol on both the micro scale, pertaining to the dynamical behavior of the evolving particle size spectrum, and the macro scale, involving the importation, buildup, and decay of aerosol over a given general geographic region. The method offers considerable possibilities for studies of aerosol physics, urban air pollution dispersion, and the identification of aerosol sources which affect the atmosphere. The method's greatest advantage is that the parameters are determined *in situ* in the free atmosphere and are integrated throughout the vertical column. There are times, perhaps even the majority of times, when the aerosol existing in the free atmosphere can have very different characteristics than those on the surface. For example, hygroscopioc aerosols, during times when humidities are about 70% or 80%, have completely different optical properties than those predicted for particles caught on filters, desiccated, and then later examined. There are also examples where the aerosol aloft has absolutely no relation to the aerosol at the surface, for instance in the case of an overpassing plume from a power plant.

These studies, though very limited in scope due to the brevity of the sampling, do suggest that the aerosol column load changes quite dramatically throughout the day, by something like a factor of two. Apparently, in the case of Fort Collins, what is happening is that in the morning the inversion lifts enough to allow Denver air pollution to convect outward and eventually into the Fort Collins area. Increases in aerosol optical thickness also seem to have preceded the onset of cumulus clouds in the late afternoon, a phenomenon probably associated with particle growth in high relative humidities.

The particle size spectrum has been deduced as being bimodal, with the small mode consisting of sulfate aerosol and the large mode, of several μm radius, due to crustal material, presumably dust injected from the local surrounding region which is relatively arid. The deduction of a bimodal distribution from the passive optical measurements, and the verification of its existence from the X-ray spectrometry analysis of collected particles, suggests that the passive optical sensing method in general is a useful one for studies of the microphysics of aerosol. It also indicates that the surface aerosol is representative of the column aerosol for this case, in which convective currents were strong; this is important to know for assessing climatic influences brought about by suspended material in the atmosphere.

It was possible to deduce from the remote sensing experiments that the aerosol in the free atmosphere above Fort Collins is very lightly absorbing, with a single scattering albedo (at 700 nm) of 0.95 to 1.0. This is reasonable if indeed the aerosols are sulfate and crustal material, since the latter is relatively light in color in this location. Another indication that strongly absorbing material is low in abundance in this aerosol mixture comes from the relatively light brown-gray color of filters exposed near the surface. The implications seem to be that carbonaceous aerosol is a minor species near Fort Collins, at least in summer (the situation may be very different in winter when fossil fuel and wood burning is going on under a capping inversion).

These optical sensing experiments, though performed with relatively modest apparatus, show an indication of being useful to delineate certain physical processes involving the dynamic evolution of particles suspended in the atmosphere.

REFERENCES

1. Parker, R. L., Understanding inverse theory. *Ann. Rev. Earth Planet. Sci. 5(35)*, 35-64 (1979)
2. Twomey, S., *Introduction to the Mathematics of Inversion in Remote Sensing and Indirect Measurements,* Elsevier Scientific Publishing, New York (1977).
3. Deirmendjian, D., *Electromagnetic Scattering on Spherical Polydispersions,* American Elsevier Publishing, New York (1969).

PASSIVE OPTICAL TECHNIQUES 355

4. Van de Hulst, H. C., *Light Scattering by Small Particles,* Wiley, New York (1957).
5. Shaw, G. E., Inversion of optical scattering and spectral extinction measurements to recover aerosol size spectra, *Appl. Opt. 18,* 988-993 (1979).
6. Shaw, G. E., and Deehr, C. S., A photoelectric coronameter for atmospheric turbidity studies, *J. Appl. Meteor. 14,* 1203-1205 (1975).
7. Deirmendjian, D., *Use of Scattering Techniques in Cloud Microphysics Research: 1. The Aureole Method,* RAND Report R-590-PR, The Rand Corporation, Santa Monica, California (1970).
8. Herman, B., Browning, R. S., and DeLusi, J., Determination of the effective imaginary term of the complex refractive index of atmospheric dust by remote sensing: The diffuse-direct method, *J. Atmos. Sci. 32(5),* 918-925 (1975).
9. Deirmendjian, D., A survey of light-scattering techniques used in the remote monitoring of atmospheric aerosols, *Rev. Geophys. and Space Phys. 18(2),* 341-360 (1980).
10. Shaw, G., and Stamnes, K., Arctic haze: Perturbation of the polar radiation budget; Aerosols: Anthropogenic and natural-sources and sinks, *Annals of the New York Academy of Science, 383,* 533-539 (1980).
11. Wiscombe, W., and Grams, G., The backscattered fraction in two-stream approximations, *J. Atmos. Sci. 33,* 2440-2451 (1976).

RESULTS ON AEROSOL LIGHT ABSORPTION

Frederic E. Volz

Air Force Geophysics Laboratory
Hanscom Air Force Base, Massachusetts

The absorption index of test aerosols and atmospheric samples collected at the First International Workshop on Light Absorption by Aerosol Particles (Fort Collins, Colorado, August 1980) was determined from 0.38- to 0.94-µm wavelength with a photometer having a wide-angle receiver, and from 2.5- to 30-µm with a Perkin Elmer spectrograph. With nonabsorbing substances, scattering indexes ≤0.001 were obtained in the visual range through a combination of index matching and high-pressure (transparent pellets) sample preparation. Apparent absorption indexes of methylene blue, soot (and mixtures with ammonium sulfate), and atmospheric aerosol are normally higher for reground pellets (smaller particle size), but may still fall short of known bulk values.

I. INTRODUCTION

The methods presented here of measuring light absorption by aerosols probably differ in several aspects from most other methods used at the workshop. These differences include using the pressed-pellet method, relating other test aerosols to samples collected at the workshop, and studying the dependence of the results on sample treatment.

For the pressed-pellet method a trace of the aerosol material is ground as fine as possible, then ground again following the addition of an absorption-free supporting and diluting agent (KBr powder). In order to eliminate or reduce scattering, this mixture is then pressed into a glassy disk for transmittance measurements. This embedding method is applied to both the visual *(0.38 to 0.94 μm)* and infrared (2.5 to 30 μm) spectral ranges. The special feature of the simple visual photometer is a wide-angle ($\sim$ 1 sr) receiver for further reduction of radiation loss by scattering. Infrared (IR) measurements are made with a standard two-beam spectrograph in the home laboratory; the instrument available at the workshop was only useful for survey spectra.

The pellet method requires at least 3 mg of aerosol substance; hence the collection method intended for the workshop (large-volume thermal precipitator) could not properly be adapted, since the aerosol mass concentrations used at the workshop were too small. Therefore, except for a very few samples (including outside air), samples for this investigation had to be prepared in the home laboratory, generally by grinding of bulk material. Regrinding of pellets generally gives higher extinction, but it is not clear which results are preferable, especially since bulk values, where known, are even higher.

II. DETAILS OF PELLET PREPARATION

About 0.6 mg to 1.5 mg (+ 0.05 mg) of test aerosol substance are well ground with pestle and mortar, and then ground into $\approx$ 150 mg of KBr (infrared quality). Pressed by tradition under vacuum (but with questionable benefit) at 140,000 psi = 9600 kg/cm^2 = 960 MPa between polished rams of a standard die, clear transparent pellets are usually obtained. Pellets have an area of 1.32 cm^2 and are about 0.8 mm thick. Their density is 0.96 to 0.99 of bulk KBr. Grinding was done forcefully for about 3 minutes; however, grinding of grains of deeply colored substances (methylene blue or soot) with a trace of KBr shows that grinding time should actually be longer. Also, a few big particles (10 to 20 μm) can normally still be seen under the microscope within the pellet.

Earlier it was found in IR work that repeated grinding and repressing of a pellet increases absorption in bands, sometimes giving absorptions close to those expected for bulk optical data (1,2). It should be stressed that this effect, which unfortunately also applies to absorption away from bands, was also found with near-spherical, spray-created particles ≤ 0.2 μm in size and pressed

AEROSOL LIGHT ABSORPTION

into pellets without grinding. Use of the workshop-produced aerosols, which on a mass basis had very wide size distributions with mass median radius always larger than 0.5 μm, would therefore have been of little interest to this method without grinding.

III. EQUIPMENT AND DATA EVALUATION

1. Photometer for Visual Measurements

Pressed pellets have been applied nearly exclusively in IR spectroscopy, but shortwave absorption measurements have also been tried by using a wide angle receiver (short of an integrating sphere) to reduce the apparent extinction by scattered light (3). Since no similar sample treatment was used at the workshop, the photometer was rebuilt as sketched in Fig. 1. The initial results

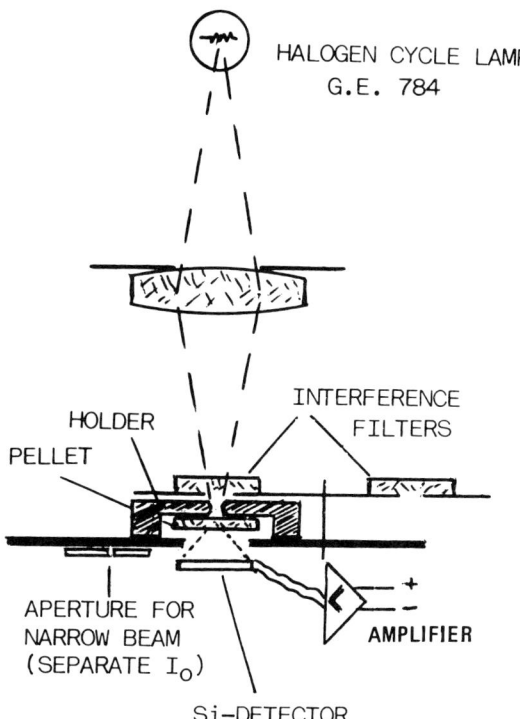

FIGURE 1. Photometer with spot illumination of pressed pellet and wide-angle receiver.

of the photometer were interesting enough to be extended and reported. Interference filters used with the photometer had half-widths of 10 nm and passed light with wavelengths of 0.38, 0.40, 0.50, 0.64, 0.68, 0.74, and 0.94 μm. All filters below 0.74 μm were secured against IR leaks by Schott BG18 filters.

2. IR Measurements

The transmittance as measured with a Perkin Elmer Model 180 double-beam spectrograph was recorded on chart and also on paper tape.

3. Data Evaluation

The relation between the mass absorption coefficient K and the imaginary part n_2 of the refractive index of a homogeneous bulk sample (e.g., a crystal plate) is

$$n_2 = \varrho \lambda \, K/4\pi \tag{1}$$

In analogy, it was assumed that measured transmittance τ of a pellet yields an extinction index

$$n_e = \frac{\varrho \lambda \, \ln[\tau(1 + 0.10\,\tau)]}{4\pi M} \tag{2}$$

The term $1 + 0.10\,\tau$ is the correction for the reflection loss of the pellet, ϱ is the density, and M (mg/cm^2) is the area weight concentration of the test substance. Since there will always be transmittance loss by scattering, $n_e = n_s + n_2$, where n_s is the "scattering index." As will be shown, n_s is about 10^{-4} for the minimum extinction observed for nonabsorbing powders in the visible range with this arrangement. To date, little attempt has been made to estimate n_s in other cases by comparing wide- and narrow-angle measurements. In the IR, n_s is probably far below the bulk background absorption (band wings) of $\approx$ 0.01 of most substances, despite the narrow spectrometer beam ($n_s < 10^{-5}$ for pure KBr pellet). However, in both spectral ranges, values of $n_2 \geqslant 0.2$ in the absorption bands may fall far short of the true values of n_2, as can be shown for substances whose bulk absorptions are known.

For clouds (or layers) of aerosols as the absorbing medium, it is known that the relation between the transmittance calculated by Mie theory (spherical particles), the size distribution of the aerosol particles, and n_2 is quite complex. However, for broad size distributions typical of atmospheric aerosols, a simple correction factor $F \approx (n_1^2 + 2)/3$ used with Eq. (2) was often found applicable. However, with $n_1 = n_1(\text{particle})/n_1(\text{medium}) \approx 1$, $F \approx 1$ in these pellet measurements except for strong bands; therefore no correction was applied. Also, the particles in question are certainly far from spherical.

IV. SOME EXPLANATIONS AND GENERAL COMMENTS

In the figures, spectra are labeled with sample number, a letter indicating pellet preparation (T = original pellet, A = reground and often diluted, B = reground twice), and sample weight (mg/pellet area = 0.76 mg/cm^2). In mixtures, sample weight and extinction index refer to the sum of both components. During all work, relative humidity was lower than 35%.

Outside aerosol was collected with a high-volume sampler by first sampling particles with r $\geq$ 0.1 μm on a polyester felt filter and then sampling the remaining smaller particles on a Delbag polystyrene filter. The felt was prewashed with distilled water and, for removal of the trace of oil applied during manufacturing, with toluene. The loaded felt was leached with distilled water, pressed and suction-dried in a funnel, leached again, and then washed under strong agitation for removal of the remaining dust and soot. The loaded Delbag filter was made wettable with ethanol before washing. Dissolution of this filter by organic solvents to recover dust and soot was not attempted. Dust was removed from all wash-fractions by sedimentation until the solution was optically clear; water solubles remained after evaporation at 60°C. Collection was from 1000 on 6 August to 1400 on 8 August (scheduled collection was on 5 August) with a total recovery of 188 ± 5 mg. With a sampling rate of 1050 cm^3/hr (± 10%), aerosol concentration was 3.4 μg/m^3 ± 15%. One-third of this was water soluble.

The density of methylene blue was obtained as 1.27 from the weight and volume of a pressed pellet. The density of the aerosol fractions and soots was assumed to be 1.8, and that of Arizona road dust was assumed to be 2.3.

At the workshop, soots were prepared by burning a high-pressure airpropane mixture with a blue flame. Since the soot used in the present measurements was deposited on a cold plated-metal surface from a weak yellow flame (propane torch with air holes closed), its properties could have been rather different from those of the workshop soots.

V. RESULTS: VISUAL DATA

1. Opal Glass.

For the study of receiver-angle effects with a stable sample, opal glass was used. The milky opal layer ($\sim$ 0.1 mm) is on one side of the 1-mm-thick glass. With the opal layer toward the detector, wide-angle extinction (shown in Fig. 2a, top) is 25% smaller than for the narrow angle. At λ = 0.94 μm, extinction/λ is 65% larger than at 0.50 μm for both wide- and narrow-beams.

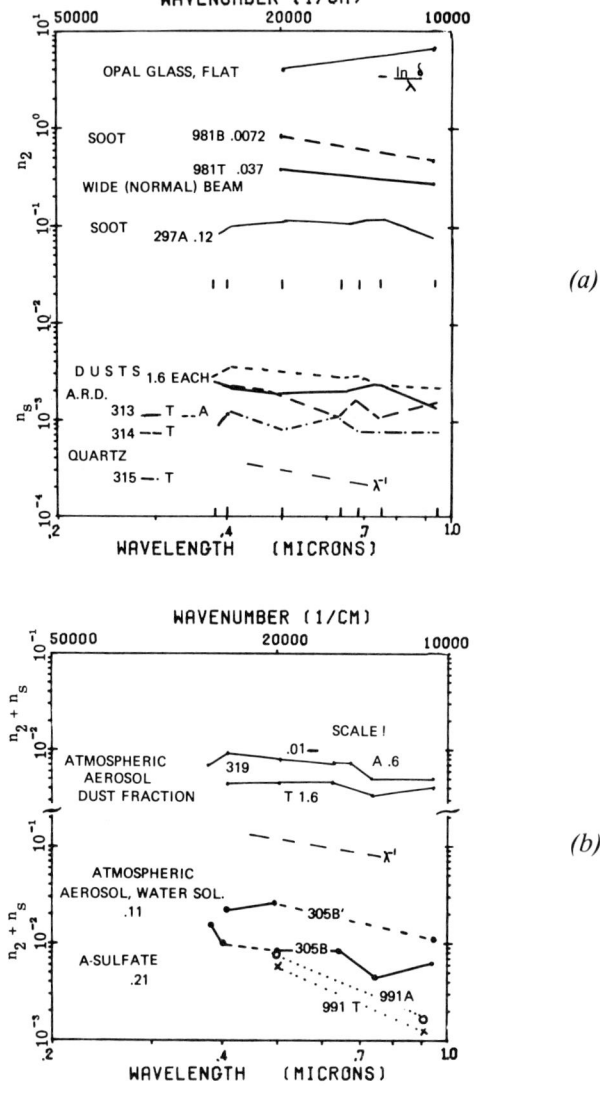

FIGURE 2. Visual and near-IR scattering and extinction index spectra, samples pressed into KBr. Spectra are labeled with sample number, a letter indicating sample preparation (T = original pellet, A = reground and often diluted, B = reground twice), and sample weight (mg/cm² of sample area). (a) Dusts and soot; $-\ln \tau/\lambda$ for opal glass at top. (b) Ammonium sulfate, dust and water solubles of aerosol.

However, extinction with the opal layer facing the light source is 4.5 times larger, suggesting a problem with pellets whose thickness varies appreciably.

2. Powders with No or Slight Absorption.

The lower part of Fig. 2a shows slightly λ-dependent extinction by Arizona road dust (ARD) and quartz powder. The extinction index of ARD is a factor 2 to 3 higher than that of the quartz, which is probably due to the absorption of iron oxides that give the dust a brownish color. Ammonium sulfate (AS) powder (Fig. 2b) shows much higher extinction at 0.50 μm, but the sample probably was contaminated with methylene blue. Reground pellets generally have higher extinction. Wide- and narrow-beam measurements were made on the AS pellet. The fraction of forward scattering to total scattering, roughly given by $(n_{sN} - n_{sW})/n_{sN}$ for wide (narrow) angle results n_{sW} (n_{sN}), is quite reasonably found to be about 0.3. The error of these low extinction indexes is on the order of 30%, but can be much reduced by higher sample concentrations and improved stability of the photometer.

Extinction of latex particles r = 0.45 μm and 1.55 μm also was about 10^{-3} with indication of different λ dependence (not shown). No clear difference between the original pellets with unground spherical particles and the pellets after regrinding could be found, although damage to ground particles was obvious under the microscope. Also, IR absorption strongly increased with grinding.

These results of minimal extinction may need more discussion. From Mie calculations for $n_2 = 0$ presented in Fig. 3 for $\lambda = 1$ μm (similar to 0.3 μm), one expects values of $n_s \ll 0.01$ only for either the Christiansen region (n/n_m close to 1 as for latex and ammonium sulfate) or for extremely small or very large particles (n = refractive index of particle, n_m of embedding medium, n_m = 1.58 for KBr). For quartz (n $\approx$ 1.55), n_s could hardly be smaller than 0.1, and it would be $\approx$ 0.3 at both 0.3 μm and 1 μm wavelength for a $N(r) \sim r^{-4}$ distribution.

3. Propane Soot

Parts of one soot sample (981, upper part of Fig. 2a) were twice diluted and reground; the extinction index increased each time. Beam width was also changed on this sample. The relative excess of narrow-beam extinction over the wide-angle value, that is, relative forward scattering, increases with regrinding from 6% to 10% in the original pellet to > 30% in the last pellet (Table 1). At this time, an explanation of these changes and a decision as to which absorption might be true cannot be given. Workshop-produced soot (sample 297) shows much less absorption, as is evident in the IR.

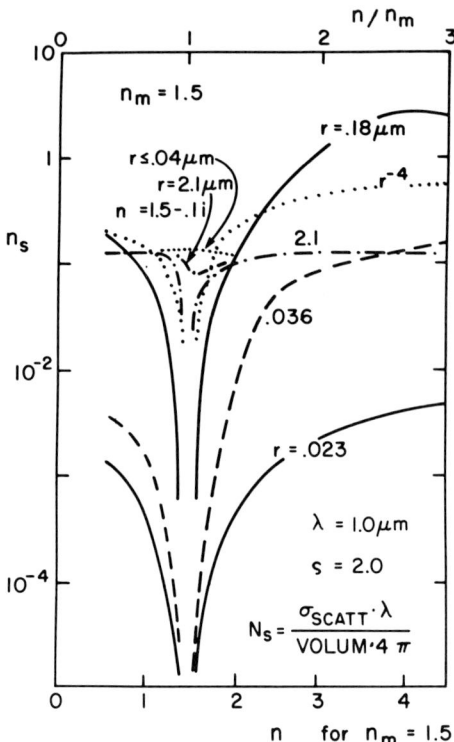

FIGURE 3. Scattering index n_s (defined like n_2) at $\lambda = 1.0$ μm of spherical nonabsorbing particles as function of ratio of refractive index of particle (n) to that of the medium (n_m). The deep Christiansen minimum disappears at moderate particle absorption. Pellet scattering index for wide-angle receiver (≈ total scattering minus forward scattering) would be expected to be 0.8 (for very small particles) to 0.2 (for large particles) for the calculated curves.

TABLE 1. Absorption Index of Soot ($\varrho = 1.8$).[a]

Sample 981 Soot	mg/pellet	λ	Wide beam 0.94	0.50	Narrow beam 0.94	0.50 μm	
T	0.0370		0.31	0.38	0.33	0.44	Original pellet
A	0.0145		0.44	0.77	0.48	0.91	1/2 reground
B	0.0072		0.68	0.99	0.90	1.34	1/2 again reground

[a]Pellet contains 20 times more aluminum sulfate (AS) than soot; for the infrared, see sample 308 in Fig. 5.

AEROSOL LIGHT ABSORPTION

4. Outside Aerosol.

The pellet with the water-soluble fraction of outside aerosol became clear for both the visual and IR range only after repeated dilution and regrinding (Fig. 2b). However, a repetition of the process (305 B ') gave higher absorption. A larger λ dependence ($\sim \lambda^{-1}$) than in most dust samples is indicated. Dust sample 324 has lower extinction and smaller regrind effect.

5. Methylene Blue (MB), Bulk and Pellets (Fig. 4a).

Bulk absorbance was determined in aqueous solution with distilled water as a reference using a Cary Mod/14 spectrophotometer. The curve (#) is a combination of results with 10-mm and 20-mm curvettes and two sets of graded solutions. The standard error is on the order of 10%. The recorded spectrum was evaluated at a few more wavelengths than the pellet measurements. While n_2 of the solution peaks at 2.2 ($\lambda = 0.65$ μm), the pellet absorption obviously is saturated at only $n_2 = 0.4$. But at $\lambda \approx 0.4$ μm the minimum absorption in pellets is two to four times higher than for the solution.

It should be mentioned that Mie calculations for a model closely resembling the MB conditions (AS optical constants, from Ref. 4, but λ reduced by a factor 0.06) would affect the peak absorption similarly for $r \approx 0.25$ μm, but background absorption would be predicted to be much higher than observed at the longer wavelength side of the main band.

Methylene blue concentrations were mostly too low for accurate pellet results beyond $\lambda = 0.7$ μm. For the results shown, rather fine particles from dried MB solution were used. A pellet of much coarser powder from the manufacturer had to be reground twice before it was reasonably clear, but absorption at 0.68 μm rose to 0.63. The real part of the refractive index derived by dispersion analysis from the n_2 spectrum of the solution is depicted in Fig. 4b. However, a perfect fit could not yet be obtained.

VI. RESULTS: INFRARED DATA

These data also require some comments on basic problems which were studied recently but which also are not yet well understood (1,2).

1. Ammonium Sulfate and Small Admixture of Soot.

Figure 5 shows results for pellets with a trace of propane soot in AS (in weight ratios of 0.049 and 0.039). Also plotted are n_2 values derived by Toon et al. (4) from transmittance and reflectance of AS crystals. That pellet absorption in band peaks does not reach the crystal values is unrelated to the small soot content, since this also is observed with pure AS in KBr. Higher band

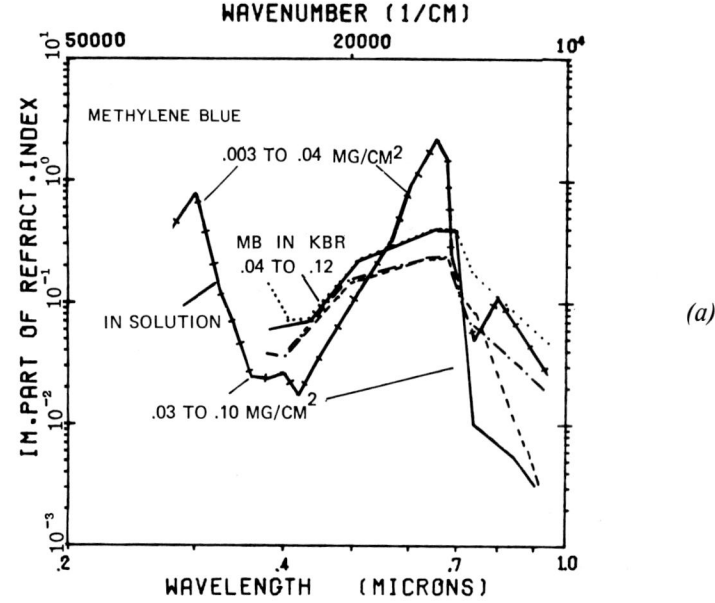

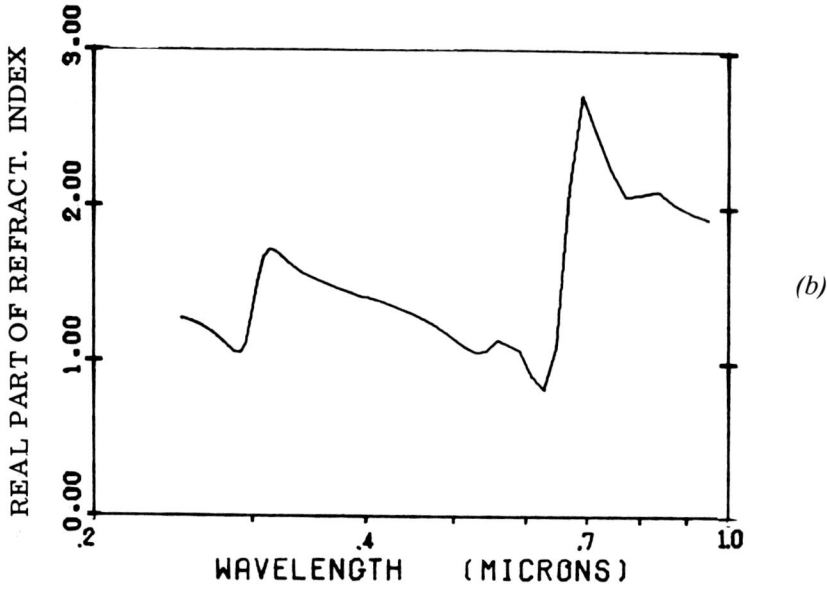

FIGURE 4. (a) Absorption index of MB in solution and in KBr pellet. (b) Real part of refractive index of MB.

absorption can often be obtained by (repeatedly) regrinding the pellet, as will be seen in several of the following data. This certainly is caused by destruction of larger particles and better homogenization (however, after too much regrinding new bands may emerge, probably caused by crystal defects and ion-exchange effects). Why background absorption (or scattering) also increases is unclear. The same conditions which apply to regrinding probably also exist for small near-spherical particles. This was shown with spray-created and thermally precipitated AS particles about 0.2 μm in size which were mixed into the KBr powder without grinding, although the degree of detrimental agglomeration of the particles was unknown. The model calculations just mentioned would require particles with $r > 1.2$ μm to cause the observed peak absorption to become smaller than the bulk value, and probably no choice of particle sizes would explain an increase of the background.

The customary way to derive optical constants in the IR is to analyze specular reflectance measurements of crystals (crystals usually properly cut, and the reflectance observed in polarized light). It was found that pressed pellets of pure AS powder gave, at near normal incidence, the same reflectance (and hence the same optical constants) as faces of AS crystals. However, reflectance in Reststrahlen bands of pressed pellets of more anisotropic substances than AS, or of hard substances (e.g., quartz), is too low. The same holds for ground mixtures of powders of soft, easily pelletable salts (e.g., AS and $NaNO_3$). This method, therefore, is also not quantitatively applicable to most samples under discussion here.

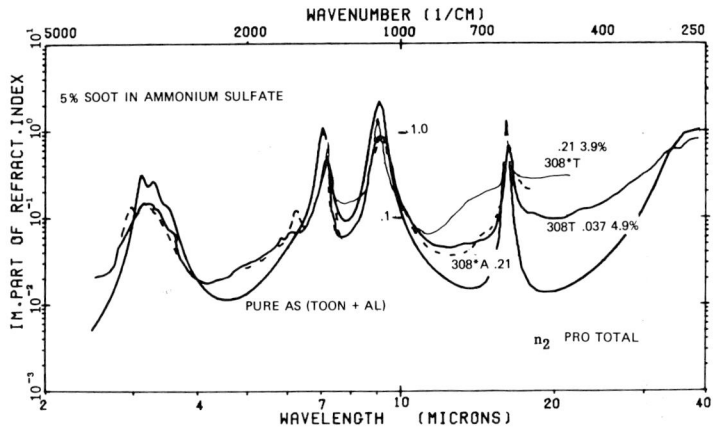

FIGURE 5. IR absorption index of low concentration soot as in AS, and of AS derived from crystaloptic measurements by Toon et al. (4).

2. Soot.

Only a tiny sample of propane soot from the workshop was available (spectrum 297 in Fig. 6). The soot showed the same weak spectral features, but as in the visible, it showed much higher absorption and a strong regrind effect. Samples 308 and 308* are unrelated.

3. Equal Mixture of Soot and AS.

As seen in Fig. 7, the AS bands are now much weaker. The background absorption caused by soot (twice the n_2 value shown) is ≈ 0.2, which is lower than for pure soot. This is noteworthy considering that those mixtures were prepared with nonworkshop soot.

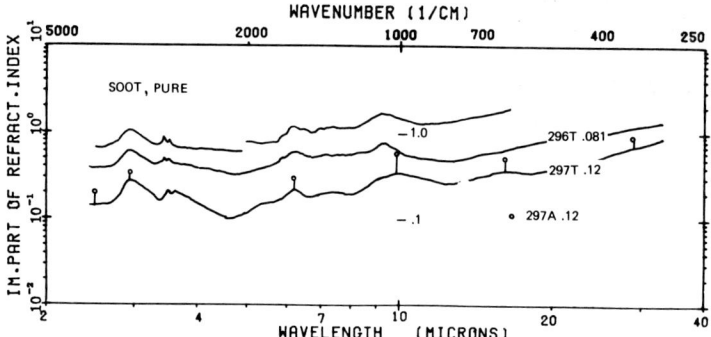

FIGURE 6. IR absorption index of soot (KBr pellet).

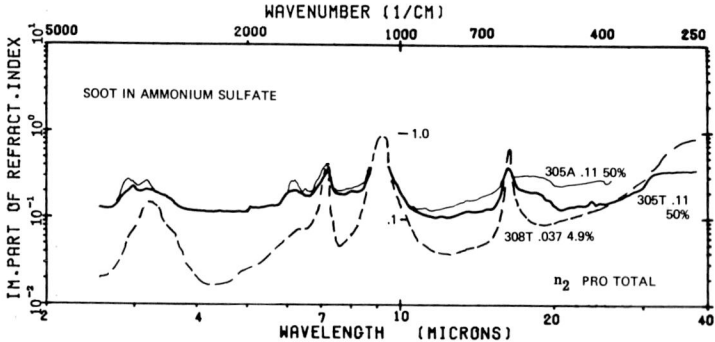

FIGURE 7. IR absorption of soot and AS (1:1 mixture) in KBr.

4. Methylene Blue.

The original pellet was very cloudy and had to be diluted. Further grinding brought no changes. Also, a reflectance spectrum of a pellet of pure MB powder was run and analyzed by dispersion theory (Fig. 8). The computer program was limited to 10 bands so that only the salient features could be iterated. Agreement between theory and measurements is good at longer wavelengths, but the method is not applicable to the weak change of reflectance at 3 μm.

5. 10% Methylene Blue in AS.

This sample was prepared with the nebulizer at the workshop but at a higher MB concentration, because the MB bands of a 5% sample run on the spectrograph at the workshop were too weak. Only significant points of the data obtained with the Perkin Elmer were evaluated and are shown connected by straight lines in Fig. 9. Again, the original pellet gave much lower n_2 values than the pellets with regrinding and dilution. The imaginary index is referred to the total sample weight.

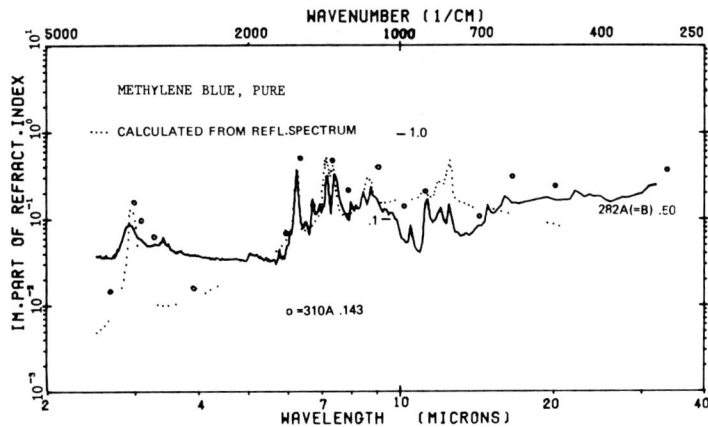

FIGURE 8. *Methylene blue (in KBr pellet) and dispersion analysis of reflectance spectrum (near-normal incidence) of pure MB pellet.*

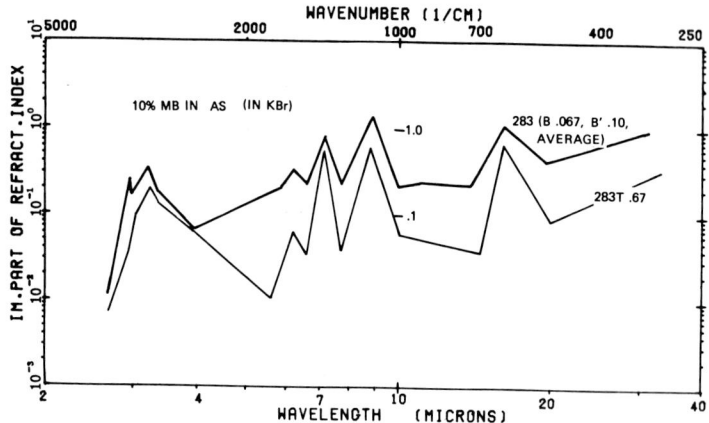

FIGURE 9. *10% methylene blue in AS; only significant points of spectrum evaluated.*

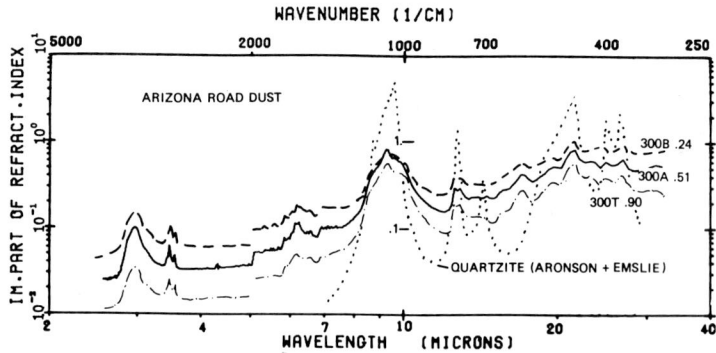

FIGURE 10. *Arizona road dust compared to result of dispersion analysis of reflectance of polished quartzite (quartzite data from Ref. 5).*

6. Arizona Road Dust.

The absorption of this silicate dust increased considerably with grinding, but the increase was least in the 9-μm band (Fig. 10). The spectrum of microcrystalline quartzite is very similar in band position except that it peaks at $n_2 = 8$ instead of 0.8 (5). Absorption by crystalline quartz is even stronger.

AEROSOL LIGHT ABSORPTION

7. *Outside Aerosol.*

The fraction of the outside aerosol which consists of dust easily removed from the felt filters ($r > 0.1$ μm, top spectrum in Fig. 11) is identical (note scale shift) to that less easily removed except for a shift in the main band and details between 11 μm and 17 μm. The superimposed ARD pellet spectrum has considerably stronger absorption. The light gray dust could not contain much soot.

The absorption spectrum of evaporation residues of the outside aerosol particles > 0.1 μm (lower spectrum in Fig. 11) is hardly different from that of larger particles (scale shift) except for a stronger 17-μm band (by sulfate, as at 9 μm). The sharp peak at 7.2 μm in the upper spectrum is typical for nitrate and ammonium ions. Absorption at 6.2 μm is caused by oxygenated water-soluble hydrocarbons which can be separated from the inorganic salts by ethanol extraction.

VII. CONCLUSIONS

Dependence on pellet treatment (regrinding) of many scattering and absorption index data reported here, and the knowledge, at least for the IR, that observed absorption spectra only marginally resemble those known from crystal-optic data (AS, Fig. 5; ARD, Fig. 10) make the results dubious for application. The pellet data on MB for the visual spectral range (Fig. 4) attest to the severe influence of particle size, even if, as in this index-matching method, particle scattering appears to be greatly reduced.

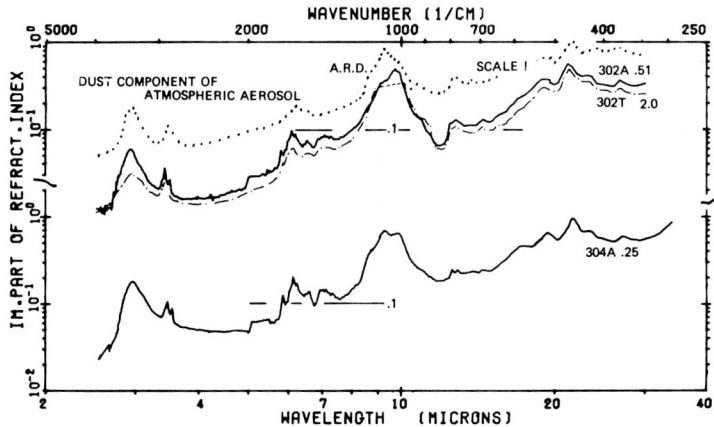

FIGURE 11. Outside aerosol, dust and soot fractions, compared to absorption of ARD (both in KBr).

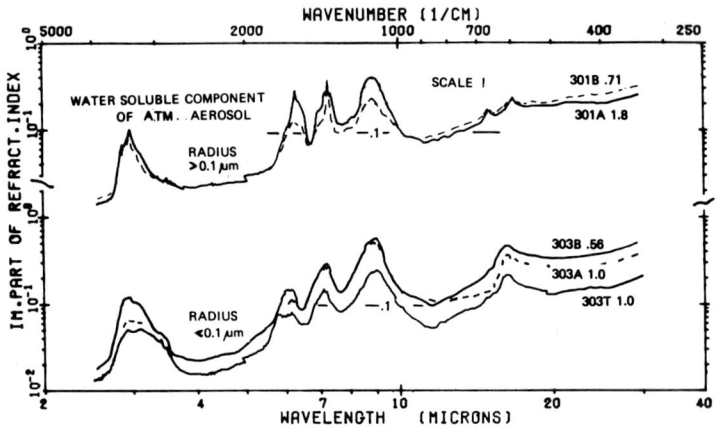

FIGURE 12. Outside aerosol, water-soluble fractions.

REFERENCES

1. Volz, F. E., Derivation of IR optical constants of aerosols and polycrystalline surfaces (specular reflection by pressed crystal powders), *Optical Phenomena Peculiar to Matter of Small Dimensions: A Digest of Technical Papers,* Optical Sciences Center, University of Arizona, Washington, Optical Society of America (1980).
2. Volz, F. E., Some basic investigations on IR optical constants of aerosol substances, paper presented at Intern. Radiation Symp., 11-16 August 1980, *Volume of Extended Abstracts,* pp. 196-198, Colorado State University, Fort Collins, Colorado (1980).
3. Volz, F. E., Infrared absorption by atmospheric aerosol substances, *J. Geophys. Res. 77,* 1017-1031 (1972).
4. Toon, O. B., Pollack, J. B., and Khare, B. N., The optical constants of several atmospheric aerosol species: Ammonium sulfate, aluminum oxide, and sodium chloride, *J. Geophys. Res. 81,* 5733-5748 (1976).
5. Aronson, J. R., and Emslie, A. G., Effective optical constants of anisotropic materials, *Appl. Opt. 19,* 4128-4129 (1980).

MEASUREMENT OF LIGHT ABSORPTION BY AEROSOLS WITH AN OPTOACOUSTIC DETECTOR[1]

Wayne M. Wright
Physics Department
Kalamazoo College
Kalamazoo, Michigan

Donald H. Stedman
Chemistry Department
University of Michigan
Ann Arbor, Michigan

Leopoldo Stefanutti
IROE of CNR
Firenze, Italy

Robert W. Terhune
Ford Motor Company
Dearborn, Michigan

The design of an optoacoustic aerosol spectrophone is described, and measurements made with this instrument at the First International Workshop by Light Absorption on Aerosol Particles, held at Ft. Collins, Colorado, August 1980, are presented.

---

[1]*Participation in the workshop was partially supported by NSF grant ATM 8005356.*

I. INTRODUCTION

Presented below are data obtained at the First International Workshop on Light Absorption by Aerosol Particles (Fort Collins, Colorado, August 1980). Measurements were made with an optoacoustic spectrophone which had been developed at the University of Michigan. Since the signal from an optoacoustic detector is essentially independent of optical scattering, data are limited to aerosols with absorbing constituents. Air containing the aerosol of interest was pumped through a small cell, such that a continous time record of the optical absorption coefficient was produced.

II. EXPERIMENTAL APPROACH

Transmission of an intensity-modulated optical beam through an absorbing gas results in localized heating of the gas and the generation of an acoustic wave. With proper alignment of the beam through a gas cell and a suitable modulation frequency, the cell can be operated in an acoustically resonant mode with a substantial increase over the free-field sound pressure. The present system uses an argon ion laser operating at either 488.0 nm or 514.5 nm wavelength. The beam is mechanically chopped at a frequency corresponding to the first-order azimuthal acoustic resonance of the cylindrical cell of approximately 5-cm inside diameter and 12.7-cm length. The average power of the chopped beam at the cell is on the order of 100 mW. A continuous flow of gas through the 0.26-liter cell, at a rate somewhat greater than 0.5 liter/min with the present pump, results in a system response time of a few minutes.

A block diagram of the experimental apparatus is provided in Fig. 1. The signal from a sensitive, low-noise condenser microphone at the cell wall is amplified, filtered, and measured with a coherent detector. Results are obtained from a strip-chart recording. The optoacoustic spectrophone was calibrated with the aid of a cylinder of compressed nitrogen containing 3 ppm (3 parts in 10^6 by volume) of nitrogen dioxide (NO_2), one of the few common gases which absorbs visible light. Values for the absorption coefficient of the NO_2 at the wavelengths of interest, obtained from transmission measurements through more concentrated samples, were taken to be $\sigma_a = 1.8 \times 10^{-3}$ m^{-1} at 488 nm and 0.9×10^{-3} m^{-1} at 514 nm. Since there was some drifting of the power in the laser beam with time, the recorded signal levels were normalized to a standard optical power before the absorption coefficient was calculated.

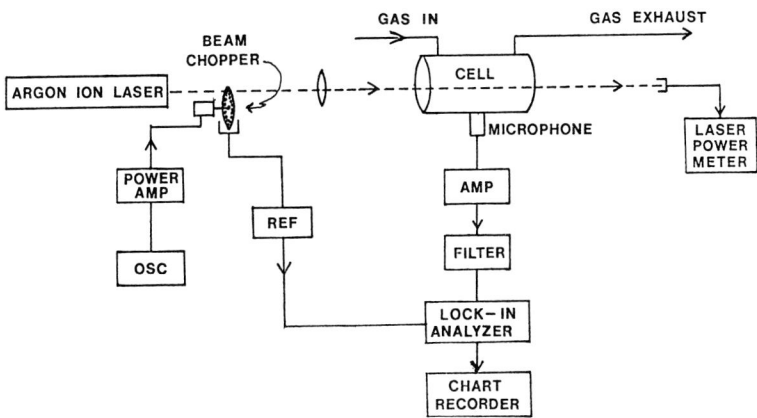

FIGURE 1.   Optoacoustic spectrophone.

III.   RESULTS

This apparatus was not designed to be portable, and all of the previous measurements had been carried out in an acoustically quiet environment. Nevertheless, recordings were obtained from this optoacoustic system over the last 7 working days of the workshop. Only about half of this chart footage was useful, however, as a result of high ambient acoustic noise levels and some technical difficulties encountered as optimization of the system was attempted.

Figure 2 presents a section of the chart record for Thursday, 31 July. Most of the long-term drift was presumably associated with changing soot concentration, while the more rapid fluctuations in signal level came from changes in the noise component that was passed by the 0.1-Hz effective bandwidth of the Ithaco model 393 lock-in analyzer. There also was a decrease in laser beam power of about 10% over this time interval. The effect on the acoustic signal of blocking the laser beam before it entered the cell is indicated here; a similar effect on the signal resulted from the insertion of a particle filter into the inlet air line.

Table 1 summarizes the measurements on steady flows of homogeneous aerosols for those circumstances in which the steady component of the signal amplitude was significantly greater than the amplitude of the short-term fluctuations. Each value for the absorption coefficient is given for an arbitrarily chosen time during an interval of relatively constant signal, which may have persisted for several minutes or for a few hours. Concentrations of soot and

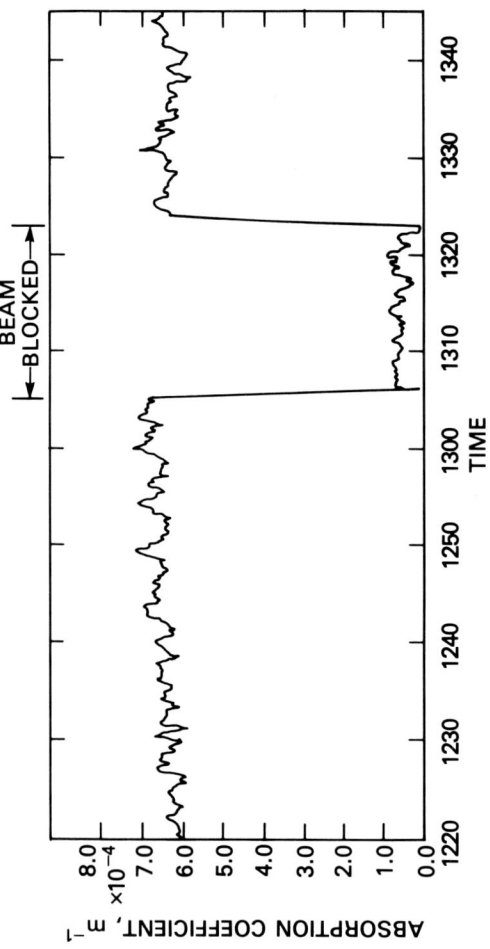

FIGURE 2. Optoacoustic signal recorded on 31 July for soot aerosol

TABLE 1. *Measured Values of the Optical Absorption Coefficient at 488 nm and 514.5 nm* for Selected Times and Various Aerosols.*

Day	Time	Aerosol	$\sigma_a (m^{-1} \times 10^{-4})$
31 July	1020	Soot	6.8
31 July	1050	Soot	6.7
31 July	1210	Soot	5.9
31 July	1300	Soot	6.8
31 July	1345	Soot	6.8
31 July	1600	Soot, salt	2.2
31 July	1650	Soot, salt	2.2*
1 August	1530	Soot	4.1*
4 August	1040	Soot	5.2
4 August	1130	Soot, salt	4.9*
4 August	1200	Soot, salt	4.7*
4 August	1300	Soot, salt	3.1*
4 August	1440	Soot	2.7*
4 August	1505	Methylene blue	1.3*
4 August	1540	Methylene blue	1.5
6 August	1100	Road dust	2.3
6 August	1130	Road dust	1.8*
6 August	1550	Soot, salt	1.2
6 August	1620	Soot, salt	1.0
6 August	1720	Soot	1.8
6 August	1750	Soot	14.0
7 August	1120	Soot, salt	1.9
7 August	1340	Soot	1.6
7 August	1500	Soot, salt, moisture	1.8

ammonium sulfate have not been included here because of large uncertainty as to their values in the low-concentration aerosol flow from which samples were extracted. No significant differences were found between results obtained with the two laser wavelengths; values corresponding to the green (514.5 nm) light are denoted with an asterisk.

The ability of an optoacoustic detector to follow time variations in the concentration of the absorbing substance is indicated in Fig. 3. This figure presents a portion of the recording obtained on the afternoon of 5 August when outdoor air was pumped into the aerosol system. It appears that this air occasionally was highly absorbing and that the aerosol particles essentially went through the system in bursts, or puffs.

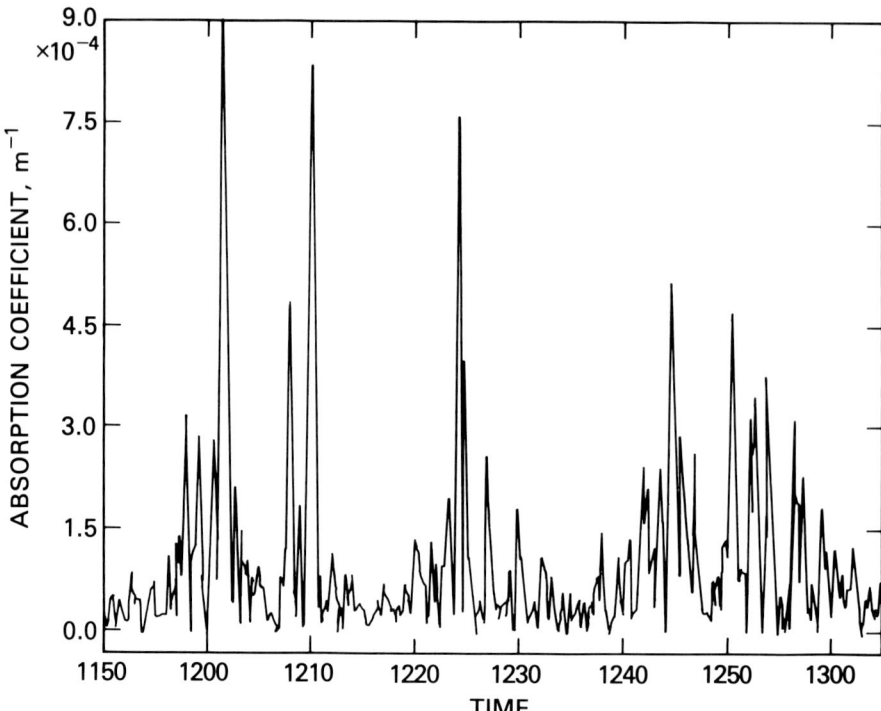

FIGURE 3. *Optoacoustic signal recorded on 5 August for ambient air.*

It was found that the addition of ammonium sulfate to a carbonaceous aerosol seemed to reduce the optical absorption, as did an increase in the humidity of the air stream. Although the signal levels for methylene blue were comparable with those of the noise for these measurements, they were considerably greater than the noise level encountered in previous work in an acoustically quieter environment.

IV. CONCLUSIONS

The optoacoustic measurements that have been described here effectively demonstrate the potential of this experimental technique. It should be possible to develop an improved apparatus that is adequately immune to environmental acoustic noise and is more suited to field use. Given that the greatest uncertainty in the present results is associated with system calibration, through determination of the value of the absorption coefficient for the calibration gas, it is important that an improved procedure be devised for use in future measurements of light absorption by aerosol particles.

DATA SUMMARY

compiled by
H. Gerber and E. Hindman

Measurements made at the First International Workshop on Light Absorption by Aerosol Particles (held at Fort Collins, Colorado, 28 July to 8 August 1980) are summarized in Table 1. Column 1 lists the investigators and assigns each group a number which is used throughout the remainder of this section and section 8. The second column gives the instrument type and the wavelengths of the measurements. The experiment date, the type of aerosol tested, and the "on" time for the CSU particle generation system are noted at the top of the table.

TABLE 1. Data Summary

Investigator	Instrument type and wavelengths of measurement	Date, particle type, and generation time		
		7/29 Salt 1114-1330	7/29 Salt 1420-1609	7/29 Salt + 1% MB 1613-1734
1 C. Bennett, Jr. R. Patty[a]	Filter spectrophone, integrating plate; 0.6328 μm (high conc. line)			
2 B. Bodhaine[b]	Integrating nephelometer; 0.450, 0.550 0.655, 0.820 μm			
3 A. Clarke A. Waggoner	Integrating plate; (also revised values *) 0.55 μm			1630-1703 $\sigma_a, \tilde{\omega}, n_2, B_a$ *
4 S. Cowen	Integrating plate; 0.56 μm (high conc. line)	1100————1600 $\sigma_a, \tilde{\omega}, B_a$		1625-1715 $\sigma_a, \tilde{\omega}, n_2, B_a$
5 W. Egan[c]	Diffuse reflectance; 0.2-1.1 μm			
6 H. Gerber	Transmissometer, nephelometer; 0.6328 μm			1620-1700 σ_e, σ_s, n_2
7 J. Gillespie[d]	Diffuse reflectance; 0.3-1.7 μm			
8 G. Grams A. Coletti	Polar nephelometer; 0.6328 μm	g,S,P	g,S,P	
9 J. Heintzenberg	Integrating plate hydrosol in sphere photometer; 0.55 μm		1529-1600 $\sigma_a, \tilde{\omega}, B_a$	1635-1733 $\sigma_a, \tilde{\omega}, B_a$
10a E. Patterson B. Marshall	Diffuse reflectance 0.3-0.7 μm		1441-1609 k, n_2, B_a	1626-1730 k, n_2, B_a (also certified standard)
10b E. Patterson B. Marshall	Laser transmission; 0.6328 μm		1438-1541 σ_a, B_a	1623-1733 σ_a, B_a (2 values)
11 R. Röhl R. Palmer W. McClenny	Filter spectrophone; 0.6-1.6 μm (high conc. line)			
12 H. Rosen T. Novakov	Laser transmission; 0.6328 μm		1529-1600 σ_a	1635-1733 σ_a
13 G. Shaw	Remote sensing; 0.70 μm			
14 F. Volz[e]	KBr pellet; 0.38-0.94 μm* 2.5-30 μm			
15 W. Wright D. Stedman L. Stefanutti R. Terhune	In situ spectrophone 0.488 μm, 0.514 μm*			

DATA SUMMARY

TABLE 1. continued

	Date, particle type, and generation time				
7/30 Salt 1101-1232	7/30 Salt + 0.3% MB 1305-1530	7/30 Salt + ~ 4% MB 1624-1800	7/31 Soot 0940-1421	7/31 Soot + salt 1433-1737	
			0940-1421 n_2, B_a		1
					2
1109-1202 $\sigma_a, \tilde{\omega}, n_2, B_a$ (2 values)*	1350-1528 $\sigma_a, \tilde{\omega}, n_2, B_a$ (2 values)*	1633-1734 $\sigma_a, \tilde{\omega}, n_2, B_a$ (2 values)*	1022-1426 $\sigma_a, \tilde{\omega}, n_2, B_a$ (6 values)	1521-1735 $\sigma_a, \tilde{\omega}, n_2, B_a$ (5 values)	3
1130-1210 $\sigma_a, \tilde{\omega}, B_a$	1350-1530 $\sigma_a, \tilde{\omega}, n_2, B_a$	1630-1730 $\sigma_a, n_2, \tilde{\omega}, B_a$	1024-1448 $\sigma_a, \tilde{\omega}, B_a$	1502-1713 $\sigma_a, \tilde{\omega}, B_a$	4
					5
					6
					7
g	g, P	g	g, P		8
	1327-1527 $\sigma_a, \tilde{\omega}, B_a$	1636-1741 $\sigma_a, \tilde{\omega}, B_a$	1015-1345 $\sigma_a, \tilde{\omega}, B_a$ (8 values)	1442-1640 $\sigma_a, \tilde{\omega}, B_a$ (3 values)	9
			1013-1346 k, n_2, B_a (3 values)	1528-1549 k, n_2, B_a	10a
	1333-1527 σ_a, B_a (2 values high&low conc.)	1633-1753 σ_a, B_a (2 values high&low conc.)	1013-1420 σ_a, B_a (3 values)	1528-1700 σ_a, B_a (2 values)	10b
			n_2, B_a (2 values)		11
1120-1232 σ_a	1327-1527 σ_a	1636-1741 σ_a	1015-1345 σ_a (7 values M_v, B_a)	1442-1640 σ_a (3 values M_v for carbon, B_a)	12
					13
					14
			1020-1345 σ_a (5 values)	1600, 1650* σ_a	15

TABLE 1. continued

Investigator	Instrument type and wavelengths of measurement	Date, particle type, and generation time		
		8/1 MB 1130-1347	8/1 Soot 1405-1600	8/4 Soot 1005-1129
1 C. Bennett, Jr. R. Patty[a]	Filter spectrophone, integrating plate; 0.6328 μm (high conc. line)			
2 B. Bodhaine[b]	Integrating nephelometer; 0.450, 0.550 0.655, 0.820 μm			
3 A. Clarke A. Waggoner	Integrating plate; (also revised values *) 0.55 μm	1133-1345 $\sigma_a, \tilde{\omega}, n_2, B_a$ (3 values)	1443-1544 $\sigma_a, \tilde{\omega}, n_2, B_a$ (2 values)	1042-1114 $\sigma_a, \tilde{\omega}, n_2, B_a$
4 S. Cowen	Integrating plate; 0.56 μm (high conc. line)	1140-1337 $\sigma_a, \tilde{\omega}, n_2, B_a$	1414-1536 $\sigma_a, \tilde{\omega}, B_a$	1032-1110 $\sigma_a, \tilde{\omega}, B_a$
5 W. Egan[c]	Diffuse reflectance; 0.2-1.1 μm			1045-1125 n_2
6 H. Gerber	Transmissometer, nephelometer; 0.6328 μm			
7 J. Gillespie[d]	Diffuse reflectance; 0.3-1.7 μm			
8 G. Grams A. Coletti	Polar nephelometer; 0.6328 μm	g, P		g, P
9 J. Heintzenberg	Integrating plate hydrosol in sphere photometer; 0.55 μm	1130-1325 $\sigma_a, \tilde{\omega}, B_a$ (4 values)	1435-1527 $\sigma_a, \tilde{\omega}, B_a$ (3 values)	1053-1113 $\sigma_a, \tilde{\omega}, B_a$
10a E. Patterson B. Marshall	Diffuse reflectance 0.3-0.7 μm			1037-1128 k, n_2
10b E. Patterson B. Marshall	Laser transmission; 0.6328 μm	1131-1345 σ_a, B_a (2 values)		
11 R. Röhl R. Palmer W. McClenny	Filter spectrophone; 0.6-1.6 μm (high conc. line)	B_a (0.71 μm, 5 values)		
12 H. Rosen T. Novakov	Laser transmission; 0.6328 μm	1300-1325 σ_a (4 values)	1435-1527 σ_a (3 values)	1053-1113 σ_a
13 G. Shaw	Remote sensing; 0.70 μm			
14 F. Volz[e]	KBr pellet; 0.38-0.94 μm* 2.5-30 μm			
15 W. Wright D. Stedman L. Stefanutti R. Terhune	In situ spectrophone 0.488 μm, 0.514 μm*		1530* σ_a	1040 σ_a

DATA SUMMARY

TABLE 1. continued

		Date, particle type, and generation time			
8/4 Soot (50%) + salt 1129-1412	8/4 MB 1454-1558	8/5 Ambient 1015-2400	8/6 Ambient 0000-0830	8/6 ARD 1017-1336	
					1
					2
1143-1400 $\sigma_a, \tilde{\omega}, n_2, B_a$ (2 values)	1536-1557 $\sigma_a, \tilde{\omega}, n_2, B_a$	1017-1146;1803-0831 $\sigma_a, \tilde{\omega}, n_2, B_a$ (for 2 size fractions)		1054-1241 $\sigma_a, \tilde{\omega}, n_2, B_a$ (2 size fractions)	3
1214-1355 $\sigma_a, \tilde{\omega}, B_a$	1519-1538 $\sigma_a, \tilde{\omega}, n_2, B_a$				4
1212-1400 n_2	1500-1545 n_2	1035————0830 n_2		1040-1300 n_2	5
				1150 σ_e, σ_s, n_2	6
				n_2, k	7
g, P	g, P	g, P			8
1140-1352 $\sigma_a, \tilde{\omega}, B_a$ (5 values)					9
	1504-1544 k, n_2, B_a (also solution transmission)			1042-1333 k, n_2, B_a (also bulk sample)	10a
1352-1402 σ_a, B_a		1025-1615: 1700-0833: 1057-1715 σ_a, B_a		1039-1334 σ_a, B_a	10b
					11
1140-1352 σ_a		1027————0827 σ_a		1044-1331 σ_a	12
			$\tilde{\omega}$		13
			1000———		14
1130*,1200*, 1300*, 1440* (soot) σ_a	1505*,1540 σ_a			1100.1130* σ_a	15

TABLE 1. Data Summary

Investigator	Instrument type and wavelengths of measurement	8/6 Soot (4%) + salt 1429-1708	8/6 Soot 1708-1807	8/7 Soot (5%) + salt 1101-1340
1 C. Bennett, Jr. R. Patty[a]	Filter spectrophone, integrating plate; 0.6328 μm (high conc. line)			
2 B. Bodhaine[b]	Integrating nephelometer; 0.450, 0.550 0.655, 0.820 μm			
3 A. Clarke A. Waggoner	Integrating plate; (also revised values *) 0.55 μm	1511-1708 $\sigma_a, \tilde{\omega}, n_2, B_a$ (2 values)	1744-1804 $\sigma_a, \tilde{\omega}, n_2, B_a$	1152-1302 $\sigma_a, \tilde{\omega}, n_2, B_a$ (2 samples)
4 S. Cowen	Integrating plate; 0.56 μm (high conc. line)			
5 W. Egan[c]	Diffuse reflectance; 0.2-1.1 μm			
6 H. Gerber	Transmissometer, nephelometer; 0.6328 μm	1430-1707 σ_e, σ_s, n_2	1740-1807 σ_s, σ_e, n_2	1106-1330 σ_e, σ_s, n_2 (also soot alone)
7 J. Gillespie[d]	Diffuse reflectance; 0.3-1.7 μm			
8 G. Grams A. Coletti	Polar nephelometer; 0.6328 μm			
9 J. Heintzenberg	Integrating plate hydrosol in sphere photometer; 0.55 μm	1455-1659 $\sigma_a, \tilde{\omega}, B_a$		
10a E. Patterson B. Marshall	Diffuse reflectance 0.3-0.7 μm	1457-1700 k, n_2, B_a		
10b E. Patterson B. Marshall	Laser transmission; 0.6328 μm	1458-1700 σ_a, B_a	1742-1803 σ_a, B_a (3 values)	1051-1338 σ_a, B_a
11 R. Röhl R. Palmer W. McClenny	Filter spectrophone; 0.6-1.6 μm (high conc. line)			n_2, B_a for soot, 7 samples
12 H. Rosen T. Novakov	Laser transmission; 0.6328 μm	1455-1659 σ_a	1741-1802 σ_a	1145-1224 σ_a
13 G. Shaw	Remote sensing; 0.70 μm			
14 F. Volz[e]	KBr pellet; 0.38-0.94 μm* 2.5-30 μm			
15 W. Wright D. Stedman L. Stefanutti R. Terhune	In situ spectrophone 0.488 μm, 0.514 μm*	1550, 1620 σ_a	1720, 1750 σ_a σ_a	1120 soot + salt 1340 soot

DATA SUMMARY

TABLE 1. continued

Date, particle type, and generation time			
8/7 Soot + salt + moisture 1455-1552	8/8 MB 0933-1115	8/8 Salt 1130-1430	
			1
			2
	1005-1104 $\sigma_a, \tilde{\omega}, n_2, B_a$ (2 values)	1217-1333 $\sigma_a, \tilde{\omega}, n_2, B_a$ *	3
			4
			5
1435-1550 σ_e, σ_s	0937-1116 σ_e, σ_s, n_2	1138-1206 σ_e, σ_s, n_2	6
			7
			8
			9
			10a
		1144-1337 σ_a, B_a	10b
			11
	0952-1051 σ_a		12
			13
			14
1500 σ_a	----1400		15

Salt—$(NH_4)_2SO_4$; soot—soot from propane flame; MB—methylene blue; ARD—Arizona road dust; ambient—aerosol from outside laboratory; σ_a—aerosol absorption coefficient; σ_e—aerosol extinction coefficient; σ_s—aerosol scattering coefficient; $\tilde{\omega}$—single scattering albedo; n_1—real part of refractive index; n_2—imaginary part of refractive index; k—absorption coefficient of bulk material; M_v—mass of aerosol particles per volume of aerosol; B_a—specific absorption coefficient (σ_a/M_v); g—scattering asymmetry parameters; S—scattering cross section; P—scattering phase function.

[a] Tested various layers and mixtures of soot and salt on filters.
[b] Continuous measurements of σ_s for duration of workshop.
[c] Measured n_2 for bulk quantities of Arizona road dust, CSU-sample and spectroscopically-pure graphite, MB, lampblack, and pure $(NH_4)_2SO_4$.
[d] Measured n_2 on bulk MB powder.
[e] Prepared own soot, $n_2{}^*$; bulk quartz, $n_2{}^*$, and ARD, n_2, $n_2{}^*$; MB, n_2, $n_2{}^*$, $n_1{}^*$, and salt powders $n_2{}^*$; MB in solution, $n_2{}^*$; bulk workshop soot, n_2; own soot and salt, n_2; MB and salt, n_2; and for ambient particles, $n_2{}^*$, n_2, for soluble and insoluble fractions removed from filter.

DATA COLLATION

compiled by
H. Gerber and E. Hindman

All data presented here are as submitted by the investigators following the workshop, except for those of Heintzenberg [9][1] who resubmitted all data with values of σ_a reduced by a factor of 0.364, and those of Clarke and Waggoner [3] who resubmitted a portion of their data following evaluation of Workshop samples with an improved method (for further explanations, see the papers of those investigators in the section entitled "Instrumentation Descriptions and Workshop Measurements"). All collation is for data in the visible spectrum due to a lack of measurements in the infrared.

The good agreement in Fig. 1 between the normalized values of σ_a, σ_s, and M_v suggests that soot particles with similar optical and physical characteristics were generated on the various dates shown. This behavior was used to extrapolate the value of M_v measured by Rosen and Novakov [12] with the total combustion method for 7/31 to the other dates shown in Fig. 1 with the aid of the light-scattering data of Bodhaine [2]. Table 1 lists the values of M_v for soot which were used in calculations of B_a (Table 2). Other measures of M_v were judged to be less reliable.

Values of M_v for methylene blue (MB) and salt in Table 1 are from the CSU filter mass measurements and the particle mass monitor (PMM) data, respectively. All data for MB in Table 2 have been wavelength corrected to $\lambda = 550$ nm using the MB spectral data of investigators [7] and [14].

[1]participant number.

TABLE 1. Mass of Soot, MB, and Salt Particles Per Volume of Aerosol $(M_v, \mu g\ m^{-3})$

Particle type	Soot					MB			Salt		
Date	7/31	8/1	8/4	8/6	8/7	8/1	8/4	8/8	7/29	7/30	8/8
M_v	177	110	128	190	37.8	236	250	425	275	136	35.0

TABLE 2. Values of B_a, $\tilde{\omega}$, and n_2 averaged for all dates for each particle type. The last row gives an overall average.

Partici-pant*	Soot (7/31,8/1,8/4, 8/6,8/7)			MB (8/1,8/4,8/8)			Salt (7/29,7/30,8/8)		
	B_a	$\tilde{\omega}$	n_2	B_a	$\tilde{\omega}$	n_2	B_a	$\tilde{\omega}$	n_2
1			0.56						
2									
3	6.45	0.180	0.883	2.11	0.442	0.163	0.065	0.988	
4	11.03	0.115		2.07	0.441	0.11	0.160	0.996	
5			0.0675			0.153			8.1×10^{-6}
6	7.00	0.180	1.04	0.976	0.626	0.03			
7						0.10			
8									
9	7.98	0.146		2.67	0.389		0.450	0.905	
10a	4.31	0.204	0.435	1.62	0.503	0.046	1.8×10^{-4}	1.00	1.6×10^{-5}
10b	5.37	0.180		1.28	0.561		0.159	0.966	
11	6.91	0.150	0.660						
12	11.37	0.090		4.46	0.290		~ 0	~ 1	~ 0
13									
14			0.422			0.203			5×10^{-3}
15	3.83	0.292		0.668	0.710				
Averages	7.14	0.171	0.581	1.98	0.495	0.115	0.139	0.971	1.26×10^{-3}

*Keyed to Table 1 in the preceding section (Section 7, Data Summary).

DATA COLLATION

TABLE 3. Values of B_a, $\tilde{\omega}$, and n_2 for various mixtures of soot and salt. Values of M_y used to calculate B_a are from CSU filter mass measurements.

Partici-pant	7/31 (~50% soot)			8/4 (~50% soot)			8/6 (~4% soot)			8/7 (~5% soot)		
	B_a	$\tilde{\omega}$	n_2	B_a	$\tilde{\omega}$	n_2	B_a	$\tilde{\omega}$	n_2	B_a	$\tilde{\omega}$	n_2
1												
2												
3	2.66	0.569	0.208	1.79	0.576	0.179	0.80	0.882	0.0406	1.10	0.675	0.0541
4	2.23	0.610		3.72	0.396							
5												
6							0.183	0.970		0.088	0.963	
7												
8												
9	2.44	0.590		1.92	0.558		0.321	0.949				
10a	0.62	0.849	0.0553				0.143	0.977	0.0128			
10b	2.66	0.568		1.77	0.578		0.433	0.933		0.665	0.774	
11												
12	4.63	0.431		3.19	0.433		0.705	0.894		1.50	0.604	
13												
14												
15	1.43	0.710		0.798	0.753		0.176	0.971		0.451	0.835	
Averages	2.38	0.618	0.132	2.20	0.549		0.394	0.939	0.027	0.761	0.770	

TABLE 4. Values of σ_a, $\tilde{\omega}$, and n_2 for ambient particles and Arizona road dust.

Participant	Ambient (8/5, 8/6)			Arizona road dust (8/6)		
	σ_a	$\tilde{\omega}$	n_2	σ_a	$\tilde{\omega}$	n_2
1						
2						
3	5.6×10^{-6}	0.742	0.38	5.4×10^{-6}	0.980	
4						
5			0.214			4.5×10^{-3}
6				1.38×10^{-4}	0.667	
7						7.0×10^{-3}
8						
9						
10a						2.1×10^{-3}
10b	7.2×10^{-6}	0.660		2.8×10^{-5}	0.908	
11						
12	8.0×10^{-6}	0.636		~ 0	~ 1.0	
13		0.95				
14						
15				2.05×10^{-4}	0.570	

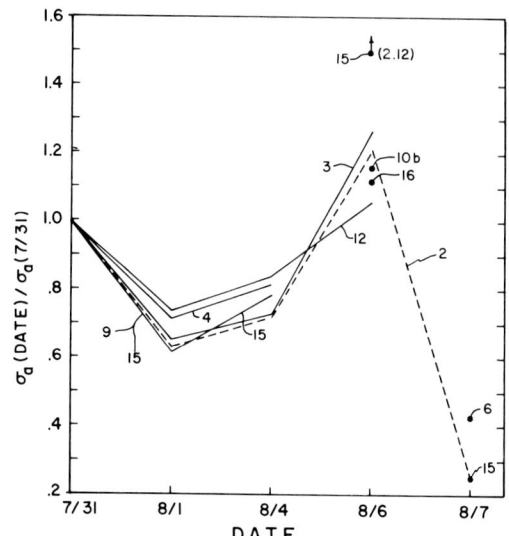

FIGURE 1. The absorption coefficient σ_a of soot normalized to the measurement on 7/31 for λ in the visible, except for data numbered 2, 6, 16. The numbers refer to different investigators (see Table 1 in section 7, Data Summary, for thekey). Data numbered 2 give the scattering coefficient σ_s normalized to the 7/31 value, $\sigma_s(DATE)/\sigma_s(7/31)$; data numbered 6 use 8/6 as a reference; and data numbered 16 give the particle mass per volume of aerosol (measured by 12) normalized to 7/31, $M_v(8/6)/M_v(7/31)$.

DATA COLLATION

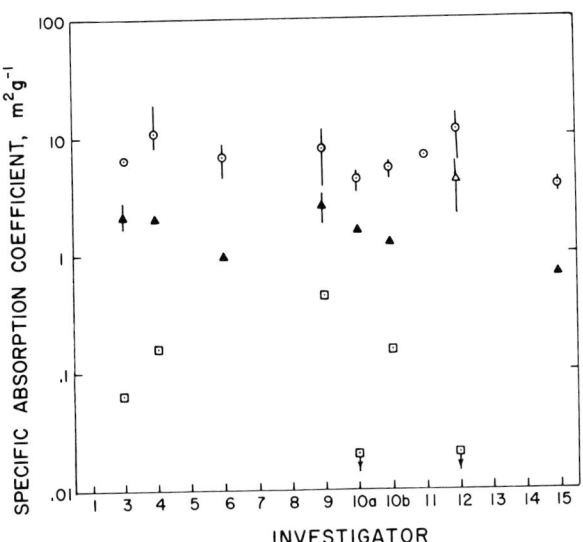

FIGURE 2. Specific absorption coefficient B_a for soot (○), MB (▲), and salt (□). Vertical lines show the range of the measurements; no lines indicate one measurement was made, or a small range existed. Table 2 gives the numerical values of B_a.

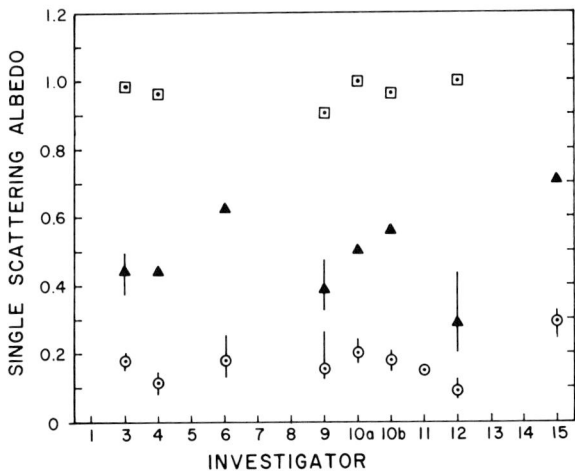

FIGURE 3. Single scattering albedo $\tilde{\omega}$ for soot (○), MB (▲), and salt (□). See Table 2 for numerical values.

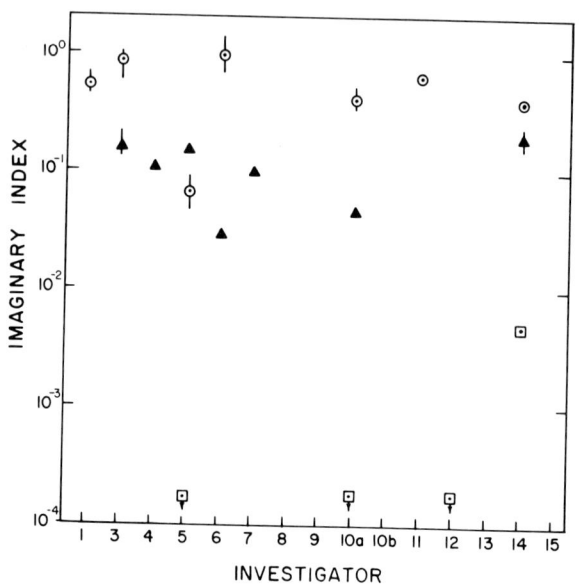

FIGURE 4. Imaginary index n_2 for soot (○), MB(▲), and salt (□). See Table 2 for numerical values.

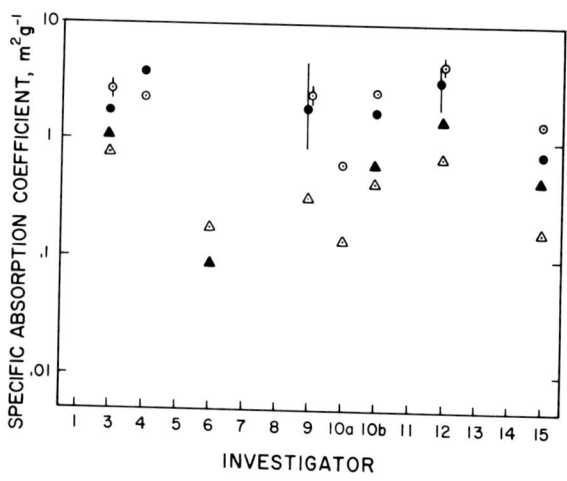

FIGURE 5. Measurements of B_a for mixtures of soot and salt on 7/31 ($\approx$ 50% soot, ○), 8/4 (50% soot, ●), 8/6 ($\approx$ 4% soot, △), and 8/7 ($\approx$ 5% soot, ▲). Numerical values are given in Table 3.

DATA COLLATION

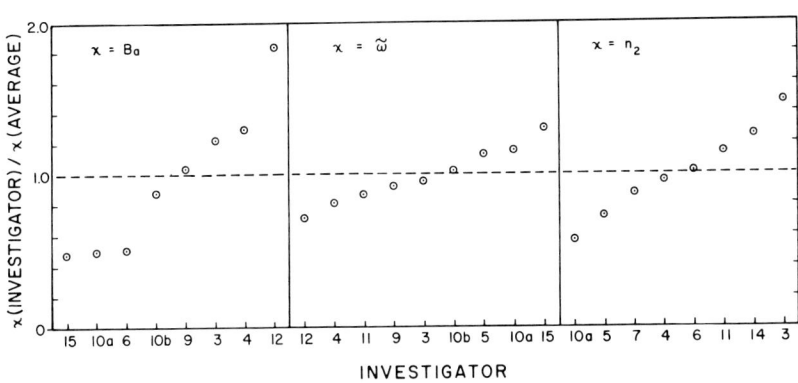

FIGURE 6. Values of B_a and $\tilde{\omega}$ for soot and MB in Table 2, and for mixtures of soot and salt in Table 3, are normalized to the overall averages given in those tables. Values of n_2 for soot and MB in Table 2 are normalized to the overall averages in Table 2.

WORKSHOP REVIEW[1]

S. Twomey

Institute of Atmospheric Physics
University of Arizona
Tucson, Arizona

Donald R. Huffman

Department of Physics
The University of Arizona
Tucson, Arizona

I. INTRODUCTION

While being appreciative of the opportunity to discuss the goals, procedures, and results of this workshop, we are conscious of the fact that, since neither of us participated in it, we are not ourselves subject to the scrutiny to which the participants have exposed themselves. At the outset, we would like to congratulate the participants for undertaking this venture and for putting their results on the line in this way, and to express the hope that they will view any negative comments which we may make in this review as an attempt to bring an intrinsically difficult topic into perspective, rather than as arbitrary criticism from insulated nonparticipants.

[1] This work was supported in part by the Atmospheric Sciences Section of the National Sciences Foundation under Contract ATM-77-11462.

II. GENERAL DISCUSSION

The title of the workshop does not, *per se*, emphasize the atmosphere or climate, but the affiliation and contributions of many of the participants suggest that atmospheric aerosol (especially its influence on climate, now and future, via radiative effects) is a prime target for the techniques being compared at this workshop. It seems desirable, therefore, to identify which parameters most directly relate to atmospheric radiation and to climate physics, as distinct from parameters which might be more relevant, for example, to investigating the composition of the aerosol.

At the present stage of climatic studies, most attention is given to aerosol particles in their role as absorbers and reflectors of solar radiation. Most of the experimental studies contained herein were carried out at wavelengths in the spectral range of solar radiation, which is in keeping with that premise. Although clearly there are severals ways for assessing the influence of aerosol particles on atmospheric energetics at solar wavelengths, the most popular seems to be the "critical single-scattering albedo" route. Values for this critical parameter from about 0.75 to about 0.95 can be found in the literature; a common value used is 0.85, from Hansen et al. (1). However, the primary parameters for climate physics are values of backscattering and of absorption by the aerosol, since these are what produce cooling and heating, respectively. Thus, the most fundamental measurement would be that of absorption and backscattering coefficients for unit volume or unit mass of air. If these were obtained directly, then size distributions, asymmetry factors, single-scattering albedos, and particulate complex refractive index would not be needed to establish the present state of the atmosphere relative to the role of the dry aerosol in climate. We therefore feel that *in situ* measurements of these two parameters would represent the optimum approach. This would be closely approximated by applying the *in situ* spectrophone technique as applied, for example, by Foot (2), with integrating nephelometry. A less desirable but still reasonable approximation to the ideal is given by the use of collection (e.g., on Nuclepore filters, etc.) to concentrate the aerosol material before measuring absorption (by spectrophone or diffusing plate methods). Some complicating factors evidently enter when collection is employed: (a) the originally widely separated particles become close-packed neighbors in the collected sample, and (b) no collection method assures the collection of all sizes of particle, and since absorption per unit particulate volume tends to a finite limit with decrease in size (unlike scattering per unit particulate volume, which tends to zero), efficient collection of particles right down through the Rayleigh size range is demanded; furthermore, disappearance of more volatile particles or drops after collection may occur. Notwithstanding these problems, since it seems that *in situ* spectrophone techniques are still insufficiently sensitive to handle cleaner atmospheric air samples (the workshop papers would suggest present detectability minimum to be around 10^{-4} m^{-1}), collection would seem to be the only alternative until the intrinsically preferable *in situ* methods become more sensitive. Most of the successful measurements at the workshop were, in fact, obtained by postcollection measurements on sampled particulate material, using either spectrophone or diffuser methods. Even those methods had problems on the ammonium sulfate aerosols, the absorption of which tended to be seriously overestimated by most of the technqiues. On the plus side of the accounting, it is encouraging that the workshop has produced a considerable improvement in basic understanding of diffusing-plate methods in the form of Rosen and Novakov's (3) analysis of their system and Clarke and Waggoner's (4) redesign of the University of Washington system in accordance with the principles enunciated in Rosen and Novakov's paper, namely, use of a diffuser with a large reflection coefficient between the light source and the sample. Application of the "sandwich" technique by Clarke and Waggoner was reported to have resulted in a substantial improvement (and incidentally reduced the inferred absorption values compared to those given by their original procedure).

Gerber's (5) method represents an alternative *in situ* approach in which absorption is obtained by differencing extinction and scattering, and even though commendable thought and planning went into the design of his instrument, it was apparently not entirely successful with aerosols of moderate absorption. That again attests to the very real difficulties which this differencing route presents.

A third approach to the problem is found in indirect, remote-sensing approaches. By their very nature, such methods could not be subjected to the intercomparison procedures of the workshop, and that in itself constitutes a serious drawback. A further serious drawback is that they involve fitting measured data via Mie theory, using imaginary refractive index as a parameter from which single-scattering albedo is calculated; an extremely indirect procedure. Involving as it does an idealization of the atmospheric aerosol (which is undoubtedly mixed in composition and shape) in order to make the Mie calculations possible, it can, at best, give what has sometimes been referred to as an "effective" or "equivalent" imaginary refractive index. We do not personally feel at all comfortable with this rationalization, especially when the values thus inferred do not resemble those for any known pure substance, as is usually the case. The perils of averaging have often been pointed out (for example, the aphorism that one foot in ice water and one foot in boiling water cannot be equated with comfort), and, in the context of particulate scattering and absorption, such perils are surely present. The contrast between a mixture of aerosol particles of two kinds (e.g., small black particles with larger white particles) and a uniform aerosol in which the same materials are intimately mixed in the proper proportions is discussed later.

III. SOME SPECIFIC POINTS

1. *Accuracy and Precision of Absorption Determinations*

Considerable work was done by the workshop organizers to produce test aerosols of known composition, yet the optical constants for these supposedly known aerosols are not presented. It seems to us that one of the great advantages of generating known aerosols is that the true values of their intrinsic optical properties should be fairly well known.

Even in the case of homogeneous solids and liquids, determination of optical constants is not necessarily an easy task in spectral regions of very high or very low absorption. Optical constants are not directly measurable quantities, but must be inferred from other measurements, such as optical transmission and reflectance, along with an appropriate calculational model such as Fresnel's equations. In the case of solid and liquid samples having smooth, plane surfaces, we have little doubt that the calculational model is both accurate and appropriate for inferring optical constants from the measurements. In the case of solids available only in the form of fine particles, the problem of optical constant determination is much more difficult, even if only one component is present in the sample. Optical measurements of transmission and reflectance or scattering can be easily made, but the calculational model used to infer the desired optical constants may not be exact (Kubelka-Munk theory, for example) or it may not be applicable (Mie theory). Although there are many techniques that have been applied to the problem of inferring optical constants of powdered materials, it is our opinion that in general the problem has not been satisfactorily solved. In some senses, that part of the workshop which aimed at determining n_2 for several pure substances is a test of the precision of such determinations. For comparison with these determinations, however, we have attempted to find optical constants for the pure materials that have been derived from bulk homogeneous materials.

In Fig. 1, we present imaginary refractive indices for carbon [taken from Williams and Arakawa (6)] and for ammonium sulfate [from Toon et al. (7)]. The optical transmission measurements on single crystals of ammonium sulfate led to an upper limit of $n_2 = 10^{-7}$ throughout the visible, as we have indicated on the diagram. Unfortunately, our brief literature search did not turn up optical constants measured on homogeneous bulk samples of solid methylene blue. We have therefore plotted n_2 from several investigators, as determined from powder samples of the methylene blue used at the workshop.

The workshop comparisons and those in Gerber's review of aerosol absorption are almost entirely concerned with consistency or precision, in the sense that results of different investigators are compared with each other or with an overall average. But this gives us very little information about accuracy, i.e.,

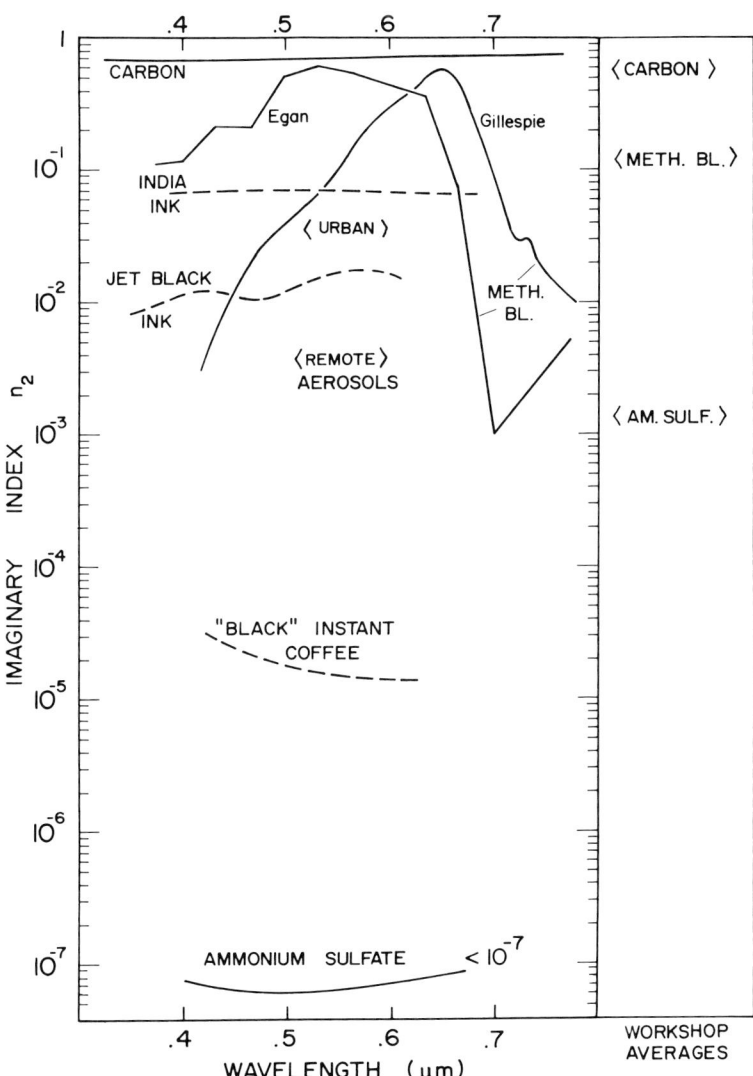

FIGURE 1. Imaginary refractive index determined from homogeneous bulk solids of carbon and ammonium sulfate, along with several measurements from workshop participants on powder samples of methylene blue. Average values of n_2 derived by Gerber (5) from workshop measurements are shown for the three pure solids and from atmospheric aerosol results reported in the literature. Results for "black" coffee, Jet Black ink, and India ink are included for perspective.

how all the measurements agree with the true values. For this comparison, we have placed on Fig. 1 the average values for carbon, methylene blue, and ammonium sulfate, taken from the data collation by Gerber and Hindman (8). Also shown are average n_2's derived by Gerber (5) from his extensive survey of the literature on optical constants inferred for natural aerosols.

The comparisons of the figure suggest that the heavily absorbing aerosols of carbon were measured fairly accurately, while the absorption by the almost nonabsorbing ammonium sulfate was generally overestimated by several orders of magnitude. In fact, there is only a factor of about 500 separating the average measured absorptions of carbon, which is one of the most highly absorbing of all common particulate materials, and ammonium sulfate, which is one of the most transparent solids. This compares with a ratio of seven orders of magnitude or more, as inferred from homogeneous solid measurements. It may also be relevant to notice that the individual investigators' results for ammonium sulfate showed a spread of more than three orders of magnitude. [The value indicated on Fig. 1 is Gerber's average value, which, as it happens, is totally dominated by the results of the investigator (9) who reported the largest value of n_2 for that substance.] The questions of averaging and reporting become crucial under such circumstances.

While the inferred value of $n_2 \approx 0.001$ for ammonium sulfate sounds quite small, we would like to emphasize with some common examples that this is, in fact, quite highly absorbing in comparison to most solids and liquids. Plotted on the figure are values of n_2 that we have determined for "black" coffee (four times the recommended strength for a popular brand of instant coffee), Sheaffer Jet Black ink, and Higgins No. 4465 India ink. Optical transmission measurements were made and converted into n_2 values by means of the equation

$$T = \exp\left(-\frac{4\pi n_2}{\lambda}\right) Cl \tag{1}$$

where C is the volume concentration of the undiluted liquid and l is the path length. In comparison to these examples, the value of n_2 from particulate ammonium sulfate measurements is about two orders of magnitude more than the "black" coffee and within about a factor of 10 of the Jet Black ink. This puts into perspective just how high the workshop average values for ammonium sulfate are. It seems that either the particulate samples of pure ammonium sulfate have been contaminated by absorbing material or the procedures for measuring and averaging to arrive at the n_2 values for nonabsorbing solids are giving rise to considerable error.

2. *Limits of Applicability of Present Methods*

The relatively large values of absorption determined for ammonium sulfate may be telling us something about the lower limit of detectability for absorption in the presence of dominant scattering, taking into account the averaging used. In view of these results, the average values of n_2 derived from measurements on atmospheric aerosols are of interest. Note that they characteristically range between about 0.001 and 0.1, with an average value for urban aerosols of 0.038, as derived by Gerber (5). Again we note that such material would be looked upon as very black; indeed, it is a little blacker than Jet Black ink.

In some sense, the viewpoint we are stressing provides a counterpoint to that presented in Gerber's review of absorption by atmospheric aerosol particles. In Gerber's average, the absorption measurements derived by Foot (2) from his spectrophone were not included because they "appeared unrealistically small." It may be, in fact, that the lowest values are the most accurate. Since we have argued that the *in situ* spectrophone may be one of the best ways to determine aerosol absorption, we must carefully consider this lowest value in Gerber's tabulation. We also point out that, even in this case, the sampling took place in the vicinity of London, certainly not the cleanest air from the standpoint of highly absorbing carbon.

3. *Effect of Particle Mixtures on the Inferred Albedo*

All of the techniques used in the workshop necessarily determine some sort of average index of refraction, in the case of mixtures of particles. Several facts indicate that the absorption derived from ambient aerosols is due to a mixture of different kinds of particles. One indication we have mentioned is that no commonly known substance has n_2 values on the order 0.001 to 0.01, as inferred. In addition, there is evidence that smaller particles in the aerosol are the more highly absorbing ones (10) and carbon in the form of soot has been identified (11) as an aerosol component. It is easy to show that the manner in which strong absorbers are mixed can make considerable difference in optical properties such as single-scattering albedo.

Consider two model systems. One model consists of an ensemble of 1% by particle volume of spherical absorbers with a single radius of 0.05 μm and a complex refractive index of $m_{bl} = 1.7 - i\, 0.7$ appropriate to carbon (12) and 99% by volume of larger particles with $m_{wh} = 1.5 - i\, 10^{-6}$, roughly approximating several possible aerosol components. Subscripts denote black (bl) and white (wh) for the appearance of the two species. As the second model, we envisage "gray" (gr) particles having similar amounts of the same kinds of strongly and weakly absorbing materials, but, in this case, homogeneously mixed into the spherical particle. Optical constants for the composite "gray" particle are properly calculated using the Maxwell-Garnett dielectric function (13). In the present case, however, the resulting average refractive index is very nearly the same as that given by the usual method of separately averaging the real and imaginary refractive indices by volume. For 1% of our black material in 99% of the white material, the result is $m_{gr} = 1.5 - i\, 0.007$. Making use of several simple approximations, we can calculate:

$$1 - \tilde{\omega} = C_{abs}/C_{ext} \tag{2}$$

for the two cases.

In the black-white mixture, the 0.05-μm black particles are approximately in the Rayleigh size limit, where absorption dominates over scattering and the absorption cross section per unit volume

$$\frac{C_{abs}}{V} = \frac{3}{4}\frac{Q_{abs}}{\alpha} = \frac{36\pi}{\lambda}\frac{n_1 n_2}{(n_1^2 - n_2^2 + 2)^2 + (2n_1 n_2)^2} \tag{3}$$

is independent of size. The large, nonabsorbing particles are approximated by $Q_{ext} \cong Q_{sca} \cong 2$, a reasonable approximation for particles above about 0.5 μm. This approximation has the added feature that complicated structure in single-size Mie calculations of extinction, which are not very realistic for broad size distributions of irregular particles, are not present. Using $C_{ext}/V = \frac{3}{4}(Q_{ext}/\alpha) = 1.5/\alpha$ and a volume fraction f of absorber

$$1 - \tilde{\omega} = \frac{fC_{bl}/V}{fC_{bl}/V + (1-f)C_{wh}/V} = \frac{1}{1 + \frac{(1-f)1.5}{f\alpha_{wh}C_{bl}/V}} = \frac{1}{1 + \frac{15.4}{\alpha_{wh}}} \tag{4}$$

In order to calculate $1 - \tilde{\omega}$ for the "gray" particles, we use the following expression derived from geometric optics (14,15)

$$1 - \tilde{\omega} = \frac{8\pi}{3}\frac{n_2 \alpha}{\lambda}\frac{1}{n_1}\left[\, n_1^3 - (n_1^2 - 1)^{3/2}\,\right] = 0.163\, \alpha_{gr} \tag{5}$$

where the numerical result follows from the use of the "average" index of refraction and $\lambda = 0.55$ μm. Equations (4) and (5) can now be used to calculate $(1 - \tilde{\omega})$ for the two methods of mixing. The approximations involved give useful estimates for large particle radii from about 0.5 μm to about 2.0 μm. The differences in the two cases are substantial. For a large particle radius of 1.5 μm, for example, the 1% of absorber gives rise to a single-scattering albedo of about 0.9 in the case of separate black and white particles and about 0.75 for the homogeneous particle.

The effect on optical properties of mixing highly absorbing and nonabsorbing components was pointed out by Bergstrom (16). Subsequently, Mie calculations were presented by several authors for black and white materials similar to those we have chosen, but with a variety of size distributions for the two components. Because of the difficulties in comparing the various results, we have simply plotted some of them as a scatter diagram in Fig. 2, with $1 - \tilde{\omega}$ for the homogeneous "gray" particle on one axis and $1 - \tilde{\omega}$ for the black-white mixture on the other. The letter symbols refer to Lindberg (17), Toon et al. (7), and Gillespie et al. (18), and the triangles to calculations of Ackerman and Toon (19). The solid curve results from our "back-of-envelope" calculations described above. Circles are some single-size Mie calculations we have made for volume fractions f from 0.001 to 0.10 and radii from 0.1- to 2.0-μm. The 45° dashed line represents independence of albedo on the method of mixing, while the horizontal dashed lines represent the critical albedo of 0.85. It is obvious that the way in which the absorber is considered to be mixed can make a large difference in the resulting single-scattering albedo. In our simple model with 1% carbon, single-scattering albedos can result which are either well into the projected global-heating domain or clearly into the cooling domain. Therefore, it is doubtful if a determination of "average" n_2 would permit one to correctly predict atmospheric effects such as temperature trends without a more complete knowledge of the optical properties of atmospheric aerosols.

4. *Can We Measure What Is Significant in the Atmosphere?*

The workshop data have provided an excellent opportunity for estimating just what levels of particulate absorption can be measured by the various techniques and perhaps allow us to gauge how much improvement may be needed to bring about the *in situ* monitoring of absorption in clean "background" air, which is felt to be desirable in the geophysical context. To achieve the latter, one must guess what levels of particulate absorption are present in the cleaner remote parts of the atmosphere and what levels would be important in climate physics. Foot, working at ground level near London, found aerosol absorption values generally somewhat less than 10^{-5} m^{-1} (2). Clean air at Mount Lemmon, Arizona (2700-m level), measured by means of samples collected over long periods by impaction on and aspiration through Nuclepore filters, was found to exhibit absorption values which typically ranged from less than 10^{-7} m^{-1} to a few times that value. This is a small amount of absorption by almost any criterion, probably too small to be geophysically significant. To obtain a rough quantitative estimate of what the latter phrase may imply, we note that Hansen et al. (1) computed a tropospheric temperature change on the order of 0.5°C and stratospheric temperature changes of about 5°C when they modeled the effects of Agung by a time-dependent optical depth perturbation which had a maximum value of 0.2 for the atmospheric column, and those numbers agreed quite well with observed temperature trends following Agung. Changes of such magnitude are substantial but clearly not catastrophic in the context of climate, whereas changes which are less than, for example, one-tenth as great would be difficult even to

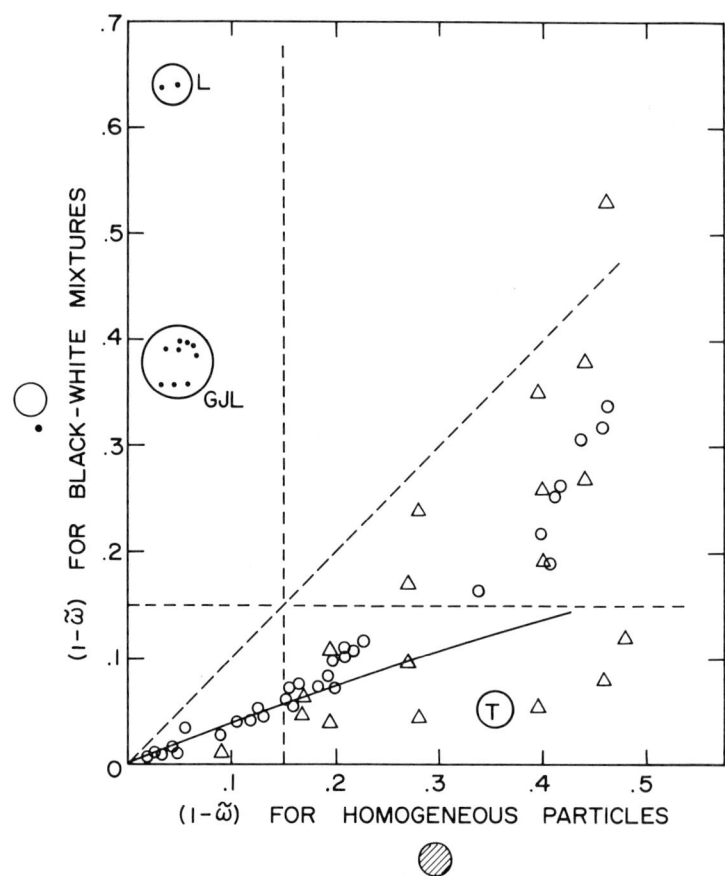

FIGURE 2. Scatter diagram comparing spherical particle calculations of $(1 - \tilde{\omega})$ for two models, one having an absorber similar to carbon and a nonabsorber as separate particles, the other having the same amount of absorber homogeneously distributed throughout the sphere. Circles are single-size Mie calculations from the present work, the solid line is from our simple approximate calculation, and the triangles are from Akerman and Toon (19). The letter symbols stand for results of Linderg (17), Gillespie et al. (18), and Toon et al. (7). The 45° dashed line represents lack of dependence of $(1 - \tilde{\omega})$ on the way in which the absorber is distributed.

detect in averaged measurements, which show undulations of one- or two-tenths of a degree having no likely connection with aerosols. We might therefore set the level of possible climate relevance at about 0.01 (in optical-depth units for the atmospheric column) or about 10^{-6} to 10^{-5} m^{-1}. The *in situ* results of the workshop seemed to suggest a detection limit of about 10^{-4} m^{-1}, and that estimate agrees substantially with comments made by Japar and Killinger (20).

The conclusion seems to be that most *in situ* techniques, while intrinsically desirable, do not at present have sufficient sensitivity for application in monitoring for climatically significant effects. Foot's technique (using the sun as a source) did, however, attain sensitivity levels comparable to those cited, which suggests that the *in situ* spectrophone technique is, in essence, capable of doing the job (2). Hopefully the next workshop will yield *in situ* measurements of absorption in ambient and diluted artificial aerosol samples down to 10^{-5} to 10^{-6} m^{-1}

5. *Reporting and Averaging Procedures*

If almost any of the techniques reported in the workshop proceedings were used at a remote climate monitoring station, it is certain that many instrumental zeros would ensue. It is also quite likely that some would sometimes give experimental single-scattering albedos exceeding unity. We feel that, even though it is obvious, we should point out explicitly the perils involved in reporting results in such circumstances. If an instrumental zero were (for example) 10^{-5} m^{-1} and one obtained such zeros except for 3 days in a month, when the results were 2×10^{-5} m^{-1}, 3×10^{-5} m^{-1}, and 2.5×10^{-4} m^{-1}, then if we reported averages, the number for the month would become 1.0×10^{-5} m^{-1}, but if we reported medians, the number would be zero; on the other hand, the figure of 2.5×10^{-4} m^{-1} might perhaps be excluded because we feel it "must" be polluted, or perhaps observation or another measurement showed it to be polluted. Similar uncertainties attend single-scattering albedo. If we obtained single-scattering albedo for a nonabsorbing aerosol with an uncertainty of ± 0.001 and then eliminated as nonphysical all values exceeding unity, the result is clearly unacceptably biased. For any trends or inferences to mean anything in such circumstances, reporting and operating procedures should be rigidly specified and rigidly adhered to. That, of course, is easy to say, but what happens when an improved technique appears? Perhaps the new method can detect down to 10^{-6} m^{-1}, and we now obtain nonzero readings for the entire month. Depending on procedures, the introduction of a better new method could produce an upward or a downward apparent trend in the long-term data.

While it may appear pessimistic, we are inclined to propose that, although the workshop activities have been valuable in assessing what can and cannot be done right now and in indicating what kinds of improvement should be looked for, they have not produced sufficient agreement between instruments, and between measured quantities and their known correct values, to justify the immediate undertaking of long-term monitoring programs using any of these techniques. Since collected samples can be preserved, the preferred monitoring strategy for the present would seem to be routine collection and archiving of aerosol material, which now (and in the future) can be subjected to nondestructive measurements while work continues on improving *in situ* techniques with respect to sensitivity and intercomparability.

6. *Contamination Problems*

Since the workshop measurements for more highly absorbing particles agreed reasonably well with known optical constants, whereas considerable disagreement occurred in the case of ammonium sulfate, two possibilities suggest themselves: (a) that the experimental techniques and procedures work well when absorption is substantial, but run into trouble when absorption is very small (compared to scattering) and (b) that the ammonium sulfate aerosol was somehow contaminated, perhaps by residual material from previous generations of absorbing particles. The disparity reported by Patterson and Marshall (21) between values obtained by them for workshop samples of ammonium sulfate and values which they obtained for "highly purified" ammonium sulfate suggest that contamination may have occurred.

REFERENCES

1. Hansen, J. E., Wang, W. C., and Lacis, A. A., Climatic effects of atmospheric aerosols, *Aerosols: Anthropogenic and Natural, Sources and Transport* (T. Kneit and P. Lioy, eds.), pp. 575-587, New York Academy of Sciences, New York (1979).
2. Foot, J. S., Spectrophone measurements of the absorption of solar radiation by aerosol, *Quart. J. Roy. Meteor. Soc. 105,* 275-283 (1979).
3. Rosen, H., and Novakov, T., Lawrence Berkeley Laboratory laser transmission method, *Light Absorption by Aerosol Particles* (H. E. Gerber and E. E. Hindman, eds.), Spectrum Press, Hampton, Virginia (1982).
4. Clarke, A. D., and Waggoner, A. P., Results from University of Washington participation in first international workshop on light absorption by aerosol particles, *Light Absorption by Aerosol Particles* (H. E. Gerber and E. E. Hindman, eds.), Spectrum Press, Hampton, Virginia (1982).

5. Gerber, H. E., Simultaneous measurements of aerosol scattering and extinction coefficients in a multi-pass cell, *Light Absorption by Aerosol Particles* (H. E. Gerber and E. E. Hindman, eds.), Spectrum Press, Hampton, Virginia (1982).
6. Williams, M. W., and Arakawa, E. T., Optical properties of glassy carbon from 0 to 82 eV, *J. Appl. Phys. 43,* 3460-3463 (1972).
7. Toon, S., Pollack, J. B., and Khare, B. N., The optical constants of several atmospheric aerosol species: Ammonium sulfate, aluminum oxide and sodium chloride, *J. Geophys. Res. 81,* 5733-5748 (1976).
8. Gerber, H. E., and Hindman, E. E., Data collation, *Light Absorption by Aerosol Particles* (H. E. Gerber and E. E. Hindman, eds.), Spectrum Press, Hampton, Virginia (1982).
9. Volz, F. E., Results on aerosol light absorption, *Light Absorption by Aerosol Particles* (H. E. Gerber and E. E. Hindman, eds.), Spectrum Press, Hampton, Virginia (1982).
10. Lindberg, J. D., and Gillespie, J. B., Relationship between particle size and imaginary refractive index in atmospheric dust, *Appl. Opt. 16,* 2628-2630 (1977).
11. Rosen, H., Hansen, A. D. A., Gundel, L, and Novakov, T., Identification of the graphitic carbon component of source and ambient particulates by Raman spectroscopy and an optical attenuation technique, *Appl. Opt. 17,* 3859-3861 (1978).
12. Pluchino, A. B., Goldberg, S. S., Dowling, J. M., and Randall, C. M., Refractive-index measurements of single micron-sized carbon particles, *Appl. Opt. 19,* 3370-3372 (1980).
13. Stroud, D., Generalized effective-medium approach to the conductivity of an inhomogeneous material, *Phys. Rev. B12,* 3368-3373 (1975).
14. Bohren, C. F., and Barkstrom, B. R., Theory of the optical properties of snow, *J. Geophys. Res. 79,* 4527-4535 (1974).
15. Twomey, S., and Bohren, C. F., Simple approximations for calculations of absorption in clouds, *J. Atmos. Sci. 37,* 2086-2094 (1980).
16. Bergstrom, R. W., Bemerkung zur Bestimmung des Absorptionskoeffizienten atmosphärischen Aerosols, *Beit. Phys. Atmos. 46,* 198-202 (1973).
17. Lindberg, J. D., The composition and optical absorption coefficient of atmospheric particulate matter, *Opt. Quant. Electr. 7,* 131-139 (1975).
18. Gillespie, J. B., Jennings, S. G., and Lindberg, J. D., Use of an average complex refractive index in atmospheric propagation calculations, *Appl. Opt. 17,* 989-991 (1978).
19. Ackerman, T. P., and Toon, O. B., Absorption of visible radiation in atmosphere containing mixtures of absorbing and nonabsorbing particles, *Appl. Opt. 20,* 3661-3668 (1981).

20. Japar, S. M., and Killinger, D. K., Photoacoustic and absorption spectrum of airborne carbon particulate using a tunable dye laser, *Chem. Phys. Lett. 66,* 207-209 (1979).
21. Patterson, E. M., and Marshall, B. T., Diffuse reflectance and transmission measurements of aerosol absorption: Report on results of the first international workshop on light absorption by aerosols, *Light Absorption by Aerosol Particles* (H. E. Gerber and E. E. Hindman, eds.), Spectrum Press, Hampton, Virginia (1982).

AUTHOR INDEX

Numbers refer to pages on which a reference is made to an author or a work of an author. *Italic numbers* refer to pages on which a complete reference to a work by the author is given. **Boldface numbers** indicate the first page of the articles in the book. For duplicate references in any one article, only the senior author is indexed.

Ackerman, T. P., 42,44,*53*,403,404,*407*
Ahlquist, N. C., 150,151,*168*
Amer, N. M., 26,*47*,137,*147*, 308,*319*,322,*333*
Anderson, J. E., 30,*50*,275,*278*
Andre, K., 26,27,33,36,44,*46*
Appleman, H. R., 33,*51*
Arakawa, E. T., 398,*407*
Aronson, J. R., 370,*372*
Auvermann, H. J., 39,*52*

Bach, W., 31,*51*
Baker, M., 25,31,37,*46,51*,52,60, *64*,130,*146,187*,174,*187*,190,*196*, 271,*273*,277,278,280,*295*,318, *319*,322,327,*333,334*
Balduin, B., 11,12,*18*
Barkstrom, B. R., 402,*407*
Benjamin, S. G., 8,10,*17*,37,*52*
Bennett, C. A., Jr., **129**
Bergstrom, R. W., 27,28,41,*48*,402,*407*
Beutell, R. G., 232,238,*241*
Bevington, P. R., 312,*319*
Blanco, A. J., 24,*46*
Blifford, I. H., 28,*48*
Bodhaine, B. A., 26,*47*,82,*96*,**149**,150, *168,*174,175,177,178,*188*,232,237, 238,240,*241*,270,*273*,322,*333*
Bohren, C. F., 402,*407*
Bolle, H. J., **1**
Bonelli, J. E., 29,*49*
Box, M. A., **169**
Braslau, N., 7,*16*
Brewer, A. W., 232,*241*
Brewer, R. J., 31,*51*
Brinkworth, B. J., 30,*50*
Browning, R. S., 29,*49*,172,*172*,298, 300,352,*355*

Bruce, C. W., 30,31,41,*50,51*,79, 275,*278*
Bullrich, K., 7,*16*
Byrne, D. M., 29,*49*,298,299,*300*

Cadle, S. H., 131,*146*
Campillo, A. J., 31,*51*
Carlon, H. R., 27,37,*48*
Carlson, T. N., 8,10,15,*17*,28,33,37,*52*
Cartwright, J., 34,*52*
Caverly, R. S., 8,*17*,28
Chang, S. C., 22,*45*
Charlock, T. P., 7,10,12,*16*
Charlson, R. J., 15,*18*,22,25,26,29, 37,41,*44,45,46,49,52,53*,60,*64*, 130,*146*,150,151,*168*,174,*187*,190, *196*,232,238,*141*,271,*273*,277,*278*, 280,287,*295,296*,318,*319*,322,327, *333,334*
Chu, L. C., 33,*51*
Chu, W. C., 15,*19*
Chylek, J. P., 9,*17*,22,44
Clark, W. E., 72,*92*
Clarke, A. D., 78,*92*,**173**,396,*406*
Coakley, J. A., 9,*17*,22,44
Coletti, A., **252**
Copp, J. D., **169**,298,*300*
Covert, D., 327,*334*
Cowen, S., **189**
Cox, S. T., 30,*50*
Cushing, K. M., 28,*49*
Cuzzi, J. N., 13,*18*,29,*49*,263,*266*

Dalzeu, W. H., 35,*52*
Dascher, A. J., 28,*48*,252,*266*
Dave, J. V., 7,*16*
Deehr, C. S., 341,*355*

Deirmendjian, D., 7,*16*,252,*266*,338, 349,352,*354*,*355*
DeLong, H. P., 24,*46*
DeLuisi, J. J., 15,*18*,29,30,33,*49*,172 *172*,298,*300*,352,*355*
Delumyea, R., 33,*51*
Dittberner, G. J., 14,*18*
Dlugi, R., 26,37,*46*,*52*
Dod, R. L., 26,*47*,322,*333*
Dodge, C. J., 31,*51*
Domoto, G. A., 7,*16*
Donoian, H. C., 271,*273*
Dowling, J. M., 28,*48*,401,*407*
Drummond, A. J., 22,30,*44*
Durham, M. D., 72,*92*
Duyckaerts, G., 27,*48*

Ebersole, J. F., 30,*50*
Eiden, R., 28,33,*48*
Egan, W. G., 26,*47*,**197**,198,199,200, 201,*230*
Emslie, A. G., 370,*372*
Ensor, D., 22,*44*,**189**,193,*196*
Evans, W. H., 28,*48*

Faxvog, F. R., 22,30,31,35,36,*45*,*50*, 52,146,*147*,275,*278*,301,303,*305*
Fegley, R. W., 15,*18*,29,*49*
Fenn, R. W., 8,10,*17*,22,24,*45*
Finnegan, W. G., **71**,174,*188*,237,*241*, 270,*273*
Fischer, K., 22,24,26,27,28,33,37,39, 44,46,60,*64*,272,*273*
Flanigan, D. F., 24,37,*46*
Foot, J. S., 30,31,32,33,*50*,396,401 403,405,*406*,*407*
Fouquart, Y., 5,*16*
Friend, J. P., 14,*18*
Fuller, W. H., 15,*19*
Furukawa, P. M., 15,*18*,29,*49*

Garvey, D. M., 72,79,*92*
Gerber, H. E., **21**,30,33,39,40,*50*,72 92,**231**,232,*241*,**379**,**387**,397,400,401, *407*
Gersho, A., 133,147,308,318,323,324
Gillespie, J. B., 27,33,44,*48*,*51*,*53*,**243**, 244,247,*250*,284,*296*,401,403,404,*407*
Gillette, D. A., 8,15,*17*,*18*,26,28,29, 47,*48*,*49*,280,*295*
Glushko, V. N., 25,*46*
Goldberg, S. S., 28,*48*,401,*407*

Goto, R., 22,*44*
Grams, G. W., 22,28,33,*45*,*48*,170,*172*, **251**,252,253,266,*266*,352,*355*
Grant, L. O., 72,*92*
Grassl, H., 37,*52*
Green, A. E. S., 28,*49*
Greenstein, J. L., 252,*266*
Gregory, G. L., 72,*92*
Grishechkin, V. S., 22,*44*
Groblicki, P. J., 131,*146*
Gundel, L., 22,26,*45*,*47*,130,*146*,280, *295*,318,*319*,322,*333*,401,*407*

Hall, J. S., 22,*45*,174,*187*,190,*196*
Hamill, P., 14,15,*18*,*19*
Hanel, G., 7,*16*,27,32,33,37,39,40, 47,*52*,**55**,239,*241*
Hansen, A. D. A., 22,26,*45*,*47*,130, 137,*146*,*147*,280,*295*,308,*319*,322,*333*, 401,*407*
Hansen, J. E., 9,12,*17*,*18*,31,*51*,253, 265,*266*,396,403,*406*
Hansen, M. Z., 28,29,33,42,*48*
Harker, A. B., 22,*45*
Harris, F. S., Jr., 72,*92*
Harshvardhan, 11,12,*17*
Hecht, G. H., 26,*47*,245,*250*
Heintzenberg, J., 26,*46*,**267**,268,*273*
Henyey, L.G., 252,253,*266*
Herman, B. M., 15,*18*,29,42,*49*,170, 172,*172*,239,*241*,**297**,298,299,*300*, 352,*355*
Hilgeman, T. W., 26,*47*,198,199,200, 201,*230*
Hindman, E. E., **71**,72,*92*,174,177, *188*,237,*241*,270,*273*,**379**,**387**,400,*407*
Hinds, B. D., 31,33,*51*
Hofmann, D. J., 15,*19*
Hoidale, G. B., 24,37,38,*46*
Horn, R. D., **71**,174,*188*,237,*241*,270, *273*
Horvath, H., 26,*46*,232,*241*
Hottel, H. C., 35,*52*
Hoyt, D. V., 3
Huffman, D. R., 23,29,36,*45*,*49*,**395**

Irvine, W. M., 5,*16*
Ivlev, L. S., 22,*44*

Jaenicke, R., 31,*51*
Japar, S. M., 28,30,31,35,*48*,*50*,*51*, **275**,275,277,*278*,284,*296*,405,*408*

AUTHOR INDEX

Jennings, S. G., 27,31,39,41,*48,51,
52,*244,*250,*284,*296,*403,*407*
Jospeh, H. H., 8,10,13,*17,18,*26,*47*
Junge, C. E., 14,*18*

Kabanov, M.V., 272,*273*
Kendall, D. N., 24,*46*
Khare, B. N., 22,28,*45,48,*228,*230,*
283,*295,*365,*372,*398,*407*
Kiang, C. S., 14,*18*
Killinger, D. K., 30,*50,* **275**,275,*278,*
405,*408*
King, M. D., 29,33,*49,*170,171,*172,*
298,299,*300*
Klimish, R. L., 34,*51*
Knollenberg, R. G., **65**,66,*69*
Kobayashi, M., 22,*44*
Kondratyev, K. Y., *3,*15,*18,*22,30,33,
44
Kortuem, G., 26,*47,*281,284,*295,*316,
319
Kuhn, P. M., 30,*50*

Lacis, A., 9,12,*17,18,*31,*51,*396,*406*
Langer, G., 72,*92*
Laude, L. S., 26,*47,*244,*250,*280,*295*
Lee, P., 9,*17,*31,*51*
Leifer, R., 14,*18*
Lenoble, J., 5,*16*
Levin, Z., 8,10,*17,*24,26,28,33,37,*46,47*
Lichtin, N. N., 284,*296*
Lin, C. I., 25,26,27,28,33,37,39,40,*46,*
60,*64,*130,131,*146,*174,175,*187,188,*
190,*196,*271,*273,*277,*278,*280,291,*295,*
318,*319,*322,323,*333*
Lin, H. B., 31,*51*
Lindberg, J. D., 24,26,27,28,33,44,
46,47,48,*51,53,*244,245,246,247,*250,*
280,284,*295,296,*401,403,404,*407*
Lippmann, M., 72,*92*
Liu, B. Y. H., 72,79,*92*
Livingston, J. M., 15,*19,*30,*50,*170,
172
Livshits, G. SH., 25,*46*
London, J., *3*
Lundgren, D. A., 72,*92*

MacCready, P. B., Jr., 150,*168*
Macias, E. S., 33,*51*
Maegley, W. J., 14,*18*
Major, G., 30,*50*
Manabe, S., 8,*18*
Marlow, W. H., 72,*92*

Marshall, B. T., 279,406,*408*
McClatchey, R., 7,*16*
McClenny, W. A., **307**
McCormick, M. P., 15,*19*
McDonald, F. A., 132,134,136,*147*
McMaster, L. R., 15,*19*
McPeters, R. D., 28,*49*
Medalia, A. I., 271,*273*
Mekler, Y., 8,10,*17,*26,*47*
Mendonca, B. G., 149,150,*168*
Middleton, W. E. K., 236,*241*
Minjares, J., 31,*51*
Mitchell, J. M., 22,*44*
Moore, J., **275**,275,*278*
Mosley, R. W., 328,*334*
Mueller, P. K., 328,*334*
Muitlbaier, J. L., 34,*52*
Murai, K., 22,30,33,43,*44*

Nagelschmidt, G., 34,*52*
Nolan, J. L., 34,*52*
Novakov, T., 22,26,34,*45,46,47,51,*130,
137,*146,147,*280,287,*295,296,*308,*319,*
321,322,*333,*397,401,*406,407*

Odencrantz, F. K., 72,*92*
Ohno, T., 284,*296*
Ohring, G., 7,9,10,*16*
Osif, T. F., 284,*296*
Palmer, K. F., 11,*18*
Palmer, R. A., **307**
Pao, Y. H., 275,*278*
Parker, R. L., 337,*354*
Patterson, E. M., 8,*17,*26,28,33,36,41,
47,*51,*247,*250,*280,281,284,*295,*
*296,*406,*408*
Patty, R. R., **129**
Pepin, T. J., 15,*19*
Perry, R. J., 29,*49*
Peterson, J. T., 14,*18*
Philips, D. T., 28,*48*
Pierce, L. B., 328,*334*
Pilat, M., 22,*44,*193,*196*
Pinnick, R. G., 27,30,31,39,*48,50,51,*
52,79,*92,*244,*250,*275,*278,*284,*296*
Pluchino, A. B., 28,*48,*401,*407*
Pollack, J., 7,9,10,11,12,13,15,*16,18,*
*19,*22,28,29,37,*45,48,49,52,*228,*230,*
263,264,*266,*283,*295,*365,*372,*398,*407*
Porch, W. M., 15,*18,*22,29,*44,49*
Pueschel, R. F., 26,30,*46,50,*232,*241*
Pui, D. Y. H., 79,*92*

Rabinoff, R. A., 15,*18*,29,*49*,239,*241*
Radcliffee, C. D., 33,*51*
Randau, C. M., 28,*48*,401,*407*
Rasool, S. I., 10,*17*
Read, A. A., 66,*69*
Reck, R. A., 7,10,*17*
Regan, J. A., 29,33,42,*49*,**297**,298,299, 300
Reynolds, D. W., 30,*50*
Riley, L. A., 22,*45*,174,*187*,190,*196*
Roach, W. T., 22,30,*44*
Robinson, G. D., 22,29,30,*44*,*49*,*50*
Robinson, N., *3*
Roessler, D. M., 22,30,31,35,36,*45*,*50*, 52,146,*147*,275,*278*,**301**,301,303,*305*
Röhl, R., **307**
Rosen, H., 22,26,27,33,35,40,*45*,*46*,*47*, 130,137,*146*,*147*,280,287,291,*295*,*296*, 308,318,*319*,**321**,322,*333*,397,401, *406*,*407*
Rosen, J. M., 15,*19*
Rosencwaig, A., 133,136,*147*,275,*278*, 308,*318*,323,*334*
Russell, P. B., 15,*19*,22,28,30,33,34, *45*,*48*,*50*,*52*,170,*172*

Sadler, M., 26,27,33,40,*46*,287,291,*296*
Sagan, C., 7,11,12,*16*,*18*,37,*52*
Sasamori, T., *3*
Sauelev, B. A., 272,*273*
SCEP, 22,*45*
Schleusener, S. A., 27,28,30,31,*48*, 66,*69*
Schnatz, G., 26,*46*
Schneider, S. H., 10,*17*
Schotland, R. M., **297**,298,*300*
Schulze, R., *3*
Schuster, B. G., 15,*18*,29,*49*,66,*69*
Sellers, W. D., 7,9,10,12,*16*,*17*
Selvidge, H., 150,*168*
Shaw, G. E., **335**,340,341,352,*355*
Sheldon, C. E., 29,*49*,247,*250*
Shettle, E. P., 8,10,*17*,22,*45*
Skidmore, J. W., 34,*52*
SMIC, 14,*18*,22,*45*
Smith, G. W., 303,*305*
Snell, F. M., 9,*17*
Snider, D. G., 244,*250*
Sparks, L. E., **189**
Spinhirne, J. D., 29,33,42,*49*,299,*300*
Stamnes, K., 352,*355*
Stanley, L., 33,*51*
Stedman, D. H., 130,*146*,**373**

Stefanutti, L., 130,*146*,**373**
Stevenson, R., 303,*305*
Stockton, B. H., 8,*17*,26,*47*,247,*250*, 280,*295*
Stroud, D., 401,*407*
Summers, A., 11,12,*18*
Swissler, T. J., 15,*19*
Szkarlat, A. C., 28,31,35,*48*,*51*,**275**, 275,277,*278*,284,287,289,*296*

Tanaka, M., 8,10,*17*
Tashenov, B. T., 25,*46*
Temkin, R. L., 9,*17*
Terhune, R. W., 30,*50*,130,*146*,275, *278*,**373**
Thorsell, D. L., 22,*45*,174,*187*,190, *196*,327,*334*
Toon, O., 7,9,10,11,12,14,*16*,*18*,22,27, 28,36,37,42,44,*45*,*48*,*52*,*53*,228, *230*,283,*295*,365,367,*372*,398, 403,404,*407*
Travis, L. D., 265,*266*
Treux, T. J., 30,31,*50*,275,*278*
Triehon, M., 14,*18*
Turco, R. P., 14,*18*
Twitty, J. T., 15,*18*,29,36,*49*,*52*
Twomey, S., 12,14,*18*,26,28,*47*,*48*,337, 338,*354*,**395**,402,*407*

Uthe, E. E., 30,*50*,170,*172*

van Camp, W., 11,12,*18*
van de Hulst, H. C., 5,*16*,24,*45*,253, *266*,338,*355*
Vassilyev, O. B., 22,*44*
Vidrine, D. W., 308,*318*
Volz, F. E., 24,27,28,37,*45*,*46*,**357**, 358,359,365,*372*,400,*407*
Vonder Harr, T. H., 30,*50*

Waggoner, A. P., 22,37,41,*45*,*52*,*53*, 78,*92*,151,*168*,**173**,174,177,*187*, *188*,190,*196*,327,*334*,397,*406*
Waldram, J. M., 22,30,43,*45*
Wang, W. C., 7,9,10,12,13,*16*,*17*,*18*, 31,*51*,396,*406*
Ward, G., 28,33,42,*49*
Weinman, J. A., 15,*18*,29,36,*49*,*52*
Weiss, R. E., 22,26,28,33,37,40,41,*45*, *53*,174,175,177,178,*187*,*188*,190, *196*,327,*334*
Wendlandt, W. W., 26,*47*,245,*250*
Wetherald, R. T., 8,*18*

AUTHOR INDEX

Wetsel, G. C., Jr., 132,*147*
Whitby, K. T., 34,*52*
White, K. O., 27,*48*
Wickramasinghe, N. C., 228,*230*
Wiedman, D., 11,12,*18*
Williams, D., 11,*18*
Williams, G. C., 35,*52*
Williams, M. W., 398,*407*
Williams, R. W., 34,*52*
Wiscombe, W. J., 170,*172*,352,*355*
Witten, R. C., 14,*18*
WMO-ICSU, 7,*16*
Wolff, G. T., 34,*51*
Woods, D. C., 15,*19*
Wootten, N. W., 30,*50*
Wright, W. M., 130,*146*,**373**
Wyatt, P. J., 28,*48*
Wyman, C. M., 28,*48*,252,*266*

Yamamoto, G., 8,10,*17*
Yamauchi, T., 22,*44*
Yasa, Z., 26,28,30,*47*,137,*147*,308,*319*, 322,*333*
Yee, Y. P., 31,*51*
Yuen, S., 327,*334*

Zuev, V. E., 272,*273*

SUBJECT INDEX

For subjects that appear on two or more consecutive pages, only the first page is listed.

Absorbance, 308
Absorption bands, infrared, 37
Absorption coefficient
 ambient measurements of, 26,30,63,389
 definition of, 23
 measurement techniques for, 24
Absorptivity, 127,308
Active Scattering Spectrometer Probe (ASSP), 91
Acoustic resonance, 276
Aerosol(s), ambient (see Particles, ambient)
 effect on climate by, 22,149,396,403
 heating or cooling of atmosphere by, 4,56
 light absorption by, 4,10,22,24,31,34,37, 56,61,298,389,396
 models of, 8,11
 monitoring of, 15,149
 production processes and rates of, 14
 stratospheric, 10
 tropospheric, 9
 volcanic, 12
Aerosol concentrator, 232
Albedo
 cloud-ice-feedback, 9
 effect of aerosols on, 4
 Lambertian surface, 7,170
 planetary, 11,13
 polar, 9
 single-scattering, 24
 surface, 4,10,298,353
Ammonium sulfate—methylene blue (MB) particles
 electron microphotographs of, 112,114
 generation technique for, 75
 mass concentrations of, 84,86
 percentge, 76
 shape of, 85
 size, mass distributions of, 94,97
 solution formulas for, 76
 summary of workshop measurements of, 379,387

Ammonium sulfate (salt) particles
 contamination of, 283,406
 electron microphotographs of, 112,121
 generation technique for, 75
 mass concentrations for, 84
 shape of, 83
 size, mass distribution of, 94,96,110
 solution formula for, 76
 summary of workshop measurements of, 379,387
Ammonium sulfate—soot particles
 aggregation of, 88
 electron microphotographs of, 115,117,119
 generation technique for, 75,88
 mass concentrations of, 84,88
 percentage of soot in, 82
 size, mass distribution of, 98,101,104,106
 summary of workshop measurements of, 379,387
Ammonium sulfate standard, 283
Ångström exponent, 151
Ångström pyrheliometer, 57
Arizona road dust
 composition of, 72
 density of, 184
 electron microphotographs of, 118
 mass concentration(s) of, 85,182
 particle shape of, 90
 resuspension technique for, 90
 size, mass distribution of, 72,104
 source of, 75
 summary of workshop measurements of, 379,387
 volume distribution of, 179
Asymmetry factor, 6,11,170,252,257,396

Backscatter fraction, 170,352,396
Becke line test, 190
Beer's law, 23
Beer-Lambert law, 191
Bernoulli's equation, 76
Bouger-Lambert law, 244,272

415

SUBJECT INDEX

Bouger's law, 23
Brewster window, 67,303
Bubbler, 75

Carbonaceous particles (*see* Soot particles)
Christiansen minimum, 364
Climate
 anthropogenic effects on, 149
 effect of aerosols on, 1,22,149,396,403
 feedback mechanisms for, 12
Climate modeling
 aerosol parameters for, 6,11
 advanced, 13
 sensitivity studies of, 9,13
 validation experiments for, 14
Cloud Simulation and Aerosol Laboratory, 72
Coefficient(s)
 aerosol absorption, scattering, extinction, 23
 aerosol to molecular scattering, 258
 bulk absorption, 23,244,280
 effective absorption, 127
 Kubelka-Munk absorption, 26,245,282
 Kubelka-Munk scattering, 245,282
 mass absorption, 23
 molecular scattering, 237
 specific absorption, 24,175,277,280
 volumetric specific absorption, 192
Column water vapor, 351
Comparisons
 instrumentation, 123
 mass concentration measurements, 83,86
 workshop measurements, 387,399
Condensation nuclei, 3,12,149
Coronameter, 342
Cosine sensor, 235,298,343
Cross section(s)
 extinction, 338
 scattering, 252,257
Cyclone, 177

Diesel exhaust particles, 34,275,305
Diffuse-direct radiation method
 ambient measurements by, 30,36
 description of, 29,125,127,169,298,351
 diffuse-direct ratio for, 170,298
 theory of, 170
 workshop measurements of, 172
Diffuse downwelling (upwelling) flux, 341,345,348,352

Diffuse-reflectance technique
 comparison with laser transmission method of, 279
 description of, 26,125,198,244,281
 error sources for, 247
 measurement consistency of, 284
 required measurements for, 245
 sample dilution for, 245,281
 saturation effects for, 284
 theory (Kubelka-Munk) of, 26,198, 200,244,281,398
Direct measurement technique(s), aerosol light absorption
 aerosol filled sphere for, 30
 difficulties with, 30,397
 flux divergence in, 29
 satellite measurements of, 30
 telephotometer and nephelometer, 30
 tower measurements of, 30
 transmission and scattering cell, 30,232
Dispersion analysis, 369

Eastman white reflectance standard, 246
Effective imaginary index, 29,41,298,397
Electrical Aerosol Analyzer (EAA), 78
Experiments
 CEANEX, 15
 GATE, 10,15
External mixture, 75
Extinction index, 360
Extinction photometer, 66

Filter(s)
 acoustic reflectivity, 136,316
 Delbag polystyrene, 281,361
 felt, 361
 fiber, 266,268
 Metricel, 77
 Millipore, 131,198,323,327,332
 Nuclepore, 77,131,174,190,287
 optical attenuation, 324,327
 optical reflectivity, 325
 particle penetration depth, 288,332
 quartz, 131,287,327
 sampling error, 268
 Teflon, 131,246,277,287,308,332
 Zefluor membrane, 289
Fluidized-bed particle generator, 75
Flux divergence, 5
Flux divergence method
 description, 29
 errors, 30

SUBJECT INDEX

417

Four-quadrant detector, 234
Fraunhofer diffraction, 338
Fresnel reflection coefficient, 265
Fresnel's equation, 244,398

Geophysical Monitoring for Climatic Change Program (GMCC), 149
Goody's theory, 351
Graphite, 199,222
Graphitic carbon, 35,284,322
Gray; particles, 401
Greenhouse effect, 4,10

Hemispheric flux sensor, 171,343
Henyey-Greenstein phase function, 253
  fit to measured, 252
Humidity
  correction for, 27,39
  effect on aerosol, 7,14,26,353
Hydrosol, 127

Indices of refraction, 23
Index of refraction, imaginary part
  ambient measurements of, 31,36,299,389
  average of mixture of, 401,403
  definition of, 23
  effective, 29,41,298,397
  ellipsometric measurement of, 200
  equivalent, 39
  formula for, 178
  humidity correction for, 39
  humidity effect on, 32
  importance of, 9
  inversions for, 28,127,266
  mean value of, 32
  measurement techniques for, 24
  relationship with $\tilde{\omega}$ of, 39,402
  spectral variation of, 36
  survey of values of, 31
  use in models of, 8,11
Index of refraction, real part
  average of mixture of, 401
  Brewster angle measurements of, 198
  definition of, 23
  oil immersion measurements of, 190
Infrared spectrograph, 127
  double-beam, 360
Instrumentation, aerosol light absorption
  cost of, 125
  description of, 24
  parameters measured by, 125
  relative error in, 125
  summary of, 123

Integrating plate method (IPM)
  comparison with other methods of, 27,39,60,277
  correction of measurements by, 185
  description of, 25,125,174,190
  effect of filter material in, 131
  effect of filter orientation in, 277
  effect of light scattering in, 58,186,194
  effect of source-detector location in, 175
  errors in, 186
  theory of, 26,60,175,193
  use with cyclone of, 185
  use with spectrophotometer of, 190
  variation of, 26,175,186,318
  workshop improvement of, 397
Integrating radiometer, 57
Integrating sphere, 25,199,246,271,281
Integrating sphere photometer, 271
International Radiation Commission (IRC), 2
Inversion(s)
  eigenvectors for, 338
  imaginary index, 28,127,266
  particle size distributions, 29,347
  single scattering albedo, 351
  methods of, 28,337
  multispectral extinction, 29,172
  optical depth and aureole, 337
  phase function, 28,127,266

Kernel functions, 339
Kramer-Kroenig dispersion analysis, 198,244
Kubelka-Munk theory, 26,198,200,244,281,398
  assumptions of, 246

Lambertain screen, 341
Lambert's cosine law, 316,341
Lambert's law, 23
Lampblack, 199,227
Langley calibration method, 340
Laser
  cavity of, 66
  Q, 67
Laser transmission method
  comparison with spectrophone of, 27,323
  correction for filter penetration in, 332
  description of, 26,126,286,322
  effect of filter material in, 287,328,333
  effect of filter orientation in, 287
  effect of particle scattering in, 326
  relationship of attenuation and absorption coefficients in, 332
  theory of, 325,327
  workshop improvements on, 397

SUBJECT INDEX

Lidar
  bistatic, 28,299
  multizenith-angle, 29
Light absorbing aerosol particles
  (see Aerosols, ambient)
  absorption properties of, 23
  composition of, 34
  size dependence of, 44
  vertical distribution of, 43
Light absorption
  measurement techniques for, 24
  monitoring of, 43,406
  wavelength dependence of, 34
Light absorption properties, 23
  relationships for, 37,402
LOWTRAN, 351

Maxwell-Garnett dielectric function, 401
Measurements, ambient
  absorption coefficient, 26,30,63,389
  heating rates of atmosphere, 63
  imaginary index, 24,28,31,389
  short wave radiation supply, 62
  single scattering albedo, 26,30,39,389
Measurement technique(s), aerosol light absorption
  calorimetric, 60
  diffuse-direct radiation, 29,125, 127,169,298,351
  diffuse-reflectance, 26,125,198,244,281
  direct, 29,123
  errors in, 43,123
  extinction to backscatter ratio, 299
  flux divergence method, 29
  bulk sample, 36
  integrating plate, 25,125,174,190
  integrating sandwich, 186
  lack of sensitivity of, 405
  laser transmission, 26,126,286,322
  limits of applicability of, 400
  optical measurement and inversion for, 28,126
  particle layer (Fischer's), 24
  particle sample, 24,123
  potassium bromide (KBr) pellet, 24,127, 358
  proposed research in, 43,405
  remote sensing, 28,298,397
  spectrophone, 30,125,129,276,302,307,323, 374
  specular reflectance, 244
  S-ratio angular scatter, 299

Methylene blue (MB) particles
  absorption by MB solution, 269,284
  color change of, 186
  electron microphotographs of, 115,117,121
  generation technique for, 75
  mass concentrations of, 84,89
  shape of, 89
  solution formula for, 76
  summary of workshop measurements of, 379,387
Mie size parameter, 37
Mie theory, 6,22,29,39,128,192,229,253,303, 338,360,397,402
Molecular scattering, 23,254
Multi-pass optical cell, 232

Nephelometer(s)
  Beutell and Brewer, 232
  calibration of, 237
  four-wavelength, 82,149
  integrating, 78,235,396
  modified MRI-1560, 174
  polar, 28,252
  reciprocal, 235
  theory of, 235
  truncation error in, 239
Nephelometer, four-wavelength
  Ångström exponent of, 164
  calibration of, 151
  comparison with reciprocal nephelometer of, 240
  description of, 150
  effective wavelengths in, 152
  sensitivity and noise in, 151
  workshop measurements for, 152
Nephelometer, polar
  angular range of, 252
  calibration of, 255
  description of, 252
  stratospheric measurement by, 254
  workshop measurements by, 255

Opal glass diffuser, 174,190
Optical depth (thickness), 6,10
  anthropogenic contribution to, 9
  formula for, 337
  global average for, 9
  measurement technique for, 340
  stratospheric aerosol, 11
  tropospheric aerosol, 8
  use with diffuse-direct radiation method of, 298

SUBJECT INDEX

Optical measurement and inversion technique(s)
  advantages of, 29
  description of, 28,126
  difficulties in, 29
Optical particle counters (OPC)
  calibration of, 79,81
  particle size channels in, 80
  PMS ASASP-X, 78
  Royco-220, 78,174
  theoretical response for, 80
Optical servo loop, 234

Particle layer (Fischer's) technique
  description of, 24
  specular reflection correction for, 26
Particle sample techniques, aerosol light absorption
  advantages of, 26
  complicating factors for, 396
  description of, 24,123
  effect of humidity on, 27
  effect of scattered light on, 27,58,186, 194,326
  use of visible vs infrared of, 27
Particles, ambient
  bimodal size distribution of, 349
  columnar size distribution of, 298
  composition of, 7,14
  horizontal homogeneity of, 299
  inhomogeneity of, 22,24,29,42,44
  organic, 34
  Raleigh size range, 396
  Raman spectra of, 35
  shape of, 6,13,24,29
  size distribution of, 6,8,11,396
  soil-derived, 7,10,12,34,37
  soluble, 7,14
  soot, 7,22,34,42,322
  variability of, 22,31,43
  volcanic, 7,10,14
Particles, workshop
  characterization of, 76
  comparisons of mass concentrations of, 86,181
  data reduction of, 79
  densities of, 79
  dilution system for, 74,76
  generation chamber for, 73
  generation techniques for, 75
  mass concentrations of, 79,83
  optical constants for, 398
  required properties of, 72
  scanning electron microphotographs of, 112
  size, mass distributions of, 93
  stability of, 82,387
  transporting system for, 74
  types of, 72,75

Phase function(s), 6,252
  appearance for ambient particles of, 42
  azimuthally integrated, 353
  dependence on backscattered light of, 170
  dependence on size distribution of, 265
  fit to workshop particle, 255,258
  Henyey-Greenstein, 252
  inversions for imaginary index of, 28,127,266
  Pollack-Cuzzi formulation for, 263
Photoacoustic cell, 276,302,304,323
Photoacoustic detection (PAD), 130
Photoacoustic effect, 302
Photoacoustic spectrometer, 309
Photoacoustic spectroscopy, 308,322
Photoacoustic techniques (see Spectrophone)
Photon counting, 268
Piezoelectric transducer, 234
Polarization planes, 253
Polarization ratio, 299
Potassium bromide (KBr) pellet technique
  description of, 24,127,358
  effect of particle size on, 371
  effect of pellet grinding on, 364,367
  pellet preparation for, 358
  sampling restrictions in, 358
  spectral ranges in, 358
Pyroelectric detector, 316

Quantum size effects, 201

Radar cross sections, 264
Radiation, atmospheric
  budget of, 3,56
  circum-global, 57
  effect of wavelength on, 4
  flux divergence of, 5,29
  shortwave supply of, 57,61
  transfer of, 5,22,28,43,229,352
Radiative transfer equation, 5
Raleigh optical thickness, 170
Raleigh scattering, 151,336
Raman spectroscopy, 34,322
Resonant absorption bands, 23
Reststrahlen bands, 367
Rotameter, 177

Scanning electron microphotographs, 83,112
Scattering efficiency factor, 253
Scattering index, 360
Scattering plane, 252
SCEP, 22
Schuster-Schwarzschild approximation, 352

SUBJECT INDEX

Single scattering albedo
  averaging and reporting procedures for, 405
  critical value of, 9,31,396
  definition of, 24
  modeling parameter for, 6,8,11,396
  particle mixture, 401
  relationship to imaginary index of, 39,402
Sky radiance, 338
SMIC, 22
Solar almucantar, 349
Solar aureole, 28,338
  diffraction pattern in, 344
  gradient in, 340
  measurement technique for, 340
  radiance for, 340
  workshop measurements of, 344
Solar constant, 62
Solar irradiance, 352
Solar photometer, 298
Solvent extraction, 322
Soot monitor, 133
Soot particles, ambient, 34,72,275,354
Soot particles, workshop
  aggregation of, 91
  Cabot Monarch, 71,270
  electron microphotographs of, 116,120
  generation technique for, 75
  mass concentrations for, 84,87
  organic fraction of, 284
  size, mass distribution(s) of, 96,98,100 104,106,108
  summary of workshop measurements of, 379,387
Soot photometer, 271
Specific absorption coefficient, 24
Specific mass, 24
Spectral solar irradiance, 338
Spectrophone, filter
  comparison with laser transmission method of, 27,323
  comparison with IPM of, 142
  description of, 30,125,129,307,323
  effect of light scattering particles on, 140
  effective filter pathlength for, 316
  errors in, 315
  filter reflectivity of, 136,316
  interference effects in, 132,136
  signal saturation in, 131,310
  theory of, 133,308
  thermal waves in, 132,134

Spectrophone, *in situ*
  calibration of, 277,303,374
  calorimetric, 60
  comparison with IPM of, 28,277
  description of, 30,126,276,302,374
  effect of volatile particles in, 31
  measurements by, 30,275,277,303
  recommendation for use of, 396,401
  resonance, 304,374
  responsivity of, 302
  theory of, 276,302
Specular reflectance measurements, 244

Thermal particle analysis, 322
Total combustion method, 387
Transmission spectroscopy, 308
Transmissometer(s), 66,233
Transmittance, 23
  accuracy of measurements of, 234

Ultrasonic nebulizer, 75

Visual range, 65

Water vapor, spectroscopic determination of, 351
Workshop
  accuracy and precision of measurements in, 398
  data logging in, 78
  evaluation of, 395
  flow measurements of, 74
  laboratory setup of, 72
  sampling stations at, 73
  schedule of experiments at, 82,380
  site of, 72
  summary of measurements at, 379,387
  supporting instrumentation at, 128
  temperature, humidity measurements at, 77
World Climate Research Program, 1

X-Ray particle spectra, 350